Statistical Methods for the Sciences

A First Course

by

Thomas W. O'Gorman

Northern Illinois University
DeKalb, Illinois

ISBN   978-0-6151-6612-4

Statistical Methods for the Sciences: A First Course

Publisher: Thomas W. O'Gorman

To my wife, Martha.

Preface

One objective in writing *Statistical Methods for the Sciences* was to develop an introductory textbook that would motivate students taking their first course in statistical methods. Many realistic examples are provided in the text and in the exercises so that students can clearly understand the importance of statistics in the modern world. Another objective was to give students greater guidance in the choice of statistical methods than is usually found in texts at this level. Consequentially, this book includes considerable discussion of which statistical methods are appropriate and it includes recommendations concerning the choice of estimators and the choice of tests. I believe that many textbooks contain stern warnings about the use of certain statistical procedures but give students few options and little guidance as to what should be done when confronted with real data. In contrast, this textbook helps the student select the most appropriate method.

This book provides a straightforward explanation of the topics most commonly found in statistical methods courses. It provides few of the frills that are included in many other introductory textbooks. The reader will not find multicolored pages, fancy graphics, or pictures. The text stays focused on the topic, which is statistical methods.

Because there are few distractions in this textbook, the teachers and students can concentrate on the understanding of the statistical methods and on the selection of the most appropriate methods. Although statistics is intellectually challenging and is quite useful, these two key motivational aspects can be missed by students who read textbooks that have many pages of unnecessary material.

Although the text and exercises are directed to the general science audience, this book could be used as an introductory biostatistics textbook. I prefer not to use statistical software in the introductory course so the book contains little software. However, if the instructor felt that software was beneficial the textbook could be supplemented with instructional material for that software. This textbook was designed for a 4 semester hour course in statistical methods. For a 3 semester hour course the optional material, which is indicated by an asterisk in the section title, could be skipped. No calculus is used, but some familiarity with algebra is necessary.

The exercises were selected to emphasize the concepts of statistics rather than the calculations. Hence, many thought provoking exercises are included which require little or no calculation but do require some thought. In addition, many of these data sets contain realistic data. Many students are motivated by the analysis of realistic data because they can see the usefulness of statistics in their lives.

There is considerable flexibility built into the book. Chapter 1 presents the usual descriptive statistics and gives some recommendations concerning the appropriateness of these methods. Probability is covered in Chapter 2 in a few brief sections that were designed to provide sufficient coverage for an introductory course. Chapter 3 presents the basic features of sampling distributions. Chapters 4, 5, and 6 introduce confidence intervals and tests of significance. In Chapters 5 and 6 non-parametric tests are used along with the usual parametric tests. In these chapters a "Guide" section is included to give students some recommendations for the selection of tests. Correlation and regression are presented in Chapter 7. The product-moment correlation coefficient is introduced first and it is then used to compute the slope and intercept. The chapter includes a discussion of the regression effect and the regression fallacy.

Chapter 8, which concerns the analysis of categorical data, includes more material on this topic than is found in most introductory textbooks. It includes the standard test for independence but also includes a test for trend and a test to be used with matched pairs data. Chapter 9 includes a parametric test and a non-parametric test for the one-way layout, along with a guide to their selection.

The book is unique in another way. It is reasonably priced. The low cost is due to the fact that the author prepared the text in an electronic format and the book is simply printed and shipped by a printer. There are no editors, text designers, or marketing staff. Although every effort was made to make the text accurate, please feel free to send any correction to me via e-mail at *ogorman@math.niu.edu*. I hope you enjoy the book.

Tom O'Gorman

DeKalb, Illinois U.S.A.

September 2007

Contents

Chapter 1. Collecting and Describing Data

1.1 Why Study Statistics?

We need statistics to understand our modern world. We may hear the results of a political poll, but may not fully understand the meaning of the results. We may see the results of a study on the effectiveness of a treatment for AIDS, but may fail to fully grasp the implications of the results. Whether we like it or not, we have facts and figures coming at us from all directions. But which facts should be noted and which fact should be ignored? The study of statistics will help us focus on the important facts, and help us make correct decisions based on those facts.

Statistical science is the science of managing information. The information is often obtained from surveys of people or from experiments on plants or animals. A politician may conduct a survey to see how much support there is for a piece of legislation. An agronomist may experiment with several varieties of corn to determine which variety produces the greatest amount of corn. Once the data are collected the next step usually consists of describing the important aspects of the data using graphs and descriptive statistics. These descriptive statistics may be presented simply with averages or may be presented in a more complex way as a three dimensional chart. However, in many circumstances, simply describing the data is insufficient, so we often need to continue our analysis in order to draw some conclusions from the data. This last step is extremely important because, after spending a considerable amount of time collecting the information, we would like to come to the correct conclusions.

Statistics are numbers which are used to summarize data. They are used in magazines, books, and on television to rapidly convey summary information about the results of surveys or scientific investigations. A television newscast may describe the political support for a candidate by quoting a statistic that a certain percentage of the voters prefer a candidate. A sports announcer may state a player's batting average, which is a statistic used to describe the overall hitting performance of that player. An article in a magazine may describe the effectiveness of a diet by quoting the average amount of weight that was lost by people on the diet. In each of these examples statistics were used to condense the information so that it could be communicated easily.

A brief overview of the steps that are typically involved in a scientific research project are displayed in Figure 1.1. In the data collection step the researcher either collects information from an observational study or from an experiment. Experimental studies, which impose treatments on the subjects of the experiment, will be discussed in section 1.3. In the next section we will discuss observational studies which are used when we want to collect information from people, but we do not need to impose a treatment or change their environment while we are collecting the data.

Figure 1.1 Steps in the Statistical Analysis of Data

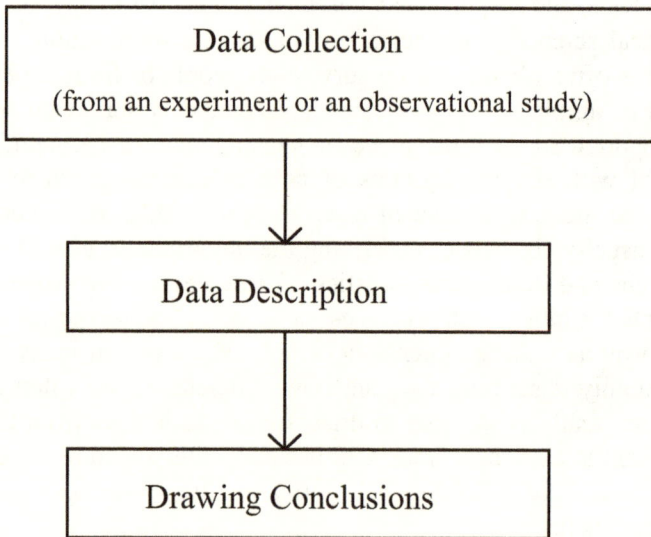

After the data has been collected, they must be properly summarized in order to effectively communicate the results of the study. Descriptive statistics can be calculated and then used to present the data concisely in a written or oral report. Graphical techniques can be used to communicate complex relationships in a clear and convincing way. These numerical and graphical methods will be presented in this chapter, starting in section 1.4.

After the data has been collected and described, the third step in a scientific study is to draw some conclusion. For example, we may decide, on the basis of results obtained from an experiment, that a particular drug really does lower blood pressure. Or we may decide, based on an opinion poll, that a certain candidate will win an election. Although it is not always easy to make correct decisions, most of the statistical methods described in this text will assist the researcher in coming to correct decisions based on the available data. In order to make an intelligent decision, we need to understand the role that chance can play. Chance, or probability, will be discussed in Chapter 2. The statistical methods used to arrive at correct decisions, which are based on chance, will be presented starting in Chapter 3 and will be the main focus of this text.

1.2 Using Surveys to Collect Data

Types of Interviews

One of the most difficult tasks in doing any statistical work is collecting the data. It is also critical that the data collection be done correctly, because no amount of mathematical manipulations can correct for improperly gathered data. A common method of obtaining information is to use a questionnaire to obtain data from people. These **surveys** are used by the government in order to estimate the unemployment rate, by businesses in marketing research, and by scientists to gather information.

There are three survey methods that are commonly used to collect data from individuals. The most popular method is the **telephone interview** which requires the interviewer to place a telephone call to an individual and record their responses. This methods is inexpensive but has several drawbacks. While almost everyone in the United States can be contacted by phone, not all can, and it is important to keep in mind that the homeless and people who are on the move cannot be contacted by phone. A more serious problem with telephone interviews is that many people will not answer the phone or will refuse to participate in the survey. These **non-respondents** pose a serious threat to the validity of most surveys because they may not be similar to the respondents with respect to the characteristic that is being measured. For example, if we are doing a survey to determine what percentage of people have trouble making payments on their debt, then the results from a telephone survey results may be misleading because people who have debt problems may not have a phone, or may be unwilling to talk to anyone on the phone about their problem. For telephone surveys, it is not unusual to have a response rate near 50%. Another problem with telephone interviews is that they are not appropriate for complex questionnaires. For example, if a person conducting a survey needs to obtain information on the dietary habits of individuals, it is important to determine the portion size of each food product. Hence, the interviewer must show a picture or object that would enable the person to estimate the correct proportion. Clearly, this sort of survey cannot be done over the telephone; it requires a **personal interview**.

Personal interviews do not have the limitations of the telephone survey, and can be used for lengthy or complex questionnaires. However, the personal interview is expensive because it usually requires that the interviewer drive to a residence or place of business to do the interview. The interviewer may also need to call the person several times to arrange the interview. The total cost may be several hundred dollars per completed interview, and because of the high costs they are not often used. Non-respondents can also be a problem with this method because some people do not want to be interviewed. The response rates vary widely but can approach 80% if the interviewer is persistent in calling back to arrange an interview.

Another commonly used method of collecting data is the **mail survey**. It is not difficult to construct a mail questionnaire and send it to hundreds of households; the big problem is convincing the individuals to return the questionnaire. As you might expect, many of these questionnaires are pitched into the trash. Unless the people who receive the questionnaire are motivated to send it back there will be few responses, especially for lengthy questionnaires. As with other interviewing methods, a high proportion of non-respondents makes it difficult to interpret the results because the people who respond to lengthy questionnaires may be quite different from those who do not respond. The respondents will generally be much more interested in the items on the questionnaire than the non-respondents will be. Consequently, the analysis of the completed questionnaires may indicate a great deal of interest from the 10% of those surveyed who responded. The opinions of the 90% who did not respond are not known and it is not correct to assume that non-responders would have answered the items on the questionnaire in the same way. A response rate of 10% is not unusual for a mail survey if the questionnaire is lengthy and there is no incentive to return the completed survey.   Sometimes, in order to increase the number of respondents, survey organizations will offer small gifts to encourage people to return questionnaires.

Regardless of the type of survey used, the non-respondents pose the most serious problem.   It is prudent to keep the following in mind when analyzing the results of surveys:

The people who choose to respond to questionnaires may be quite different from those who choose not to respond. Thus, results of surveys having a high percentage of non-respondents may be seriously biased.

Researchers sometimes do follow-up surveys of the non-respondents to determine how they might have responded.   Also, it is important to remember than a large number of returned questionnaires does not guarantee validity if the percentage of non-respondents is large.

Methods of selecting a sample

The major objective in selecting a sample is to select it in such a way that it reflects the population that it came from.   For example, if we need to estimate the amount of support for a candidate for a state office and we want to take a sample of 500 voters, we will want that sample to reflect all voters in the state.   We do not want 500 males or 500 females or 500 senior citizens.   We also do not want 500 voters from one part of the state.   A good selection method is to take a **simple random sample** of 500 voters.   A sample of size 500 is said to be a simple random sample if the selection is done in such a way that all combinations of 500 voters have the same chance of being selected.   To select a simple random sample we would use a computer program that generates random numbers corresponding to the voter on the list of all voters.   For example, if we had a million voters in the state, and we wanted to select 500 voters at random, we would use a computer program that would generate 500 random numbers, each of which would be between 1 and 1,000,000.   Suppose the first few numbers were:

387962
796872
121211
etc.

We would then look on our list of all voters in the state and select the 387962nd voter, the 796872nd voter, the 121211th voter, etc.. If the computer has the complete list of every voter in the state, it is relatively easy to select a simple random sample. The advantage of the simple random sample is that it should represent the voters in the state, with respect to gender, age, and race.

Simple random sampling is a **probability method** of sample selection. Probability methods are sound scientific methods in which everyone in the sample has some chance of being selected and the exact method of selecting the sample is known. There are many probability methods of sample selection which are used by professional polling organizations that are more complex that simple random sampling, but we will not deal with these complex methods in this text. Just remember that all probability methods give everyone in the population some chance of being selected, and the chances must be known.

In contrast, methods that are not based on probability do not give everyone a chance of being selected in the study, and consequently cannot be used to produce scientifically valid estimates. One example of a method that is not based on probability is the **convenience** or **judgment method** of sample selection. This methods relies on the interviewer's judgment to select individuals for the sample. The problem with judgment sampling is that the interviewer may select too many older voters, or too many female voters, or too many voters who are similar to what he or she believes the typical voter looks like. It is difficult to determine just how much bias the judgment selection introduces into the estimates; the estimates could be accurate or they could be quite misleading.

Another method of collecting information that is not based on probability is the **voluntary response survey**. These samples, unfortunately, are quite popular because they are quick and inexpensive. A radio station may ask callers to call in and register their opinion on a certain issue. A magazine may ask readers to return a survey on sexual practices. These are examples of voluntary response surveys because the listener or the reader decides if he or she wants to participate. Naturally, people who feel strongly about an issue are more likely to respond and it should be no great surprise that voluntary response surveys have the potential of greatly underestimating the more moderate responses that would be observed in the population had a probability method been used.

To summarize, a great deal of time and effort goes into the proper collection of data. If the data is not properly collected no amount of sophisticated analysis can fix it. A well designed survey:

- Uses a simple random sampling method or other probability method for selecting the sample.

- Uses a questionnaire that has been carefully designed.

- Makes a great effort to keep the non-response rate low, preferably less than 20%.

Professional polling organizations follow these general guidelines. The unprofessional researchers who ignore these points often produce estimates that are misleading. The most common mistake that unprofessional researchers make is to assume that a large sample size can compensate for serious flaws in the sampling procedures. If there are biases in the selection method, the bias will not be reduced by increasing the sample size. Large samples do not always produce valid estimates.

As an example of an improperly conducted survey, consider the survey reported in *Women and Love*, by Shere Hite (Knopf, 1987). To collect the data Hite sent the questionnaires to organization rather than to individuals. The author described the distribution of questionnaires as follows:

Clubs and organizations through which questionnaires were distributed included church groups in thirty-four states, women's voting and political groups in nine states, women's rights organization in thirty-nine state, professional women's groups in twenty-two states, counseling and walk-in centers for women or families in forty-three states, and a wide range of other organizations, such as senior citizens' homes and disabled people's organizations, in various states.

The survey was also sent to an unspecified number of women who wrote to the author to request that a questionnaire be sent to them. In the survey 100,000 questionnaires were distributed, but only 4,500 were returned by mail, which is a 4.5% response rate or a 95.5% non-response rate. One of the results of the survey was that "70% of women married five years or more are having sex outside of their marriages." In order to evaluate the validity of this results we need to discuss the study design.

There were several problems with the design of this study. First, a probability method was not used to select the sample. This is important because the millions of women who were not members of one of those organizations had little or no chance of being included in the sample. Second, of those individuals who did receive the questionnaire, only 4.5% responded. Since we do not know anything about the 95.5% who did not to respond, it is impossible to say anything definitive about the 95,500 women who received the survey but chose not to respond. If the women who were having extra-marital sex were more inclined to return the questionnaire then the estimate could be seriously biased. These problems do not prove that the extra-marital sex estimate is inaccurate, but they do show that the estimate may be inaccurate.

Exercise Set A

1. Many scientific calculators can produce random numbers that fall between 0 and 1. Suppose that a students uses a calculator to produce the following random numbers :

    0.571   0.072   0.077   0.747   0.298

    In practice, we often multiply these random numbers by 1000 to produce the integers

571   72   77   747   298

These random numbers are produced in a manner such that each 3 digit number has the same chance of being produced and the value of one random numbers does not influence the chances for the next number. Thus, the chance of obtaining these five random numbers is equal to the chance of producing any other five numbers.

a) Suppose you are enrolled in a college having 1000 students and you wish to take a simple random sample of 10 students. Describe in detail how you could use the calculator's random number generator to select a simple random sample.

b) Describe in detail how you might take a simple random sample of 10 students in a college of 500 students. There are several acceptable ways of doing this; you need to describe only one way.

2. A used car dealer wanted to estimate how satisfied the customers were with the cars they purchased in the last month. The dealer wanted to take a simple random sample of 10 cars from a list containing the 52 individuals who purchased a car in the last month. The dealer did not have a random number table or a scientific calculator, but he did have an ordinary deck of cards. Describe how the card deck could be used to take a simple random sample of 10 recent buyers.

3. The National Institute of Mental Health sponsored a series of personal interview surveys to estimate the prevalence of certain mental disorders. One of these Epidemiologic Catchment Area (ECA) surveys was done in New Haven, Connecticut using probability methods. In this survey 4045 households were selected for study and one member of each household was selected for the survey. There were 3058 completed interviews.

a) What was the percentage of non-respondents?

b) Considering that this was a survey of mental health, do you believe that the respondents would, on average, be more healthy or less healthy, than the non-respondents? Explain.

4. The survey reported by Shere Hite in *Women and Love*, (Knopf, 1987) had a high proportion of non-responders. In the book the author reported that "the final statistical breakdown of those participating according to age, occupation, religion, and other variables known for the U.S. population at large in most cases quite closely mirrors that of the U.S. female population." Does this support a belief that the results of the study can be generalized to the U.S. female population?

5. A baking products company wanted to estimate the percentage of women who bake cookies for their families. They used a telephone survey and found that 47 of the 100 housewives they contacted did bake cookies for their families. To conduct the survey they called 600 households during the day on the first Tuesday and Wednesday of October. The households were selected by a simple random sample procedure. Do you believe that approximately 47% of all housewives bake cookies for their families? Why or why not?

6. A graduate student wanted to determine how many alcoholic beverages were consumed per week by the typical college student. To investigate the matter he conducted personal interviews with 50 students. To select the students, the graduate student stood in the lobby of the library and asked passerby's if they would to participate in the survey.

    a) In your opinion would this selection method produce valid estimates of alcohol consumption? Why or why not?

    b) If the graduate student selected 500 students, instead of 50 students, in the lobby of the library would the results be improved? Why or why not?

    c) How could this study be improved? If your method requires that a list be obtained, specify how that list might be obtained.

7. (Challenging) Consider again the ECA study that was described in exercise 3. Assume that the households were selected by using a simple random sample and that within the household one individual was selected to be interviewed from all individuals in the household

18 years of age or older. Did everyone in the community over the age of 18 have the same chance of being selected? Explain.

1.3 Conducting an Experiment to Collect Data

Many researchers in the sciences use experiments, rather than observation studies, to collect data. The problem with observational studies is that they are not too useful in determine if a scientific theory is true or false. For example, suppose we wanted to find out if a certain drug reduces the amount of back pain in people who had recently developed back pain. We could give all patients who had recently developed back pain a supply of these back pain pills and we could take a sample of some of those patients. Now suppose we notice that 80% of the patients who took the back pain drug recovered after two months. What could we conclude? We could not conclude that the drug was effective in reducing pain because it is possible that these patients would have gotten better without any treatment. The only way to find out if the drug was effective is to compare people who took the drug to those who did not.

However, if we do use two groups of patients, those who took the drug and those who did not, and if we let patients decide if they wanted to take the drug, we will have difficulties interpreting the results. The problem is that people who chose to take the drug may have greater pain than those who decided not to take the drug and, if this occurred, the results of the experiment cannot be attributed solely to the influence of the drug. Consequently, it may be impossible to determine the effectiveness of the drug.

A better approach is to randomly assign patients to a drug group or to a group that does not receive a drug. A group that receives no treatment or a standard treatment is called a **control group**. A **placebo** is a dummy treatment which, in this example, is a fake pill. In order to randomly assign a patient we flip a coin and if it turns up heads we will assign the patient to a treatment group, otherwise we will assign the patient to the control group. This will insure that the two groups will have, on average, the same severity of illness. If, after two months, the researcher observes a difference between the two groups, it is reasonable to conclude that the difference was due to the drug and not to any other factor. This is an example of a **randomized experiment**.

Randomized experiments are popular because it is possible to draw firm conclusions from the results. In a simple experiment the **experimental units**, which are the people, animals, or plants that are used in the study, are assigned at random to one of several groups. The groups are defined by the levels of the treatment factor. In our back pain example, if a difference is observed between the two groups, it must first be determined that the observed difference was probably not due to chance. This point will be taken up in later chapters. For now, let as assume that we observe a difference and that the difference could not have been due to chance. If the observed difference was not due to chance, then it must be due to the drug because, if the randomization was effective, the groups are comparable in all other ways. The basic rules for randomized experiments can be summarized as follows:

- Use a control group so that you can fairly evaluate the effectiveness of the treatment.

- Randomly assign units to the control and treatment groups.

- Use a sufficient number of units so that you can detect a difference between groups if the treatment really does make a difference.

In many studies it is important that the experimental subjects know which treatment they are receiving. If the subjects do not know what treatment they are receiving we say that they are **blind** to the treatment. It is often desirable to use a **double-blind** study, where the people who administer the treatment also do not know what treatment is being given. Of course, someone must know which treatment was administered to each experimental unit. Typically, the researchers who supervise the study have access to the codes that enable them to figure out, after the study, which patients received which treatments. In the back pain experiment the back pain pills and the placebo pills would be packaged in a way that would make it difficult for the patients to figure out which was the real pill. Thus, the patients would be blind to the treatment, and if the people working directly with the patient also did not know which pills were real and which were fake, it would be called a double-blind study. Double-blind experiments make it more difficult for the subject to figure out what treatment is being given and makes it less likely that the people taking the measurements will consciously or unconsciously bias the results. This reduces the chance of obtaining a false conclusion based on biased reporting of results.

Experimenters make a great effort to reduce the influence of other variables on the results. These outside variables, if not properly controlled, can jeopardize the validity of the results. For example, an experimenter who randomly assigns rats to two groups may decide to house the rats in the same room in order to assure that the temperature and the lighting are the same for both groups. When human subjects are used in a randomized experiment, it is usually difficult, but not impossible, to remove the effects of these outside variables.

To illustrate the possible effects of outside variables, consider an experiment to determine if children are more active if they have a large amount of sugar in their diet. You could randomly assign them to a sugar diet or to a sugar substitute diet, and have them eat their three meals a day in a special kitchen where they would be given a meal prepared for one of those diets. Would this really be a good experiment? What about snacks and soft drinks that they might consume when they are away from the special kitchen? Would they also eat the same foods that their family is eating at home? Before reading further, take a scrap of paper and jot down how you would design an experiment to see if sugar increase activity in children.

The results of an experiment to investigate this question was published in a paper entitled "Effects of Diets High in Sucrose or Aspartame on the Behavior and Cognitive Performance of Children" by M. L. Wolraich, *et al.* (*The New England Journal of Medicine*, 1994, p.301-307). In this study the researchers randomly assigned 48 children who were 3 to 10 years of age to a sucrose diet, an aspartame diet, or to a saccharin diet. Behavioral measurements were taken after each week in this nine week experimental period. The children stayed on the diets they were assigned for 3 weeks after which they were switched to one of the other two diets for three weeks. After six weeks on the study they were switched to the remaining diet. Thus, each child was randomly assigned to a sequence where they experienced each of the three diets for three weeks each. At this point we need not be concerned with all the details of this study; it is sufficient for our purposes to state that all children were randomly assigned to one of the diet sequences.

We want to focus on how the experimenters controlled the outside variables. The following procedures were used to reduce or eliminate the possible effects of these outside variables:

- Before the study, all food was removed from the home.
- During the study period all foods were prepared by the researchers and were delivered by van to the home.
- Care was taken to make the sweetened foods look normal, even though the sugar content had been altered.
- Soft drinks were specially prepared by three national bottlers in unmarked bottled that were coded so that only the researchers knew which sweetener was in the bottles.
- The families were told that the sweeteners would change each week. In fact, the diets prepared for each of the three weeks were carefully prepared to look different from week to week, although the sweetener did not change over the three week period. This was done to keep the participants guessing about what kind of sweetener was actually used.

You can see that a great deal of time and money was spent to control for these outside variables. However, even with this degree of control, it is possible that a child may be tempted to depart from the prescribed diet. The researchers anticipated this problem by adding chemical markers to the food sweetened by sucrose and aspartame. They then obtained urine samples weekly to ensure that the children were eating only the prepared foods. The conclusion of the research was that neither sucrose nor aspartame affects children's behavior. You may or may not believe the results, but it is clear that the researchers made a great effort to control many of the outside variables. This makes the research more valuable because there are fewer alternative explanations for the results.

If experimental results are so conclusive, why aren't they always used? One reason is that the results from experiments cannot always be generalized to real world situations. For example, research on the effectiveness of safety devices in automobiles may rely on experimental evidence obtained under strictly controlled circumstances. That is, all results may have been obtained from automobiles that were traveling at 35 miles per hour before they crashed straight into a concrete wall. Under actual driving conditions, automobile crashes that are exactly like these experimental crashes are rare, so it is not entirely clear if the experimental results would accurately predict the real world experience. It would be prudent to use observational data, in addition to experimental data, to evaluate the actual crash worthiness of a vehicle.

Another problem with experiments is that they are sometimes impractical. A researcher, investigating the effect of high fat food on heart disease, may randomly assign half of the volunteers to a very low fat diet for one year. However, it is likely that most of those assigned to a low fat diet would fail to follow the low fat diet. The failure of many subjects to comply with the experimental plan will produce misleading results.

There are also many ethical problems associated with experiments on animals and humans. It has now been pretty well established that cigarette smoking causes cancer. Would it have been ethical, 50 years ago, to randomly assign half of the volunteers to a smoking group, when the researchers strongly suspected that smoking caused cancer? Because experiments using humans were unethical and impractical, research in this area has relied on observational studies that compare the cancer rate of smokers to the cancer rate of non-smokers.

Exercise Set B

1. Suppose you want to randomly assign animals to treatment groups but have only one coin and one die.

   a) If you want to randomly assign an approximately equal number of animals to four treatment groups, how do you do the assignment?

   b) If you want to randomly assign an approximately equal number of animals to three treatment groups, how do you do the assignment?

2. Take a fair coin and flip it 10 times. Suppose each flip of the coin determines the assignment of an experimental unit to one of two treatments.

a) Would this method of assignment be acceptable for a randomized experiment?

b) With your coin flips how many units were assigned to the two treatments?

c) If 1000 experimental units are assigned by using coin flips, what is your guess about the chance that exactly 500 units will be assigned to each treatment?

   i) There is a small chance of obtaining exactly 500 heads.

   ii) It is likely that exactly 500 heads would be flipped.

   iii) It is certain that exactly 500 heads would be flipped.

3. A college professor wanted to determine if daily ingestion of a multivitamin pill would increase student performance. About half of the students in his class volunteered to take the vitamin pills every day for the entire semester. At the end of the semester the grades of those who volunteered to take the vitamins were better than the grades of those who did not volunteer to take the vitamins. For this exercise we can assume that the difference in grades was probably not due to chance. The professor concluded that the daily ingestion of multivitamins did improve student performance.

a) Describe any flaws that you see with this study.

b) Do you believe that the professor's conclusions were justified? Why or why not?

c) If you believe the study could be improved, describe how it could be improved.

4. A surgeon developed a new surgical procedure which he claimed would reduce the time needed for recuperation from surgery. He used the new procedure on some of his patients and used the standard surgical procedure on the other patients. He encouraged his colleagues to use the new procedure because the patients who had the new procedure needed less time for recuperation.

His colleagues disputed his claim that the patients who had the new procedure needed less time for recuperation. They believed that he may have selected patients for his new treatment who were generally healthier than the other patients. The surgeon plans to do another experiment. How would you suggest that he modify his procedure to convince his colleagues of the effectiveness of the new treatment?

5. A psychologist wanted to determine if marijuana increased the ability of students to write a composition on a given topic. A double-blind randomized experimental design was used. The marijuana was given as an ingredient in brownies that were given to students in the marijuana group. Individuals in the control group were given brownies that were similar in taste, texture, and appearance to the marijuana brownies. An English professor who did not know what treatment was given, graded the compositions. The results indicated that there was little difference in average writing ability between the two groups.

This results was somewhat surprising because a prior study showed that marijuana ingestion decreased writing ability. The prior study was identical to this study except, in the prior study, the English professor knew which students ate the marijuana brownies. Give one possible explanation for the different conclusions.

6. The long term health effects of cigarette smoking are well known. However, suppose a researcher wanted to determine the average amount of impairment of lung function that non-smokers would experience after two years of smoking. He made arrangements to do a randomized experiment at a prison where 30 prisoners agreed to participate in the study. Fifteen of the prisoners were randomly assigned to smoke a pack of cigarettes every day while the other prisoners were assigned to chew a pack of gum each day. After two years the lung function of all prisoners was recorded.

a) What this a double-blind study? Explain.

b) Did the control treatment appear to be reasonable considering the objectives of the experiment?

c) Assuming that the prisoners complied with the treatment, do you believe that the experiment was ethical? Explain.

d) If the activity of the prisoners was not closely monitored, do you believe the prisoners would comply with the treatments? Explain.

7. A group of physicians decided that a new treatment for bladder cancer appeared promising. They decided to evaluate the relative effectiveness of the new treatment by comparing the proportion of patients who survived more than 5 years under the new treatment to the proportion who survived more than 5 years under the standard treatment. A double-blind randomized experiment was used with approximately 50 patients assigned to each group.

a) In this study the new treatment is compared to the standard treatment. What practical and ethical problems would be encountered if a fake treatment was used as the control treatment?

b) Many cancer treatments have serious adverse effects that are well known to the nurses and physicians. In your opinion will it be practical to do a double-blind study ? Explain.

1.4   Using Summation Notation to Compute the Average

In this section we will describe how to use subscripted variables and summation notation to compute the average. Summation notation is commonly used in statistics because it makes it easy to quickly and precisely state what calculations should be done. In this section we will assume that we have obtained n measurements and that these measurements can be listed as $x_1, x_2, ..., x_n$. Suppose we have recorded the height of four students, in inches, as {66, 70, 69, 65}. Thus, $n = 4$ and the subscripted variables are $x_1 = 66$, $x_2 = 70$, $x_3 = 69$, and $x_4 = 65$. We know that the average, or **mean**, equals the sum of the measurements divided by the number of measurements. The mean is written as $\bar{x}$, which in this example is

$$\bar{x} = \frac{x_1 + x_2 + x_3 + x_4}{4} = \frac{66 + 70 + 69 + 65}{4} = 67.5 \text{ inches.}$$

It can become very tedious to list every term in the expression for $\bar{x}$ if n is large. Summation notation can greatly reduce the length of these expressions. Suppose we obtain measurements from a sample of 47 students and we use x_i to indicate the measurement obtained from the ith member of the set of measurements. Then the sum of these 47 measurements can be written as

$$\sum_{i=1}^{47} x_i = x_1 + x_2 + \ldots + x_{47},$$

where $\sum$ is the Greek letter sigma which indicates that the values following the summation must be added. The "i" that appears below the summation sign indicates that we must add the values of x_i, where the "i" will vary. The number that appears below the summation sign indicates the lower limit of the summation and the number that appears above the summation sign indicates the upper limit of the summation. For example, if we wanted to sum the 3rd, 4th, and 5th measurements we could write that sum as

$$\sum_{i=3}^{5} x_i = x_3 + x_4 + x_5.$$

Using summation notation we can write the general formula for the mean, or average, as

$$\bar{x} = \frac{\sum_{i=1}^{n} x_i}{n} = \frac{x_1 + \ldots + x_n}{n}.$$

If we have 47 measurements, we can now write the mean of the 47 measurements as

$$\bar{x} = \frac{\sum_{i=1}^{47} x_i}{n} = \frac{x_1 + \ldots + x_{47}}{47}.$$

The mean is one of the most commonly used descriptive statistics. It is easy to calculate and states succinctly the middle of the measurements.

A data set can also be described by making a **dot diagram.** To understand how these are constructed, consider the height data, in inches, for five students : $x_1 = 64, x_2 = 65, x_3 = 67, x_4 = 69, x_5 = 75$.   A dot diagram is shown in Figure 1.2 for this data, where each dot represents one observation.   The mean, which is $\bar{x} = 68$, is indicated by a dark triangle below the line.   It can be shown that the mean is the point that balances the dots on the dot diagram.

Figure 1.2 Dot Diagram

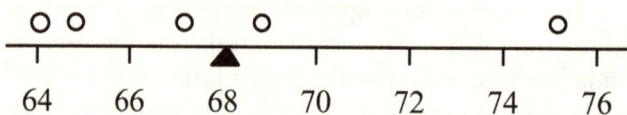

Showing a dot diagram has several advantages over showing an average value. The dot diagram in Figure 1.2 shows that there is a great deal of variability in height, and that one student is much taller than the average. In the next section we will investigate the variability in a sample.

Exercise Set C

1. Let $x_1 = 3.0$, $x_2 = 4.2$, $x_3 = 8.7$, $x_4 = 10.9$.

    a) Compute $\sum_{i=1}^{4} x_i$.

    b) Compute $\sum_{i=2}^{3} x_i$.

2. Using summation notation, write the sum $x_2 + x_3 + x_4 + x_5$.

3. A researcher selects a sample of 6 students from a population of all students on campus. The number of cups of coffee ingested per day during finals week is recorded for these 6 students. The measurements are

$$x_1 = 2, \; x_2 = 1, \; x_3 = 12, \; x_4 = 0, \; x_5 = 4, \; x_6 = 2 \; .$$

a) Using summation notation, write the definition of the mean and compute it.

b) Make a dot diagram and indicate the mean on the diagram.

4. Suppose in the previous exercise the data for the third student could not be obtained. What is the mean of the other five measurements? Compare the mean to the mean obtained in the previous exercise. Did the mean change greatly? Why did the mean change?

5. A quality control inspector in a soft drink bottler selects 20 cans of pop at random and empties the contents into a container. He determines that the 20 cans contain a total of 224 ounces of pop. What is the average amount of pop in these cans?

6. Fifteen people get on an elevator which has a maximum capacity of 2500 pounds. Their average weight is 180 pounds. Have they exceeded the maximum capacity of the elevator?

7. Five students are in a room. Their average height is 70 inches. After one student leaves the room the average height becomes 68 inches.

a) Was the person who left the room taller than 70 inches? No computations should be necessary.

b) Compute the height of the person who left.

1.5 The Sample Standard Deviation

The mean is the most popular measure of the location of the center of the measurements. Unfortunately, a **measure of central tendency** such as the mean does not always provide a true description of the measurements in the data set. To provide a more complete description of the measurements, we need to describe the **variability** in the data set.

An example may help to illustrate the importance of variability. Suppose you lived near two lakes and you wanted to learn how to swim. Before attempting to swim in either lake you row a boat around both lakes in order to take a sample of 7 measurements from each lake. The data for both lakes are summarized in Figure 1.3 in a dot diagram.

Figure 1.3

Lake 1

O O O O O O O

| | | | | | | |
1   2   3   4   5   6   7

x=depth in feet

Lake 2

O O O   O   O O O

| | | | | | | |
1   2   3   4   5   6   7

x=depth in feet

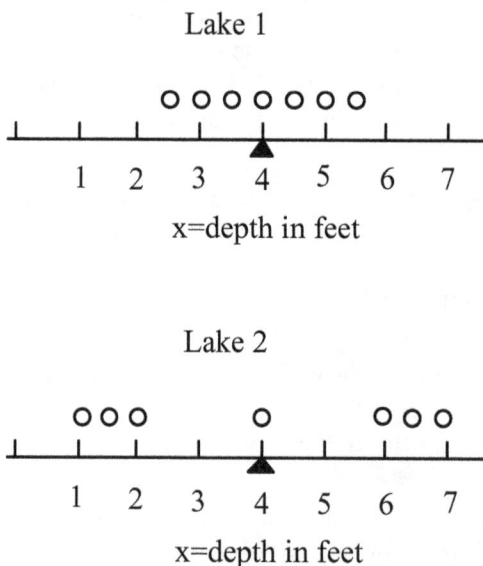

Note that the average depth equals four feet for both lakes. Although the average depths are equal, most people would prefer to learn to swim in Lake 1. It appears that Lake 2 is too shallow in some areas and too deep in other areas. The mean was not useful in making the decision because we needed to know something about the **variability** of the depth measurements.

Several **measures of variability** have been proposed.   The **range**, which is the difference between the largest and smallest measurement, is not a good measure of variability because it is difficult to interpret. The problem with the range is that it depends only on the smallest and the largest value. If a single measurement is unusually large or small the range will be large. Consider the dot diagram of the measurements of the depths of two lakes shown in Figure 1.4. Note that the range of the measurements in Lake A is greater than the range in Lake B. However, it is not clear that there is greater variability in Lake A, and a reasonable person may prefer to learn to swim in Lake A, realizing that is as easy to drown in 7 feet as in 8 feet of water. Consequently, the range is not a useful measure of variability and is seldom used to indicate the overall variability.

Figure 1.4 The difficulties associated with using the range as a measure of variability

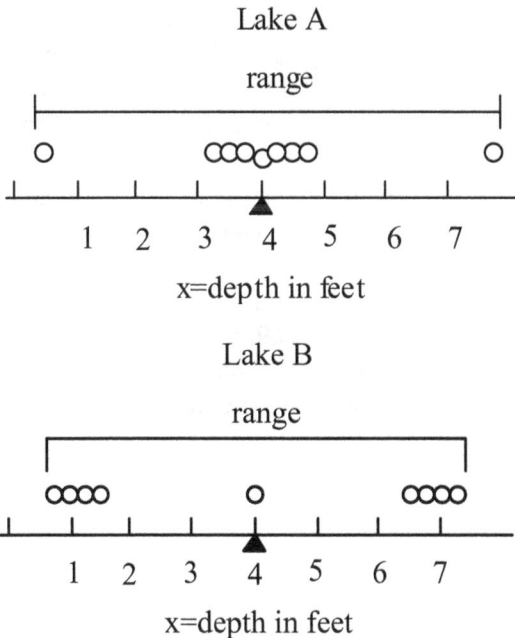

Lake A

range

x=depth in feet

Lake B

range

x=depth in feet

The most popular measure of variability is the **standard deviation**, which we will denote by s. To compute the standard deviation we first calculate the deviations between each measurement and the mean of the measurements so that we obtain $(x_1 - \bar{x}), (x_2 - \bar{x}), \ldots, (x_n - \bar{x})$. If we try to obtain a measure of variability by summing these deviations we will obtain a sum of zero. Instead, we square the deviations, which will remove the negative signs, before summing. The formula for the standard deviation is

$$s = \sqrt{\frac{\sum_{i=1}^{n}(x_i - \bar{x})^2}{n-1}} = \sqrt{\frac{(x_1 - \bar{x})^2 + \ldots + (x_n - \bar{x})^2}{n-1}}$$

It may appear strange that in the computation of s we sum the squares of the n deviations $(x_1 - \bar{x}), (x_2 - \bar{x}), \ldots, (x_n - \bar{x})$ and divide by $n-1$ rather than by n. However, it can be shown that using n in the denominator produces a standard deviation that tends to underestimate the standard deviation in the population that was sampled. Therefore, by using $n-1$ in the denominator instead of n we introduce less bias into the estimate. The formula containing $n-1$ in the denominator is the most common formula for the standard deviation and most scientific calculators use this formula to calculate the standard deviation. If you plan to use the standard deviation function on a calculator you should carefully check the results to ensure that it uses $n-1$ in the denominator.

When computing the standard deviation by hand it is helpful to compute the deviations   because they can tell us which observations contributed the most to the standard deviation. We can illustrate the calculations by computing the standard deviation of the heights of four students who had heights of 66, 70, 69, and 65 inches. This data set is shown in Figure 1.5 along with the mean, which is indicated by the triangle at 67.5. The deviations from the mean values are indicated in the figure as distances from the observations to the mean.

Figure 1.5 Deviations from the Mean

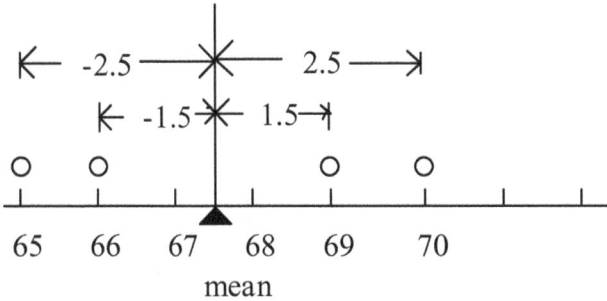

mean

The standard deviation, which is a function of the deviations from the mean, is

$$s = \sqrt{\frac{\sum\limits_{i=1}^{4}(x_i - \bar{x})^2}{4-1}} =$$

$$= \sqrt{\frac{(66-67.5)^2 + (70-67.5)^2 + (69-67.5)^2 + (65-67.5)^2}{4-1}}$$

$$= \sqrt{\frac{(-1.5)^2 + (2.5)^2 + (1.5)^2 + (-2.5)^2}{3}} = \sqrt{5.66} = 2.38 \text{ inches.}$$

It is important to note that the standard deviation is expressed in the original units, which in this case was inches.

We will also need to develop some intuitive understanding of the standard deviation. At this point the standard deviation s=2.38 inches should be seen as a measure of the amount of variability in the sample. If someone else had a set of measurements with a standard deviation of $s = 4.03$, we could say that our sample was less variable than theirs.

One important property of the standard deviation can be illustrated with the following example. Suppose that the student who took the height measurements of 66, 70, 69, and 65 inches did not notice that the ruler started at 2 inches instead of 0 inches. That is, the ruler was defective and overestimated the true height by 2 inches. Thus, the true heights were 64, 68, 67, and 63 inches. Can we correct our sample mean and standard deviation without redoing the calculations?

Since each x_i will be decreased by 2 inches to correct the measurement we can easily verify that $\bar{x}$ will decrease by 2 inches. This makes sense because $\bar{x}$ is a measure of location and the location has decreased by 2 inches. However, the standard deviation should not change since it is a measure of variability and the variability has not changed. This can be verified by recoding the measurements in the dot diagram in Figure 1.5 and noting that the derivations $(x_i - \bar{x})$ do not change with a change in location. Therefore the mean would change to $\bar{x} = 65.5$ and the standard deviation would remain at $s = 2.38$.

The **variance** (s^2) is another popular measure of variability. As the notation implies the variance is simply the square of the standard deviation. It is somewhat more difficult to work with than the standard deviation and is expressed in the original units squared. We will use it occasionally in the text. The standard deviation, which can always be computed as the square root of the variance, will be our main measure of variability.

At this point, the meaning of the standard deviation may not be perfectly clear. However, we will use the standard deviation in the next section and offer some further explanation of it as a measure of variability. You may wonder why it is necessary to square the deviations in the formula for s. If the deviations were not squared, they would always sum to zero. It is possible to sum the absolute value of the deviations, and a measure of variability has been proposed based on absolute values, but it has not been as popular as the standard deviation.

Exercise Set D

1. An aerobics instructor wanted to describe the ages of the four students in her class. The ages were $x_1 = 22$, $x_2 = 23$, $x_3 = 20$, and $x_4 = 31$ years of age.

   a) Compute the mean.

   b) Compute the standard deviation.

2. Consider the dot diagrams for the entrance exam scores for 9 students at three universities.

Entrance Exam Scores

University A

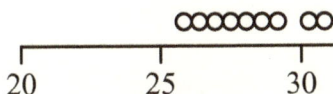

20	25	30

University B

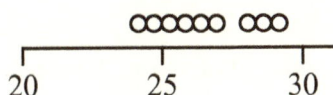

20	25	30

University C

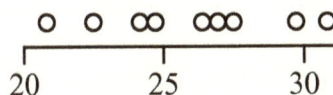

20	25	30

a) Which university had the greatest average entrance exam score?

b) Which university had the greatest variability in entrance exam scores?

3. Two students, who live in separate dormitories, obtained the weights of 5 students in their dormitories. The measurements, after sorting from lowest to highest, were:

Dormitory   1:   140,   150,   155,   160,   170

Dormitory   2:   130,   150,   155,   160,   180

Which sample had the larger standard deviation? No calculations should be necessary.

4. Compute the standard deviations for the weights listed in the previous exercise for both dormitories. Compute the variances.

5. In the previous exercise, which observations contributed most to the sum of the squared deviations. For dormitory 2 how much did the squared deviations for the smallest and largest observations contribute to the total sum of squared deviations?

6. Consider the following eight observations:

   9,   11,   9,   11,   9,   11,   9,   11

   a) What is the mean?

   b) Describe the squared deviations from the mean.

   c) What is the standard deviation?

   d) If the data set consisted of 500 "9"s and 500 "11"s, what would be the approximate value of the standard deviation.

7. A researcher recorded the body temperature of 12 cyclists after one hour of cycling. To simplify the calculation of the standard deviation she subtracted 90° F from each measurement. Will she obtain the correct standard deviation? Why or why not?

8. Suppose a contest requires you to correctly pick 6 numbers between 1 and 50 in order to win the grand prize.

   a) Without repeating any numbers, how would you pick numbers to minimize the standard deviation? There may be more than one correct answer.

   b) Which numbers would you pick to maximize the standard deviation?

9. The ages of the five students who graduated from a very small high school were:

   18, 16, 17, 18, 21

a) Compute the average age and the standard deviation of the ages.

b) At their 10 year reunion their ages were:

28, 26, 27, 28, 31

State their average age and the standard deviation of their ages at the time of the reunion.

c) You should have been able to quickly state the standard deviation of their ages at the time of the reunion. Explain why it is easy to calculate the standard deviation at the time of the reunion if the standard deviation at graduation is available.

1.6 Histograms and the Empirical Rule

The mean describes the center of the measurements and the standard deviation describes the variability of the measurements. However, sometimes these two statistics don't really tell the whole story about the measurements. The **frequency histogram** provides a simple graphical method of describing the measurements.

To illustrate how we construct a frequency histogram, suppose we obtain the mid-term exam scores for 50 students who took a mid-term exam. We can obtain a very good picture of how these students performed by constructing a frequency histogram. To construct this type of histogram we first must decide on the class intervals that would produce an attractive and informative histogram. Class intervals are defined by the boundaries, which are usually determined by the person who is analyzing the data. If too many intervals are used, the resulting histogram will have an irregular appearance. If too few intervals are used, the resulting histogram will not convey much information to the reader. In our example, we would determine the class boundaries and then count the number of measurements that fall into each interval. The frequency histogram is constructed by making the height of the vertical bars above each class represent the class frequencies.

This process is illustrated in Table 1.1 for the 50 exam scores. The "2" under "Class Frequencies" indicates that 2 students had scores in the range of 20-29. The frequency histogram is constructed by using the class intervals on the horizontal axis and the class frequencies determine the heights of the vertical bars. The frequency histogram for the exam scores is given in Figure 1.6.

Table 1.1 Class Frequencies And Class Relative Frequencies
For Mid-Term Exam Results

Class	Intervals	Frequencies	Relative Frequencies
1	20-29	2	.04
2	30-39	0	.00
3	40-49	1	.02
4	50-59	4	.08
5	60-69	6	.12
6	70-79	13	.26
7	80-89	16	.32
8	90-100	8	.16

This frequency histogram shows several features of the data. It shows that almost all students scored above 50 on the exam, that a few students had very low scores, and that many students had scores between 70 and 89. The mean and the standard deviation would not indicate these features as clearly as does the histogram. The histogram is the best method of displaying the measurements in a large data set if you want to see how the measurements are really distributed.

Figure 1.6 Frequency Histogram for Exam Scores

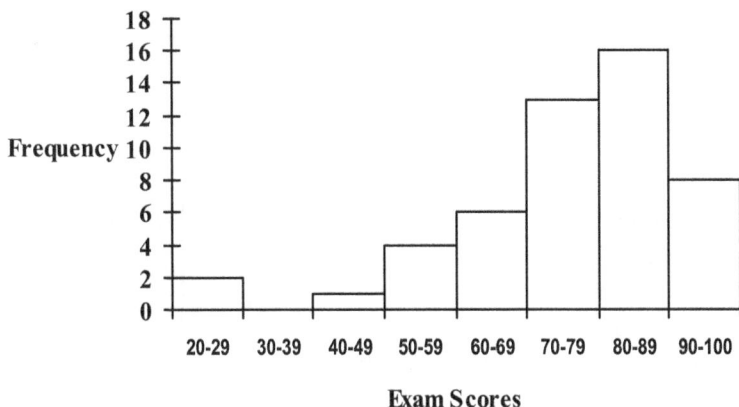

It is often more convenient to plot the proportion of students who fall into these classes instead of plotting the frequencies. A **relative frequency histogram** is a graphical display of these proportions. If we have n measurements in the data set, then the relative frequencies are obtained by

$$\text{relative frequency} = \frac{\text{class frequency}}{n}.$$

For example, in the first class the class frequency was 2, the total number of measurements was 50, so the relative frequency was $2/50 = 0.04$. The relative frequencies in our example are tabulated in Table 1.1 and are plotted in Figure 1.7. Note that the relative frequency histogram has the same appearance as the frequency histogram in Figure 1.6 except that the vertical axis is changed to reflect the proportion in each class.

Figure 1.7 Relative Frequency Histogram for Exam Scores

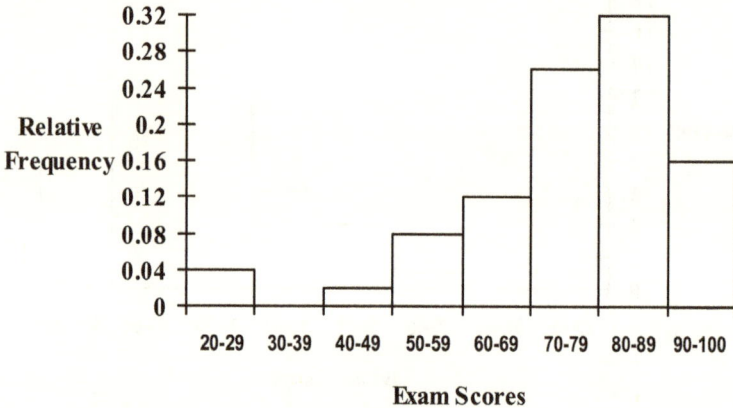

We can learn a great deal about the measurements by the careful study of a histogram. Consider the relative frequency histogram of cigarettes smoked per day as shown in Figure 1.8. To determine the proportion of adults who smoke 20 or fewer cigarettes we could add the proportion for the "16-20" interval to those to the left of the "16-20" interval to obtain $.50 + .11 + .08 + .10 = .79$. To obtain the proportion of adults who smoke more than 5 cigarettes per day we add the proportion for the "6-10" interval to the proportions for the intervals to the right of the "6-10" interval. The sum of these intervals would equal .5 .

A much quicker method of calculating this probability uses the fact that the sum of these proportions must equal 1.00 . Since half of the adults smoke 5 cigarettes or fewer per day we know immediately that half must smoke more than 5 cigarettes per day. The fact that the proportions must sum to 1.00 is often used in this text. If we wanted to roughly compute the proportion who smoke fewer than 35 cigarettes, we would note that the proportion who smoke more than 35 cigarettes is approximately .03, so the proportion who smoke fewer than 35 cigarettes must be approximately 1.00-.03=.97 .

Figure 1.8 Relative Frequency Histogram of Cigarettes Smoked per day

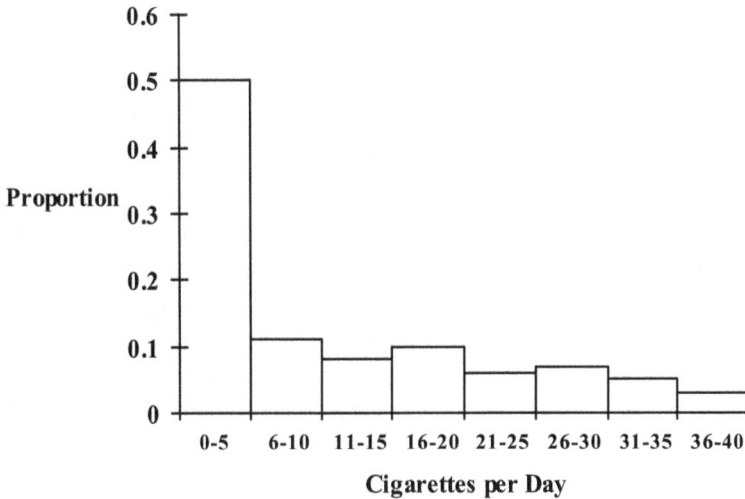

The histogram shown in Figure 1.8 has a long right tail which shows the proportion of individuals who smoke many cigarettes per day. This histogram, which has a long right tail and no left tail is said to be **skewed to the right**. A histogram that has a long left tail and a short right tail is said to be **skewed to the left**. If a histogram is not skewed to the right or to the left we say that it is **symmetric**. Figure 1.9 gives examples of skewed and symmetric distributions. Many real data sets have histograms that are skewed to the right. Symmetric, or nearly symmetric, histograms are not uncommon, but real data sets rarely have histograms that are skewed to the left.

Figure 1.9 Skewed and Symmetric Histograms

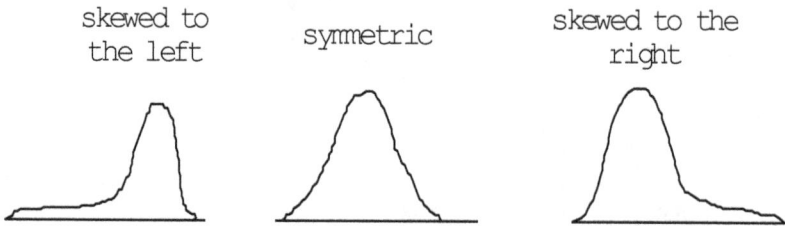

skewed to
the left

symmetric

skewed to the
right

Sometimes we will need to describe the lengths of the tails of the distribution. In Figure 1.10 we observe three distributions, one which has shorter tails than the normal and one which has longer tails than the normal. The proportion of the observations that are found in the tails is a measure of the **kurtosis** of the histogram. A long tailed distribution has a high degree of kurtosis while a distribution that has shorter tails than the normal has a small degree of kurtosis. The distributions need not be symmetric, any distribution with long tails has a great deal of kurtosis.

Figure 1.10 Kurtosis in Three Distributions

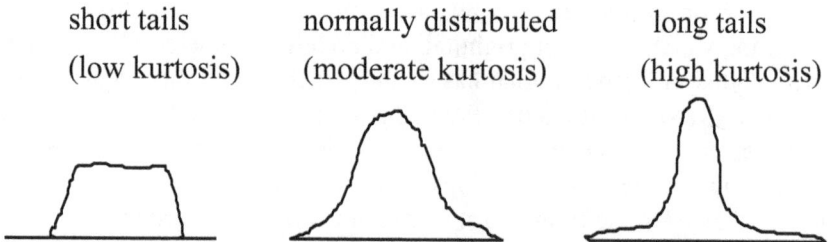

short tails
(low kurtosis)

normally distributed
(moderate kurtosis)

long tails
(high kurtosis)

An example of a symmetric histogram having moderate kurtosis is given in Figure 1.11. Suppose this histogram describes the histogram of heights of adult females in the United States. If the histogram of the data is symmetric, the mean will approximately equal the median. Furthermore, if the histogram has the **bell shape** of the histogram in Figure 1.11 we say the data is approximately **normally distributed**. For normally distributed data it can be shown that approximately 68% of the measurements fall within one standard deviation of the mean and that approximately 95% fall within two standard deviations of the mean.

Figure 1.11 Relative Frequency Histogram for the Heights of Adult Females

mean=64 inches, standard deviation=2.5 inches

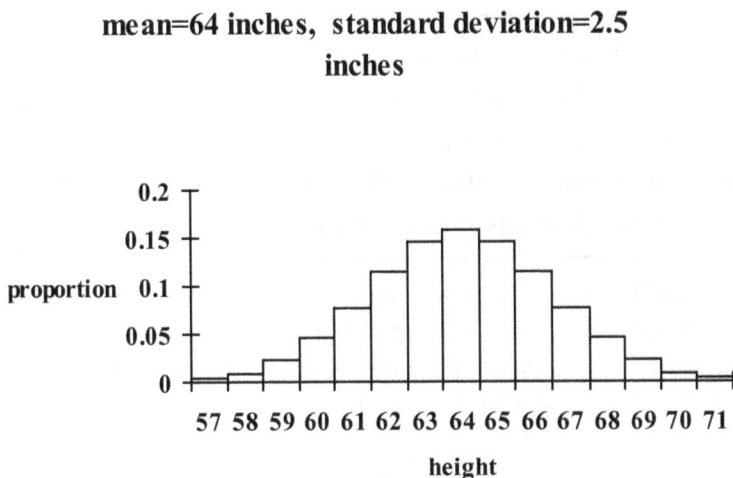

Thus, in our example, which has $\bar{x} = 64$ and $s = 2.5$, the measurements within one standard deviation of the mean are those in the interval $(64 - 2.5, 64 + 2.5) = (61.5, 66.5)$, and we should find 68% of the adult females to have heights in that interval. You should verify this by summing the proportions associated with the appropriate intervals. Similarly, we should find 95% of adult women to have heights in the interval $\left[64 - 2(2.5), 64 + 2(2.5)\right] = (59, 69)$. The interpretation of the standard deviation applies to bell shaped or normal histograms and applies approximately to histograms that are roughly normal in shape. The interpretation of the standard deviation does not apply to highly skewed histograms or to histograms that have very long or short tails.

These guidelines allow us to interpret the standard deviation for histograms that are approximately normal. That is, if the distribution of weight is roughly normal and the standard deviation is approximately 20 pounds, then we know that 95% of the measurements will be within about 40 pounds of the mean, and that about 70% will be within 20 pounds of the mean. To determine these percentages we only need to know the standard deviation and the shape of the distribution. We will often use these guidelines for nearly normal distributions and will call them the **empirical rule.**

> The empirical rule states that if the histogram is approximately normal in appearance, then approximately 68% of the measurements fall within one standard deviation of the mean, and approximately 95% of the measurements fall within two standard deviations of the mean.

It should also be noted that the empirical rule implies, for an approximately normal histogram, that approximately $68\%/2 = 34\%$ of the observations are between the mean and one standard deviation above the mean. For the measurements described with the histogram shown in Figure 1.10, we should find approximately 34% between 64 inches and 66.5 inches. Similarly, we should find about $95\%/2 = 47.5\%$ between the mean and two standard deviations above the mean. The normal histogram often appears in statistical work, and we shall make extensive use of it in this text. We will consider the normal distribution in greater detail in Chapter 3.

Exercise Set E

1. A sociologist took a sample of 20 students and asked them to estimate how many hours of television they watched in the last week. The measurements were recorded as:

   8, 2, 2, 5, 56, 7, 9, 4, 5, 6, 11, 1, 0, 47, 5, 0, 8, 7, 5, 9

   a) Select some class intervals and construct a relative frequency histogram.

   b) Compute the mean of the measurements.

   c) How many measurements are below the mean? How many above?

d) In your opinion does the mean value reflect the viewing time of the typical students in this sample? Why or why not?

2. A relative frequency histogram of the number of children under 18 in a family is given in the figure below.

Relative Frequency Histogram of the Number of Children under 18

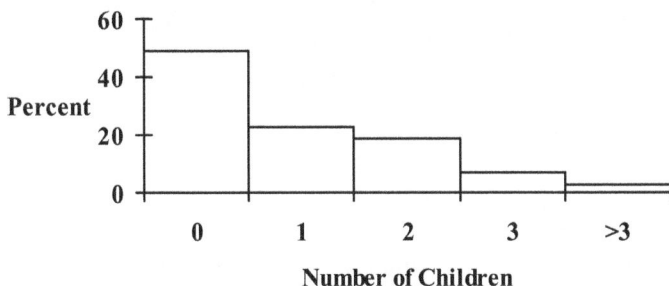

a) Approximately what proportion of families had exactly one child under the age of 18?

b) Approximately what proportion of families had 3 or more children under the age of 18?

    3%   10%   50%   90%

c) Approximately what proportion of families had 1 or 2 children under the age of 18?

    10%   20%   30%   40%

3. The histogram of the number of children under 18 displayed in the previous exercise is

skewed to the left   symmetric   skewed to the right

4. Consider the relative frequency histogram of the weights of 7th grade boys. The midpoints of the intervals are specified in this histogram. Assume that the mean is 100 pounds and the standard deviation is 10 pounds.

proportion

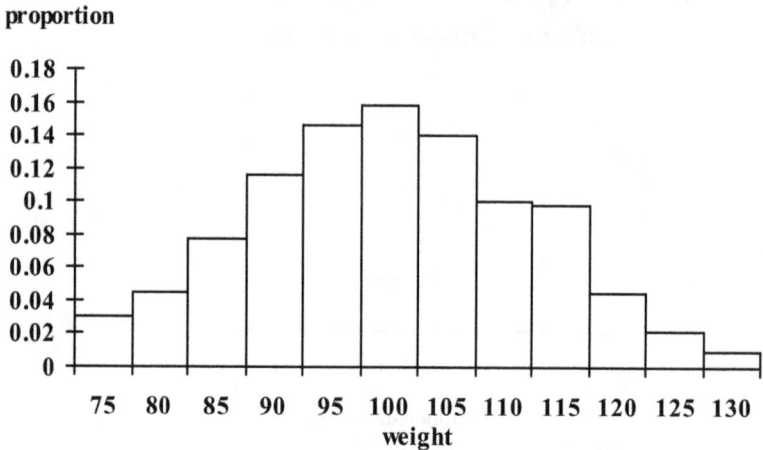

a) Approximately what proportion of boys weigh between 88 and 108 pounds?

b) Approximately what proportion of boys weigh between 78 and 118 pounds?

5. Suppose the scores on a college entrance examination had a relative frequency histogram that was approximately normal in appearance. The scores had a mean of 500 and a standard deviation of 100.

a) Draw a rough sketch of the histogram of these scores.

b) Approximately what proportion of scores fell between 400 and 600?

c) Approximately what proportion of scores fell between 400 and 500?

d) Approximately what proportion of scores fell between 300 and 700?

e) Approximately what proportion of scores were greater than 700?

6. This relative frequency histogram shows weights for high school soccer players. The midpoints of the intervals are specified in this histogram.

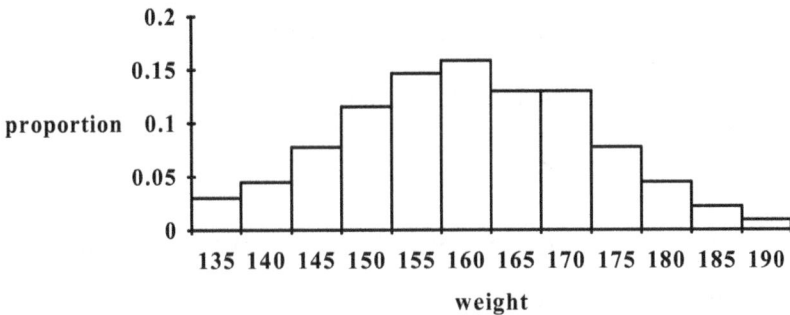

a) Is this histogram roughly normal in shape, or is it skewed to the right?

b) The standard deviation can be roughly estimated from the figure by using the empirical rule. It is approximately equal to:

    2.5 pound   5 pounds   12 pounds   25 pounds

7. For the histogram

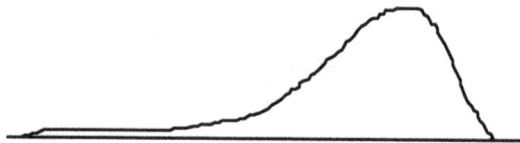

the distribution is:

skewed to the left

approximately symmetric

skewed to the right

1.7 Sample Percentiles and the Interquartile Range

Percentiles and the Median

Percentiles are often used to report the results of standardized tests. For example, a student who took a test and scored at the 70th percentile received a grade that exceeded 70% of the other students who took the test. A student who scored at the 98th percentile received a score that exceeded 98% of the other students and, consequently, was lower than the score received by 2% of the students.

To roughly calculate a percentile we take the set of measurements $x_1, x_2, \ldots x_n$, and we sort them to obtain the **order statistics** $x_{(1)}, x_{(2)}, \ldots, x_{(n)}$. For example, if we had the $n = 4$ observations $[x_1 = 7,\ x_2 = 9,\ x_3 = 4,\ x_4 = 11]$ we would sort these to obtain the order statistics $[x_{(1)} = 4,\ x_{(2)} = 7,\ x_{(3)} = 9,\ x_{(4)} = 11]$. If we have $n=100$ measurements and we want to approximate the 90th percentile we find the value that exceeds 90% of the measurements. After sorting we find that the 90th percentile is $x_{(90)}$, because $x_{(90)}$ must be greater than or equal to 90 of the 100 measurements. This rough method of calculating percentiles will work adequately if n is large and if we are satisfied with a rough estimate of the percentile.

In most cases we will want to use a more precise definition of percentiles. Some notation will be convenient. Let p equal the proportion such that the $100p$th percentile is to be computed. For example, if we want the 90th percentile we use $p = .9$. To compute the $100p$th percentile we compute the value $p(n+1)$. If $p(n+1)$ is an integer we use the order statistic $x_{(p(n+1))}$.   Thus, if $n = 9$ the 90th percentile is $x_{(p(n+1))} = x_{(.9(9+1))} = x_{(9)}$.   If $p(n+1)$ is not an integer we must interpolate between the order statistics having indexes closest to $p(n+1)$. For example, if $n = 4$ and we want the 50th percentile, we need to interpolate between order statistics closest to $p(n+1) = .5(4+1) = 2.5$.   That is, we need to take the average of $x_{(2)}$ and $x_{(3)}$.

The **median** is defined as the 50th percentile so that roughly 50% of the observation are less than the median and 50% are greater than the median. If $n = 7$ the median is $x_{(p(n+1))} = x_{(.5(7+1))} = x_{(4)}$ which makes sense because it is the middle observation. The **first quartile** is the 25th percentile and the **third quartile** is the 75th percentile. If $n = 8$ and we want to compute the 75th percentile, we first compute $p(n+1) = .75(9) = 6.75$, so that the 75th percentile is between $x_{(6)}$ and $x_{(7)}$. Since $p(n+1) = 6.75$ we need to interpolate a value 75% of the distance between $x_{(6)}$ and $x_{(7)}$. Consequently, the exact value of the 75th percentile would be $x_{(6)} + .75(x_{(7)} - x_{(6)})$.

Both the median and the mean are measures of the center of location of the histogram, and are often approximately equal. However, with skewed data the median may not approximate the mean. In Figure 1.12 we see a relative frequency histogram that is skewed to the right. This indicates that some measurements are much greater than the median but no measurements are much smaller than the median. Recall that the mean, which is marked by a small triangle, is the balancing point of the histogram.   Generally, the mean is greater than the median if the histogram is skewed to the right. The mean is larger than the median because it is greatly influenced by the measurements in the right tail having very large values whereas the median is not affected by the magnitude of these large measurements.

Figure 1.12 Histogram for data that are skewed to the right

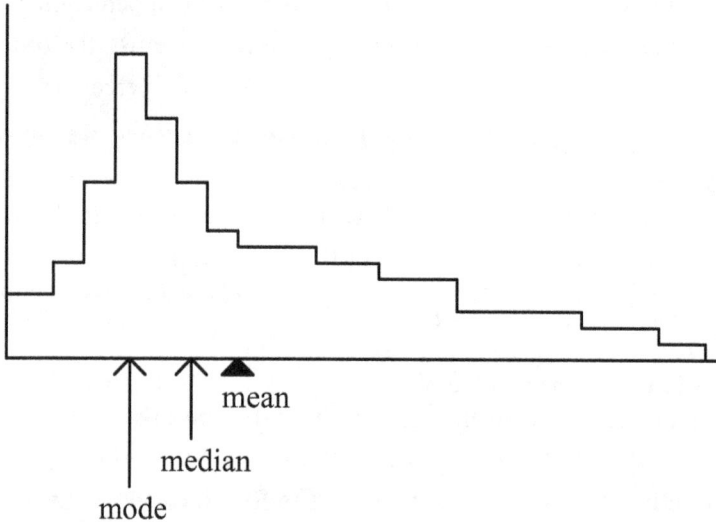

The Trimmed Mean

Means and medians are the most popular measures of the location of the histogram. The mean is easy to compute and can be used to compute an estimate of a population total. The median has the advantage of being insensitive to very large or small values. However, there are other measures of location that combine some of the best features of the mean and median.

We define the 10% trimmed mean as the mean of the observations that remain after we throw out the largest 10% and the smallest 10% of the measurements. That is, we average all order statistics having indexes between $.1n$ and $.9n$. For example, if $n = 25$ so that $.1n = 2.5$, the two smallest values, $x_{(1)}$, and $x_{(2)}$, are removed and the two largest values, $x_{(24)}$ and $x_{(25)}$, are removed. We then average the remaining 21 measurements. This statistic has the advantage of not being influenced by a few unusual observations. Although it is also possible to define a 20% trimmed mean in a similar way by throwing out the largest 20% and smallest 20% of the observations, we will use only the 10% trimmed mean because it appears to be the most popular of the trimmed means.

The Mode

The **mode** is occasionally used to describe the data. The mode is the most frequent value in the distribution that, on a histogram, will be indicated by a peak. The mode is indicated on Figure 1.12 for a histogram. Note that the mode is the value that is at the tallest point on this histogram and is slightly less than the median. However, with many data sets the mode is not too useful as a measure of central tendency. Take another look at Figure 1.8, which is the relative frequency histogram of cigarettes smoked per day. The mode would be in the range 0-5 cigarettes because that is the most common response. If we allowed for a response of zero on the questionnaire about smoking we would obtain a mode of zero because most people do not smoke. In this case the mode of zero is not very informative. Indeed, for many data sets the mode is not as useful as the mean or the median, so we will not use it in the remainder of this text. However, it is necessary to understand the definition because the mode is occasionally used as a descriptive statistic.

The Interquartile Range

The standard deviation is the most popular measure of variability, but another popular measure of variability is the **interquartile range (*IQR*)**, which is defined as

IQR=75th percentile - 25th percentile.

To illustrate this measure of variability consider the following data set of systolic blood pressure measurements obtained from a sample of 9 individuals. After the data are sorted we obtain the order statistics:

106, 108, 114, 115, 118, 120, 121, 125, 170

In order to compute the IQR we need the 25th and 75th percentiles. To compute the 25th percentile we first compute $p(n+1) = .25(9+1) = 2.5$ which implies that the 25th percentile is the average of the 2nd and 3rd order statistics, which is $\frac{108+114}{2} = 111$.

For the 75th percentile $p(n+1) = .75(9+1) = 7.5$ so the 75th percentile is the average of the 7th and 8th observations, which is 123.   Therefore $IQR = 123 - 111 = 12$.

Consider the histogram for the weights of female undergraduates shown in Figure 1.13. The 5th, 25th, 50th, 75th, and 95th percentiles are indicated on the histogram. We can roughly estimate the IQR to be approximately $IQR = 165 - 110 = 55$ pounds. The IQR, like the standard deviation, is a measure of the variability in the data. Usually, either the IQR or the standard deviation is used as a measure of variability. Generally, if the median is used as the measure of the center of location, the IQR is usually used as a measure of variability. If the mean is usually used as the measure of the center of location, the standard deviation is usually used as the measure of variability.

Figure 1.13 Some Percentiles for a Histogram of the Weights

of Female Undergraduates

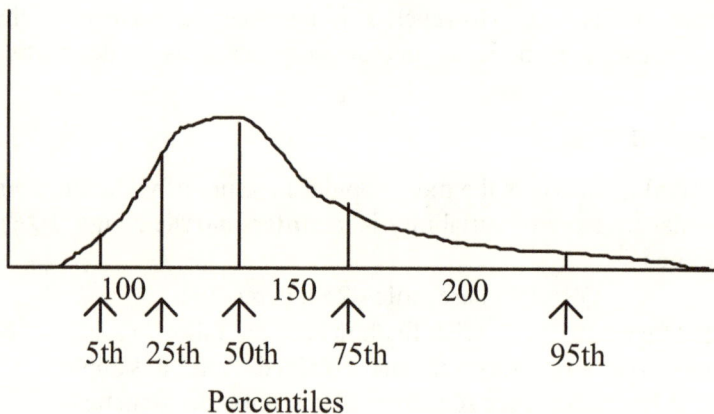

Exercise Set F

1. A researcher obtained the weights of 4799 adult males between the ages of 35 and 45. The researcher used a computer program to sort the data to obtain the order statistics $x_{(1)}, x_{(2)}, \ldots, x_{(4799)}$. Express the following as order statistics. (For example, the median is $x_{(2400)}$.)

a) The 25th percentile.

b) The 75th percentile.

c) The IQR.

2. A sociologist wanted to determine the typical duration of long distance telephone calls between teenagers. A sample of 8 teenagers was selected from the population and one call was obtained from each teenager. The duration of the calls were:

$$3, 16, 19, 180, 11, 14, 2, 8$$

a) What was the median duration of the calls?

b) What was the mean duration of the calls?

c) In your opinion should you use the mean or the median to report the typical duration. Explain your reasoning.

3. Refer to the data on the duration of telephone calls in the previous exercise. Compute the following percentiles using the method given in the text.

a) The 25th percentile.

b) The 75th percentile.

c) The IQR .

4. A study was conducted at a state university to determine the amount of coffee consumed per week. One thousand students were selected and their consumption per week was displayed with the histogram shown below.

**Relative Frequency Histogram of Coffee
Consumption in 6 Ounce Servings per Week**

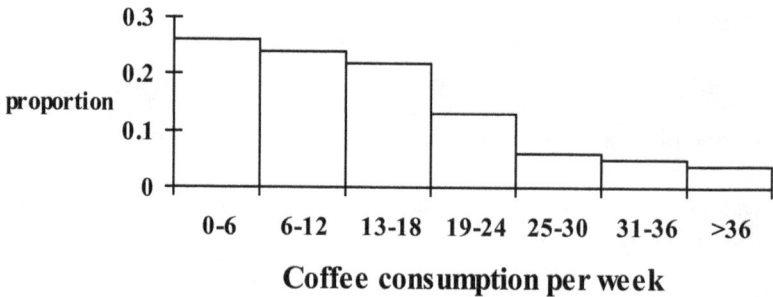

a) Use the histogram to roughly estimate the median.

b) Use the histogram to roughly estimate the 90th percentile.

c) Use the histogram to roughly estimate the interquartile range.

5. An ice skater received the following scores from ten judges at an ice skating competition:

4.1	5.8	5.8	5.9	6.0
5.7	6.0	5.9	5.7	6.0

Compute the mean and the 10% trimmed mean. Which measure best describes the location of the histogram?

6. Suppose we have obtained the following sample percentiles for weights from a large sample of males.

> 5th percentile=130 pounds
> 25th percentile=150 pounds
> 50th percentile=165 pounds
> 75th percentile=190 pounds
> 95th percentile=230 pounds

a) Compute the IQR.

b) The 80th percentile would be around:

> 80 pounds   180 pounds   195 pounds

1.8 Box Plots and Stem-and-Leaf plots

Box Plots

   Although the histogram is one of the best ways to construct a graphical display of a set of measurements, two other techniques will be described in this section that may be more suitable for certain data sets. The **box plot**, which is sometimes called a box-and-whisker diagram, is shown in Figure 1.14. This box plot is for the height data of adult women that was displayed in the histogram shown in Figure 1.11. The left side of the box is placed at the 25th percentile of the data and the right side is placed at the 75th percentile. The vertical line in the box is placed at the median. Just by looking quickly at the box we can see that the median is about 65 inches, the 25th percentile is about 62 inches, and the 75th percentile is about 67 inches. The vertical line for the median happens to be at the center of this box because the histogram is symmetric. If the distribution were skewed, the line for the median would not be in the center. Since the ends of the box are at the 25th and 75th percentiles, the IQR is represented by the width of the box. The horizontal lines on each side of the box, which are sometimes called the whiskers, extend to the most extreme points or to a distance of at most 1.5 times the IQR beyond the ends of the box, whichever is shorter. Box plots can be drawn horizontally or vertically and are included in most statistical software packages. However, there does not appear to be a standard method of displaying the outliers on box plots. If software is used to make the plot, the documentation should be consulted concerning how the values that extend beyond the whiskers are shown on the plot.

Figure 1.14 Box Plot for the Height of Adult Women

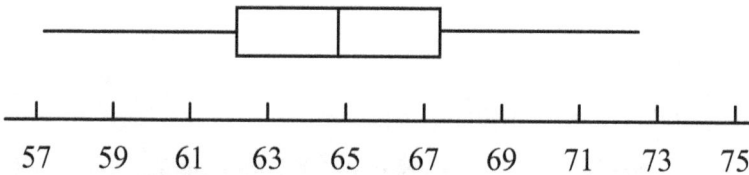

Box plots are often placed next to each other to compare several groups. For example, suppose we want to compare the entrance scores for freshman at three universities. The box plots shown in Figure 1.15 indicate that the scores at the first university are generally higher than those at the other universities. They also indicate that the scores at the second university are much more variable than those at the third university. The entrance scores at these universities could have been compared using medians and IQRs, but the box plot gives a more complete description of the entrance scores.

**Figure 1.15 Box Plots for Entrance Scores to
Three Universities**

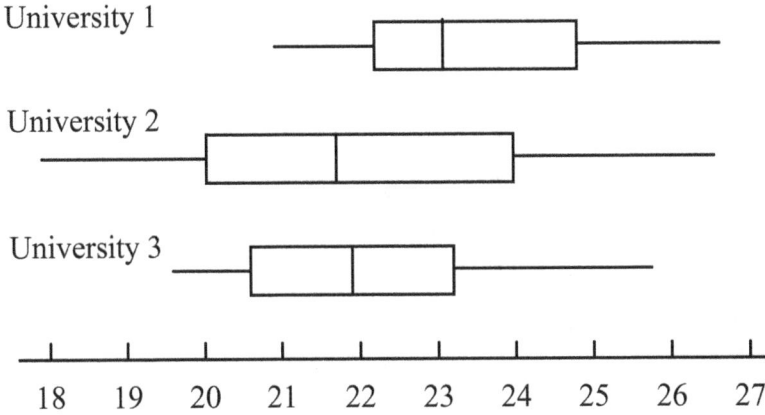

University 1

University 2

University 3

| 18 | 19 | 20 | 21 | 22 | 23 | 24 | 25 | 26 | 27 |

Stem-and-leaf plots

The **stem-and-leaf plot** is another graphical display which is suitable for certain kinds of data sets. The procedure is best illustrated with an example. Suppose we have the weights from a sample of 20 males as follows:

165, 184, 171, 141, 214, 157, 126, 162, 158, 191,
172, 157, 168, 143, 169, 174, 182, 187, 147, 156

The basic idea is to list the first two digits on the left of the vertical line shown in Figure 1.16 and then to list the last digit of all observations that have the same first digits on the right side. Since the minimum is 126 and the maximum is 214 we list the first two digits from 12 to 21. We begin our tabulation by recording the first weight of 165 by putting a 5 to the right of the vertical line on the row that has 16 to the left of the vertical line. Our second observation will be recorded as a 4 to the right of the vertical line in the row having 18 to the left. We continue to tabulate in this manner to complete the stem-and-leaf plot.

Figure 1.16 Stem and Leaf Plot for Weight Data

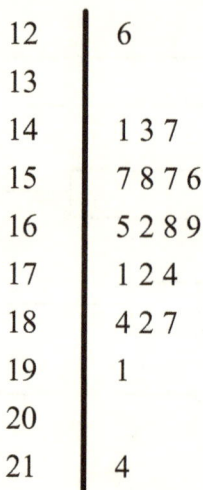

12	6
13	
14	1 3 7
15	7 8 7 6
16	5 2 8 9
17	1 2 4
18	4 2 7
19	1
20	
21	4

The stem-and-leaf plot provides a quick way to display the distribution of the data and, when turned on its side, gives the appearance of a histogram. It is an excellent graphical method for $10 \leq n \leq 50$ if the first digits can be used to define several categories. However, the stem-and-leaf plot has serious limitations. Sometimes the digits chosen to go on the left provide too few categories to make an informative plot. This would occur if we had decided to make a plot of the heights in inches that would have a 6 or a 7 as the first digit. In this situation some modification of the basic stem-and-leaf would be required. Also, the stem-and-leaf plot does not work well for large data sets having thousands of observations because the digits on the right of the vertical line quickly exceed the margin of the paper. For large data sets the basic stem-and-leaf must be modified by making each digit on the right of the vertical line represent several observations. Despite these shortcomings, the stem-and-leaf plot is a popular method of displaying data.

Exercise Set G

1. The box plots for the heights of 13 year old girls and boys are:

Box Plots for the heights of 13 year old girls and boys.

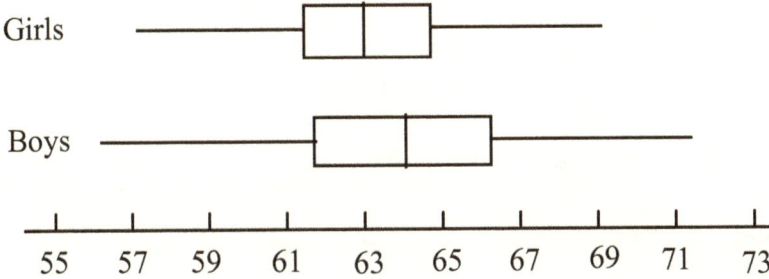

Girls

Boys

55	57	59	61	63	65	67	69	71	73

a) Is a girl of median height taller or shorter than a boy of median height?

b) Are some girls taller than some boys. Explain.

c) Is a girl who is at the 25th percentile taller or shorter than a boy who is at the 25th percentile?

d) Approximately what percentage of 13 year old boys are shorter than a 13 year old girl who is at the 75th percentile of height?

20%   40%   60%   80%

e) Compute a rough estimate of the IQR for the heights of the boys and girls.

f) As measured by the IQR, are boys heights more variable than girls?

2. The Environmental Protection Agency (EPA) tests cars to determine the number of miles they will travel on a gallon of gasoline. The order statistics of these mileage estimates for 19 cars are :

    19.1, 22.1, 22.6, 23.3, 23.5, 23.8, 23.9, 24.4, 25.1, 25.6,
    25.8, 26.3, 26.3, 28.1, 29.3, 29.7, 30.6, 32.5, 36.4

    Construct a stem-and-leaf plot for the data.

3. Using the data from the previous problem

    a) Calculate the 25th percentile, the 50th percentile, and the 75th percentile.

    b) Construct a box plot for the data.

4. A college conducted a survey of the starting salaries of their most recent graduates. They classified the graduates as business graduates, science graduates, or other graduates. The results of the survey were displayed in box plots as follows:

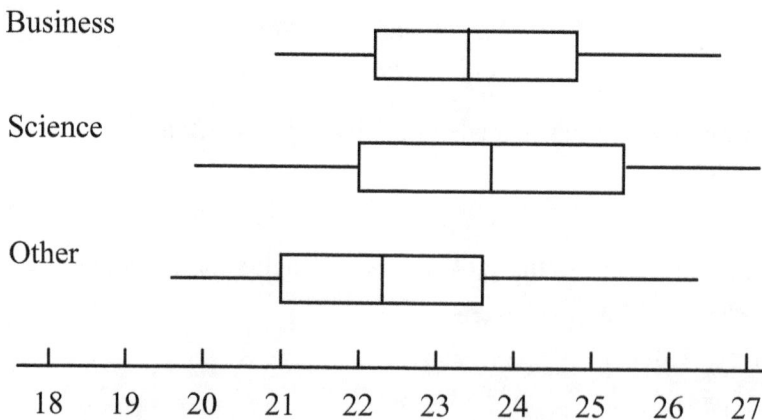

Business

Science

Other

18   19   20   21   22   23   24   25   26   27

    a) Which graduates tended to receive the lowest starting salaries?

    b) Which graduates had the most variable salaries?

c) A graduate classified in the "other" category and making the median starting salary for that category would have a salary that is at the ____th percentile of the business graduates.

1.9 Guide to Descriptive Statistics

Descriptive statistics are very useful tools that make it easy to communicate the results of surveys and experiments in a concise way. However, it pays to be cautious when interpreting these statistics. Manufacturers have been known to use statistics to overstate the advantages of their products. Politicians have been known to choose their statistics very carefully in order to promote their views. Therefore, before you attempt to interpret descriptive statistics ask yourself the following questions:

1. If the data were obtained by using a sample, how was the sampling done and what was the non-response rate? If the sampling method was not specified, or if the sampling method was not based on probability, then you should be suspicious of the results. The results could be correct with non-probability sampling, but you have no assurance that they are correct. If the survey was done using a probability method, but the response rate is low, say less than 80%, then the results should be interpreted with caution because the non-responders may have responded differently to the survey. It is wise to ignore the results from a voluntary response survey, since they may be far from the truth.

2. If the data came from an experiment, was it a randomized experiment? Were the subjects blind to the treatments that were given? If randomization was not used the results may simply reflect the difference between those people who volunteered for the treatment and those who did not. Randomized experiments are commonly used in the agricultural and biological sciences. Caution should be exercised if the usual randomization procedures were not followed.

3. Was some measure of variability, either the standard deviation or the interquartile range, published so that the variability could be determined? The results may be valid even if the variability was not stated, but is does help to have some idea of the precision of the results. Sometimes estimates of precision are stated in terms of standard errors or confidence intervals which will be studied later in this text.

4. Could a graphical method, such as a histogram or a box plot, present the data more effectively? These methods are especially helpful when two or more groups are compared in an experiment because they allow the reader to judge how much overlap was present in the experimental groups.

The sum up, it is vital that you understand how the data was collected and analyzed. You need to be aware of the difficulties of interpretation associated with observational data so that you are not so easily fooled by bogus statistical arguments.

Chapter Review Exercises

1. A researcher, who wanted to estimate the average amount of money spent per month on entertainment, sent questionnaires to 1000 residences that were selected at random. The questionnaires requested information about family members, family income, and detailed information about expenditures for recreational items. There were 87 questions on the five page questionnaire. The results for the 42 people who returned the five page questionnaire were published. The researcher would like to improve the survey next year. In a brief paragraph describe a change that you believe may improve the survey.

2. A sociologist wanted to determine the role that religion plays in the lives of the people who live in a certain community. She used a telephone survey to obtain the information from a simple random sample of people who were listed as active members of religious organization in the community. In your opinion, would the survey yield valid estimates? If not, what changes should be made to the design of the study?

3. A nutritionist used rats in an experiment to determine if a low calorie diet would increase the life expectancy of rats. The 20 lab rats that he ordered from a science supply company were delivered in two batches. The first batch of 10 rats were randomly assigned to the low calorie diet. The second batch of 10 rats were then assigned to the standard diet.

    a) What were the experimental units?

    b) Were the experimental units randomized?

    c) What could be done to improve the experiment?

4. One hundred middle aged men who were experiencing some hair loss volunteer to participate in a trial of a drug designed to reduce hair loss. The men were assigned at random to either a treatment group or to a control groups. Of the 48 men who were assigned to the treatment group, 22 reported that they felt that they experienced some growth of new hair over the three month period that the study was conducted. Would it be fair to conclude that the drug is effective in reducing hair loss? Why or why not?

5. A sociologist wanted to investigate the relationship between criminal behavior and alcohol in male teenagers. He interviewed 800 teenagers and found that 50 had an arrest record and that 750 had no arrest record. The 50 who had an arrest record reported an average alcohol consumption that was about twice that of the 750 males who had no arrest record. The sociologist concluded that the arrest of young men causes them to consume more alcohol.

    a) Was this an experimental or an observational study?

    b) Do you agree with the conclusion? Why or why not?

6. An anatomy professor wants to investigate the role of a certain organ in the regulation of blood pressure in rats. He obtains 20 rats and assigns 10 at random to a treatment group and 10 rats to a control group. The experimental design specifies that rats in the treatment group will have their blood pressure taken before and after the surgery to remove the organ.

   a) Should the rats in the treatment group be housed in a separate cage or should the treatment rats be marked for identification and housed in the same cage? Explain.

   b) Should control rats also undergo a surgical procedure that is similar to that used on the treatment rats except that no organ is removed? Why or why not?

7. Fourteen people are on an elevator that has a maximum capacity of 2500 pounds. The average weight of these fourteen people is 140 pounds. A man who weighs about 350 pounds attempts to get on the elevator. Will this additional weight cause the elevator to be over-loaded ?

8. You graduate from college and receive an offer for a job in a distant city. Before accepting the position you travel to the city to determine the cost of purchasing a home. You take a sample of 5 homes and obtain the following home values:

   $172,000 $181,000 $186,000 $189,000 $672,000

   a) Compute the mean and the median of these home values.

   b) Does the mean home value describe the value of the typical house? Why or why not?

9. Using the data in the previous exercise compute the standard deviation of the home values. [Hint: It may be easier to work in units of thousands of dollars.]

10. Consider the following dot diagrams that represent the reading scores on standardized tests for 5th graders at three school.

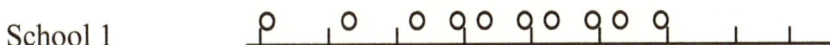

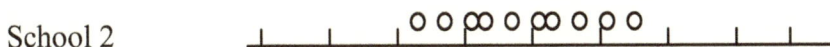

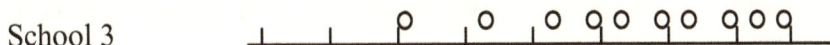

School 1

School 2

School 3

   a)   Which school appears to have the highest average reading score?

   b)   Which school appears to have the least variability in scores?

11. Consider the measurements shown in the following dot diagrams.

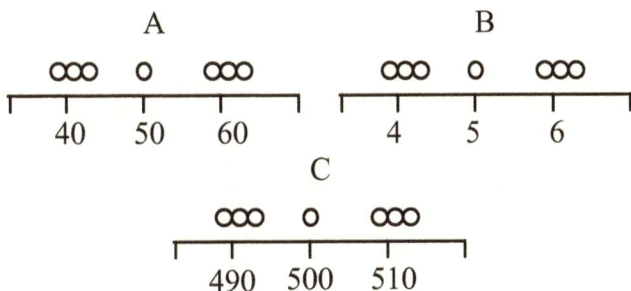

A

40   50   60

B

4   5   6

C

490   500   510

   a)   The standard deviation of the measurements in diagram A is approximately

          1   10   100

   b)   The standard deviation of the measurements in diagram B is approximately

          1   10   100

   c)   The standard deviation of the measurements in diagram C is approximately

          1   10   100

12. a) Can the standard deviation ever be zero? If it can, give an example of a data set with s=0.

    b) Can the standard deviation every be negative? If it can, give an example of a data set with s<0.

13. The National Institutes of Health obtained blood pressure data on 965 males and 945 females aged 15 through 20. A computer program determined the following percentiles for the systolic blood pressure for males and females:

Percentile	Males	Females
5th percentile	100	90
25th percentile	110	100
50th percentile	120	110
75th percentile	128	120
95th percentile	142	130

a) Construct side-by-side box plots for males and females.

b) Compute the interquartile range for the males and the females.

c) Compare the variability in blood pressure for the males to the variability for the females.

14. The relative frequency histogram displays the distribution of the weekly consumption of chocolate candy bars by college students.

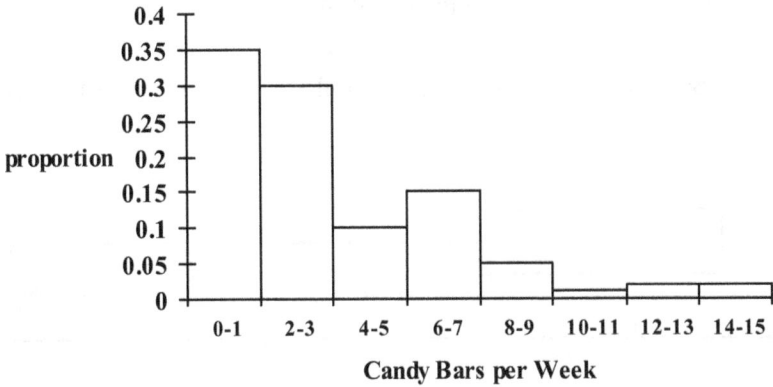

Candy Bars per Week

a) What proportion of students consume 10 candy bars or more per week?

b) What proportion of students consume 3 or fewer candy bars per week?

15. The dot diagram shows the heights of 20 grade school students who are 8 years old. You may assume that height has a normal histogram in this population.

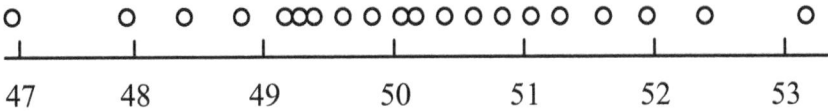

a) The median height is approximately

$$49 \qquad\qquad 50 \qquad\qquad 53$$

b) The standard deviation of the height is approximately

$$.5 \qquad 1.5 \qquad 3 \qquad 6$$

Hint: Refer to the Empirical Rule.

16. The box plot shows the number of miles people commute to work or to school.

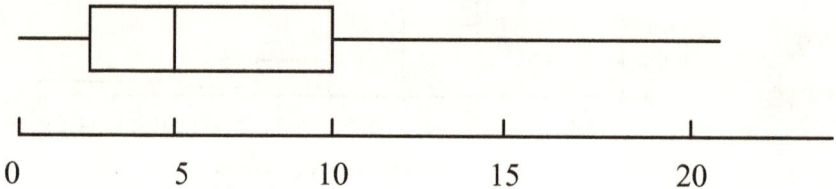

a)   What is the median?

b)   What is the IQR?

c)   If you made a histogram would it be symmetric or skewed?

17. Assume that the heights of males has a normal histogram and that the average is 69 inches and the standard deviation is 3.

a)   Draw a rough sketch of a histogram.

b)   What percentage of males have heights in the interval (66, 72)?

c)   What percentage of males have heights in the interval (63,75)?

18. The histogram of total family income (indicated in thousands of dollars per year) for two communities is displayed in the following box plots:

Community 1

Community 2

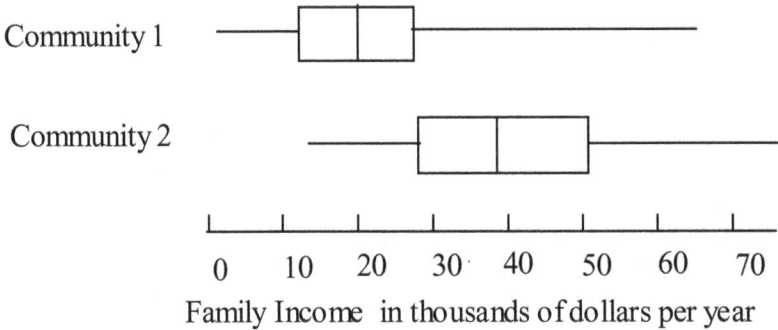

    0   10   20   30   40   50   60   70

    Family Income in thousands of dollars per year

  a) As there individuals in Community 1 who have a greater family income   than   some   individuals   in   Community   2?

  b) Suppose a family in Community 1 has an income of $28,000. Roughly estimate what proportion of families in that Community have   incomes   that   are   less   than $28,000 ?

  c) What proportion of families in Community 2 have incomes that are less than $28,000 ?

Chapter 2. Probability

2.1 Approaches to Probability

If we roll a die or flip a coin we cannot be certain of the outcome. This uncertainty can be evaluated by the science of probability, which has developed over the last few hundred years. Although some probabilities for dice had been calculated some years earlier, Blaise Pascal (1623-1662) and Pierre Fermat (1601-1665) developed methods for calculating probabilities for some games of chance. We have no need to analyze games of chance in detail, but we do want to become sufficiently familiar with the basic rule of probability that we can apply the rules to statistical problems.

Before discussing probability it is important to understand the distinction between probability and statistics. Suppose we have a large bowl with 100 red marbles and 200 blue marbles. If we select 10 marbles at random from the bowl we would use probability methods to compute the chance of getting 5 red marbles and 5 blue marbles. Probability methods are used when we know the proportion of red marbles in the bowl and want to compute the probability of obtaining a certain number of red marbles in a sample.

Statistical methods are used when we have obtained a sample and want to estimate the population proportion. For example, statistical methods would be used if we took a sample from a large bowl that had an unknown number of red and blue marbles. Suppose we selected 10 marbles at random from the bowl and obtained 5 red and 5 blue marbles. We would use statistical methods to estimate the proportion of red marbles in the bowl.

We use statistical methods, rather than probability methods, in political polling because we need to estimate the proportion of voters who will vote for a candidate, based on the results from a random sample of a few hundred voters. But to fully grasp the statistical methods, we first need to understand probability so that we can see the relationship between the population proportion and the sample proportion. Consequently, some understanding of probability will clarify the development and the use of statistical methods.

There are several approaches to estimating the probability of an event. The **frequency approach** is based on the long run relative frequency of the event. Suppose you to roll a die 6000 times and keep track of the number of times you roll a "1", "2", "3", "4", "5", or "6". What do you expect the results to be? If you have a fair die with six sides then each side should be equally likely and you should expect to obtain about 1000 "1"s, about 1000 "2"s, and so forth. The expected outcomes for this experiment are displayed in Figure 2.1.

Figure 2.1 Expected Frequencies for the Six Outcomes in the Die Experiment

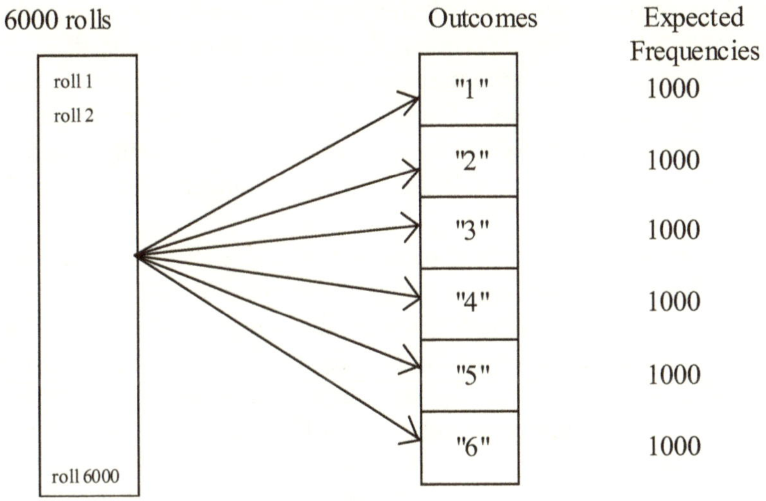

6000 rolls	Outcomes	Expected Frequencies
roll 1	"1"	1000
roll 2	"2"	1000
	"3"	1000
	"4"	1000
	"5"	1000
roll 6000	"6"	1000

We say that the probability, or chance, or rolling a "4" is $\frac{1000}{6000} = \frac{1}{6}$ because, with a large number of rolls, the proportion of these rolls that we are "4" should be near $\frac{1}{6}$. This is an example of the frequency approach to probability.

For many games of chance we can use the **classical approach** to probability instead of the frequency approach. The classical approach is used when we have good reason to believe that the outcomes of an experiment are equally likely. For example, suppose we inspect a die very carefully and decide that it is cubical and that it is not "loaded" in any way. It seems reasonable to assume that the six outcomes are equally likely, and since the probabilities must sum to one, the probability of rolling a "4" must equal $\frac{1}{6}$. This classical approach is useful for games of chance but cannot be used when the assumption of equally likely outcomes cannot reasonably be made. For example, if we want to know the probability of rolling a "4" with a defective or "loaded" die, the classical approach could not be used because the outcomes are not equally probable. For this die the frequency method would need to be employed. Fortunately the methods developed in this chapter are valid for probabilities that were estimated with either the frequency or the classical approaches.

In order to develop methods for calculating probabilities some definitions will be helpful. We will call the list of all possible outcomes the **sample space**. In the die experiment the sample space consisted of six outcomes. Sometimes we are interested in a subset of the sample space which we will call an **event**. For example, we may be interested in calculating the chance of rolling an even number. The event, which is rolling an even number, is a collection of three outcomes {rolling a 2, rolling a 4, rolling a 6}.

To determine the probability of an event using the frequency approach we find the proportion of times the event will occur when the process is repeated many times under the same conditions. The probability of rolling an even number equals the limit, as the total number of rolls increases, of the number of rolls of even numbers divided by the total number of rolls. We will assume that such a limit exists. Let A represent any event and let $n(A)$ be the number of times that the event A occurs in n repetitions of an experiment. Using this notation we can define the probability of A, $P(A)$, as the limit of $\frac{n(A)}{n}$ as the number of repetitions increase without bound. Consider the event of rolling a "4" with a fair die. If we roll $n=6000$ times we would expect to obtain $n(A)=1000$ rolls to be "4"s, if we roll $n=600,000$ times we expect 100,000 rolls to be "4"s. Therefore, we expect the proportions of "4"s will be the limit of $\frac{n(A)}{n}$ as n increases, which will equal $\frac{1}{6}$.

For many games of chance, either the frequency approach or the classical approach can be used to determine the chance that a single event will occur. However, we often need probabilities of more complex events. We may want to determine the chance of rolling a "4" or a "5" on one roll of a die or we may interested in determine the chance of rolling a "4" on the first roll and an even number on the second roll. The probability rules given in the next two sections will enable us to calculate probabilities for these more complex events.

2.2 The Addition Rule

The addition rule is used to compute the probability that either one of two events will happen when they both cannot happen at the same time. To illustrate the rule, consider again the die experiment and its outcomes. In Figure 2.2 we display the outcomes and probabilities associated with those outcomes in a **tree diagram**. If we are interested in the event of rolling an even number, we can compute its probability as

P(roll an even number) = P(roll a 2) + P(roll a 4) + P(roll a 6) .

That is, in this situation we can simply add the probabilities of the individual outcomes to obtain the probabilities of this event. However, suppose we are interested in the events:

Event A: roll an even number
Event B: roll a number ≤ 2.

We can see from Figure 2.2 that:

Probability of rolling an even number = $P(A) = \frac{3}{6}$.
Probability of rolling a number ≤ 2 = $P(B) = \frac{2}{6}$.

We can also see directly from Figure 2.2 that

$P(A\ or\ B) = P$(roll a 1) + P(roll a 2) + P(roll a 4) + P(roll a 6) = $\frac{4}{6}$.

We note that in this situation that we cannot simply add $P(A)$ to $P(B)$ in order to obtain the correct answer, since $P(A) = \frac{3}{6}$ and $P(B) = \frac{2}{6}$. However, we observed that in our first situation we could add the probabilities to obtain the correct answer. We need to know when we are justified in adding the probabilities.

Figure 2.2 Probabilities for the six outcomes in the die experiment.

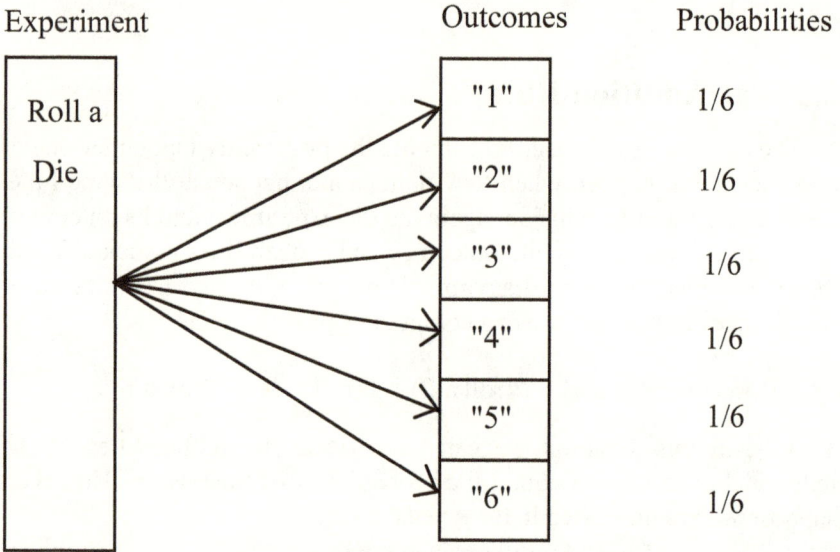

Experiment   Outcomes   Probabilities

| Roll a | | |
| Die | | |

"1"	1/6
"2"	1/6
"3"	1/6
"4"	1/6
"5"	1/6
"6"	1/6

From Figure 2.2 we can see that if two events have no outcomes in common the probabilities can be added. If two events have no outcomes in common they are called **mutually exclusive events**. If the events are mutually exclusive the computation of probabilities is simplified because there is no possibility of counting the same event twice. The **addition rule** states:

> **If two events are mutually exclusive the probability that either will occur equals the sum of their probabilities.**

An example may illustrate this point. Suppose we shuffle an ordinary deck of cards containing 13 spades, 13 hearts, 13 clubs, and 13 diamonds, and we draw one card off the top of the deck. There are 52 outcomes that are displayed in Figure 2.3.

Figure 2.3 Outcomes of drawing one card from an ordinary card deck.

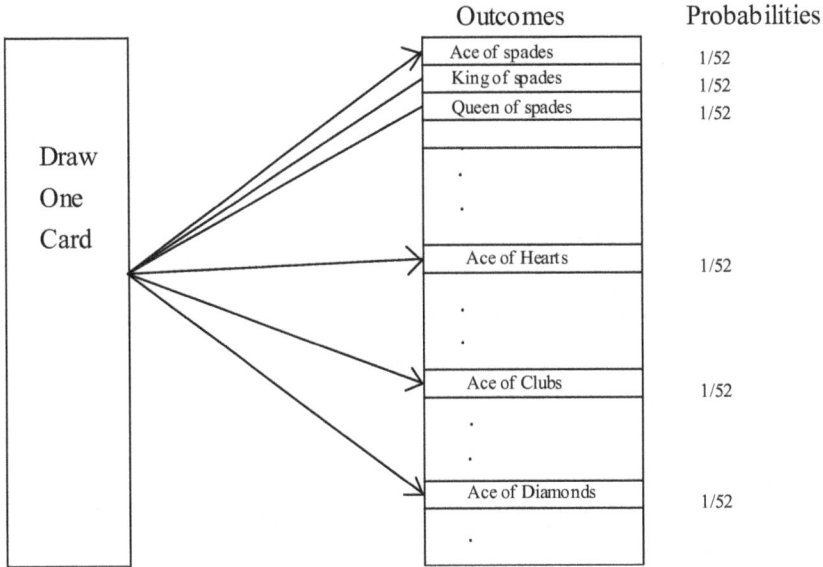

	Outcomes	Probabilities
	Ace of spades	1/52
	King of spades	1/52
	Queen of spades	1/52
Draw One Card	.	
	.	
	Ace of Hearts	1/52
	.	
	.	
	Ace of Clubs	1/52
	.	
	.	
	Ace of Diamonds	1/52
	.	

From Figure 2.3 it can be seen that $P(\text{spade}) = \frac{1}{52} + \ldots + \frac{1}{52} = \frac{13}{52} = \frac{1}{4}$. Since drawing a spade and drawing a heart are mutually exclusive events, $P(\text{spade}) + P(\text{heart}) = \frac{1}{4} + \frac{1}{4} = \frac{1}{2}$. However, since drawing a spade and drawing an ace are not mutually exclusive, we are not justified in adding probabilities to compute $P(\text{spade or ace})$. If we did try to add the probabilities we would be counting the ace of spades twice. To correctly compute $P(\text{spade or ace})$ we can simply list all outcomes that are either aces or spades. Since 16 cards are either spades or aces we find that $P(\text{spade or ace}) = \frac{16}{52}$.

In general, when we want to compute the probability of A or B occurring, we first need to determine if the two events are mutually exclusive. If they are mutually exclusive, we can add the probabilities to obtain the correct answer. However, if they not mutually exclusive we must identify all of the outcomes that are either A or B, being careful to avoid counting some outcomes twice.

Exercise Set A

1. Suppose a coin has been bent so badly that it no longer can be used in vending machines. You suspect that the probability of getting a head does not equal ½. Briefly describe an experiment that would allow you to estimate the probability of getting a head with this coin.

2. We shuffle a deck of cards. Consider the following events:

    Event A: draw a King
    Event B: draw a Jack

    a) Are the two events mutually exclusive?

    b) What is P(A or B)?

3. A bowl has 30 red marbles, 20 white marbles, and 50 blue marbles. If the marbles are mixed thoroughly, what is the chance of drawing a blue or a white marble from the bowl?

4. A roulette wheel has 38 slots. Two are green, 18 are black, and 18 are red. The wheel is made so that the chance that the ball will land in any slot is equal to the chance that it will land in any other slot.

    a) What is the chance that the ball will land in a red slot?

    b) What is the chance that the ball will land in a green slot?

    c) What is the chance that the ball will land in a red or a green slot?

    c) What is the chance that the ball will not land in a black slot?

5. Suppose you play roulette, which is described in the previous exercise, by betting one dollar on a red number. If the ball lands in a red slot you win one dollar, otherwise you lose your dollar. Suppose you have $1000 dollars and play roulette 1000 times, making a one dollar bet each time.

    a) How many times would you expect to win?

    b) How many times would you expect to lose?

c) How many dollars would you expect to take home?

d) Suppose you bet $10 instead of $1 each time. How much would you expect to take home?

6. A box contains 10 tickets numbered 1 through 10. The four tickets having a number less than or equal to 4 are black, the other tickets are white. Consider the events:

> Event A: Draw a ticket with a "9" on it.
> Event B: Draw a white ticket.

a) Are the two events mutually exclusive?

b) What is the chance of drawing a 9 or drawing a white ticket?

7. Alan is quite particular about his soft drinks. He will only drink a regular (non-diet) Brand A cola. Suppose he selects, at random, a soft drink out of a cooler that has:

> 6 cans of regular Brand A cola
> 6 cans of diet Brand A cola
> 12 cans of regular Brand B cola
> 12 cans of diet Brand B cola

a) What is the chance that he selects a regular Brand A cola?

b) What is the chance that he selects a diet cola?

c) What is the chance that he selects a diet cola or Brand B cola?

8. Angela has a fair 20 sided die that has the numbers 1 through 20 inscribed on the faces. These twenty sided die are called icosahedrons or D20 die. The die is constructed in a way that makes each of the 20 outcomes equally likely.

a) What is the chance of rolling a "4" with this die?

b) What is the chance of rolling an even number?

c) What is the chance of rolling an even number or a number less than or equal to 9?

2.3 The Multiplication Rule with Independent Events

The multiplication rule is often used to compute the probability that several events occur in a sequence of independent trials. For example, suppose we roll a die twice and we want to compute the chance of rolling a "1" on both rolls. For each of the six outcomes of the first roll there are six outcomes for the second roll. Consequently, there are 36 outcomes, which are displayed in the tree diagram in Figure 2.4.

Figure 2.4 Probabilities for the outcomes of rolling a die twice.

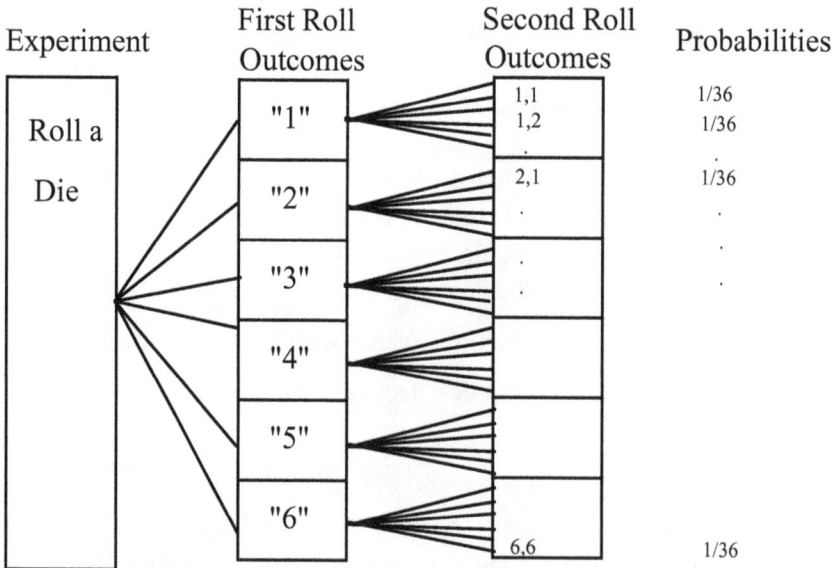

Note that, if the die is shook well after the first roll, the outcome of the second roll should be unrelated to the outcome of the first role. To compute the probability of rolling a "1" on the first and second rolls, we repeat this experiment 3600 times. We would expect to get about 600 "1"s on the first roll, and on about 100 of these 600 rolls we would get a "1" on the second roll. Therefore the probability of rolling a "1" on the first and second rolls would be $\frac{100}{3600} = \frac{1}{36}$. Thus, the probability of rolling a "1" on both rolls can be calculated by taking the probability of rolling a "1" on the first roll times the probability of rolling a "1" on the second roll, or $P(1,1) = \frac{1}{6} \times \frac{1}{6} = \frac{1}{36}$.

Now suppose we roll a die three times and we want to compute the probability of rolling a "4" on each roll. Since the outcomes of subsequent rolls are unrelated to the outcome of the previous rolls, we can simply multiply the probabilities. The probability of rolling "4" on each of the rolls is $\frac{1}{6}$ so that $P(4,4,4) = \frac{1}{6} \times \frac{1}{6} \times \frac{1}{6} = \frac{1}{216}$. We can also use this method to compute the probabilities of obtaining a sum. For example, if we roll a die four times, what is the probability that the sum is 24? We solve this by noting that we need to roll a "6" on every roll to obtain a sum of 24, so that $P(\text{Sum} = 24) = P(6,6,6,6) = \frac{1}{6} \times \frac{1}{6} \times \frac{1}{6} \times \frac{1}{6} = \frac{1}{1296}$.

For many problems involving dice or coins the outcome of one event does not change the probability associated with a subsequent event. We say that two events are **independent** if knowledge of the first event does not change the probability of the second event. The multiplication rule for independent events says :

If A and B are independent then
$$P(A \text{ and } B) = P(A) \times P(B).$$

We can also use the multiplication rule to calculate probabilities for coin flips. Suppose we flip a coin 5 times and wish to determine the probability of flipping five heads. Coin flips are independent because the chance of getting a head on the second flip does not depend on the outcome of the first flip. Thus, the probability that the first two flips are heads is $P(H,H) = \frac{1}{2} \times \frac{1}{2} = \frac{1}{4}$. We can extend the multiplication rule to calculate the chance that the first three flips are heads as $P(H,H,H) = \frac{1}{2} \times \frac{1}{2} \times \frac{1}{2} = \frac{1}{8}$. By continuing in this fashion we see the chance of getting five heads on five flips is

$$P(H,H,H,H,H) = \frac{1}{2} \times \frac{1}{2} \times \frac{1}{2} \times \frac{1}{2} \times \frac{1}{2} = \left(\frac{1}{2}\right)^5 = \frac{1}{32}.$$

Often the calculation of probabilities can be simplified by using the fact that the probabilities must add to 1.0. Suppose we want to calculate the chance of obtaining at least one head in five tosses of a coin. This is not too easy to calculate directly because we need to calculate the chance of getting exactly one head, exactly two heads, exactly three heads, and so forth. Then we need to add those probabilities. This would be a great deal of work. Fortunately, it is possible to greatly simplify the work by recognizing that, in five tosses, we can either get no heads or we can get at least one head. Consequently,

$$P(\text{no heads}) + P(\text{at least one head}) = 1$$

and it follows that

$$P(\text{at least one head}) = 1 - P(\text{no heads}).$$

But the probability of getting no heads is the same as the probability of getting five tails which is $P(T,T,T,T,T) = \left(\frac{1}{2}\right)^5 = \frac{1}{32}$. Therefore, the probability of getting at least one head is $1 - \frac{1}{32} = \frac{31}{32}$.

Exercise Set B

1. Two ordinary dice are rolled. What is the chance of rolling a "2" on the first roll and a "4" on the second roll?

2. A coin is flipped twice. Compute the chances of obtaining:

    a) a head on the first flip and a tail on the second flip.

    b) a tail on the first flip and a head on the second flip.

    c) exactly one head.

3. Two dice are rolled.

    a) What is the chance that the sum will be 12?

    b) What is the chance that the sum will be 11?

4. Five dice are rolled.

    a) What is the chance that all five dice will show "6"?

    b) What is the chance that all five dice will show the same number of dots?

5. A large bowl has 400 red marbles and 600 blue marbles. Suppose we draw a marble at random and note its color before replacing it and drawing another.

   a) What is the chance that both marbles will be red?

   b) What is the chance that both marbles will be the same color?

6. Alan will only drink a certain brand of soft drink. A cooler has 6 cans of his favorite soft drink and has 6 cans of another soft drink.

   a) He selects a can from the cooler, discovers that it is his favorite kind, and drinks it. He then selects another can at random. What is the chance that the second can will also be his favorite?

   b) Are the following events independent?

   Event A: Selecting his favorite soft drink on the first attempt.

   Event B: Selecting his favorite soft drink on the second attempt.

7. A roulette wheel has 38 slots. Two are green, 18 are black, and 18 are red. The wheel is made so that the chance that the ball will land in any slot is equal to the chance that it will land in any other slot. The wheel is spun twice and the results are recorded.

   a) What is the chance of getting a red on both spins?

   b) What is the chance of getting a red on the first spin and a black on the second spin?

   c) What is the chance of getting a black on the first spin and a red on the second spin?

   d) What is the chance of getting a black on both spins?

   e) Sum the chances you found in (a) to (d). If the sum does not equal 1.0 explain why the sum does not need to equal 1.0.

2.4 The Multiplication Rule with Dependent Events.

The multiplication rule for independent events, which was developed in the last section, allows us to compute the probabilities for many outcomes involving games of chance. If we roll a die several times, the events defined for the rolls are independent events because the probabilities of the outcomes do not depend on the prior outcomes. The same could be said for coin flips. However, card games are generally more difficult to analyze because the events may not be independent. In this section we will develop methods for computing probabilities when the events are dependent.

Suppose we deal a heart off the top of a well-shuffled card deck. What is the probability that the second card is also a heart? This problem is more difficult that those considered in the last section because the deck will change after one card is removed from the deck. The probability that the first card is a heart is $\frac{13}{52} = \frac{1}{4}$. After the heart is dealt there are only 12 hearts remaining in the deck of 51 cards. Thus, the **conditional probability** of dealing a heart as the second card, given that a heart was dealt as the first card, is $\frac{12}{51}$. The probability that both are hearts is

P(first is a heart and the second is a heart)

=P(first is a heart) × P(second is a heart given that the first was a heart)

$$= \frac{13}{52} \times \frac{12}{51} = \frac{1}{17} \ .$$

In general, the chance that two events will happen equals the chance that the first will happen times the chance that the second will happen. This is called the **multiplication rule** and it can be stated as follows:

For any two events A and B,

P(A and B)=P(A) x P(B given that A has occurred).

The multiplication rule can be generalized to three or more events. For example, if we deal four cards off the top of a well-shuffled deck, the chance that all four will be Hearts is $\frac{13}{52} \times \frac{12}{51} \times \frac{11}{50} \times \frac{10}{49} = \frac{11}{4165}$. The chance that the first four cards will be Queens is $\frac{4}{52} \times \frac{3}{51} \times \frac{2}{50} \times \frac{1}{49} = \frac{1}{270725}$.

We usually do not shuffle the deck after we deal a card. However, if we deal a card off the top of a well-shuffled deck, replace the card in the deck, and shuffle the deck again before we deal the second card then

P(first is a heart and second is a heart)

$= P$(first is a heart) $\times P$(second is a heart)

$= \frac{13}{52} \times \frac{13}{52} = \frac{1}{16}.$

In this example we replaced the first card before drawing the next card so that the probabilities for the second draw are identical to those for the first draw. In this unusual card problem the conditional probabilities equal the unconditional probabilities so the events are independent and we are justified in using the multiplication rule for independent events. However, in most card problems we use conditional probabilities because we usually do not replace the cards before we draw the next card.

Recall that in the examples using dice we did not concern ourselves with conditional probabilities. We could ignore conditional probabilities because the probabilities of the outcomes for the second role were not changed by the outcome of the first role. We said that two events are independent if knowledge of the occurrence of the first event does not change the probability that the second event will occur. Thus, if we want to compute the probability that two independent events both occur, we can multiply the unconditional probabilities as we did in the previous section. However, if the two events are dependent the multiplication rule must be used with conditional probabilities.

Exercise Set C

1. Two cards are dealt of the top of a well-shuffled deck of cards.

    a) Compute the probability that the first card is a Jack and the second card is a Jack.

    b) Compute the probability that the first card is a Jack and the second card is a King.

2. Four cards are dealt off the top of a well-shuffled deck.

    a) Compute the probability that the four cards are hearts.

b) Compute the probability that the first and second cards are hearts and the third and fourth cards are clubs.

3. A bowl has 6 red and 4 blue marbles.

   a) If two marbles are drawn without replacement, what is the chance that both will be red?

   b) If two marbles are drawn with replacement, what is the chance that both will be red?

4. A card deck is shuffled. The top card is turned over and it is a Jack of Diamonds.   What is the probability that the second card is a Queen of Clubs given that the first card was a Jack of Diamonds?

5. A card deck is shuffled. The top card is not observed, but it is set aside face down. What is the probability that the second card is the Queen of Clubs?

6. At a small college there were 30 female faculty members and 20 male faculty members. A hiring committee, consisting of 3 faculty members, will be selected to interview the applicants for the position.

   a) What is the probability that all members on the committee will be females?

   b) What is the probability that all members on the committee will be males?

7. (Challenging) Three cards are dealt of the top of a well-shuffled card deck.

   a) What is the chance of getting three hearts?

   b) What is the chance that the first two cards are hearts and the third card is not a heart?

c)   What is the chance of getting exactly two hearts?

d)   What is the chance of getting no hearts?

e)   What is the chance of getting one or more hearts?

* 2.5 Bayes' Rule

In the previous section we used conditional probabilities in the multi-plication rule. In this section we will use Bayes' Rule to compute certain probabilities based on known conditional probabilities. Bayes' Rule has many medical applications in screening tests, which are used by physicians to give some indication that a patient has a disease. One example of a screening test is a blood test for the early detection of prostate cancer. A positive screening test result does not mean that the patient has cancer, but it does indicate that further testing should be done.

A perfect screening test would always indicate that the person has a disease when that person actually has the disease and would always indicate that the person does not have the disease when the person actually does not have the disease. However, not all screening tests are perfect. A positive test result indicates that the person may have the disease, but does not, with absolute certainty, determine that the person has the disease. Similarly, a negative test results does not, with absolute certainty, determine that the person does not have the disease.

Now suppose a young man has tested positive for a disease. He will probably be interested in knowing the chance that he actually has the disease given that he has tested positive for the disease. This is called the **predictive value positive**. In this situation the physician may know the conditional probability that the test is positive given that a person has the disease and the conditional probability that the test is negative given that a person does not have the disease. But the patient is not interested in these probabilities. The patient is interested in the conditional probability that he has the disease given that the test was positive.

Bayes' rule can be used to calculate this conditional probability. The following notation will be used:

T^+: the person has received a positive test

D: the person has the disease

$\overline{D}$:   the person does not have the disease.

We will use the vertical slash to indicate conditional probability. Using this notation $P(D|T^+)$ is the probability that the person has the disease given that the test was positive. We can use the multiplication rule to compute the probability that the person has the disease and the test result is positive

$$P(D \text{ and } T^+) = P(T^+)P(D|T^+) = P(D)P(T^+|D). \qquad \textbf{(2.1)}$$

We are interested in finding the conditional probability of having the disease given that the test was positive. From the middle part of equation (2.1) we see that

$$P(D|T^+) = \frac{P(D \text{ and } T^+)}{P(T^+)} \qquad \textbf{(2.2)}$$

Since a person who has tested positive either has the disease or does not have the disease we can write the denominator of equation (2.2) as

$$P(T^+) = P(D \text{ and } T^+) + P(\overline{D} \text{ and } T^+).$$

We can now substitute this expression into equation (2.2) to obtain

$$P(D|T^+) = \frac{P(D \text{ and } T^+)}{P(D \text{ and } T^+) + P(\overline{D} \text{ and } T^+)}.$$

Finally, we rewrite this expression using the last part of equation (2.1) to obtain Bayes Rule

$$P(D|T^+) = \frac{P(D)P(T^+|D)}{P(D)P(T^+|D) + P(\overline{D})P(T^+|\overline{D})}. \qquad (2.3)$$

Thus, we can compute $P(D|T^+)$, which is the predicted value positive, if we know $P(T^+|D)$ and $P(T^+|\overline{D})$. For a person who has a disease, the chance of getting a positive test result is called the **sensitivity** of the test, which is $P(T^+|D)$. The chance of a disease-free individual getting a positive test is $P(T^+|\overline{D})$, so the chance of a disease-free individual getting a negative test is called the **selectivity** of the test, which is $1 - P(T^+|\overline{D})$. For a screening test to be effective, the sensitivity and the selectivity should be close to 100%.

For example, consider a hypothetical screening procedure for HIV infection. Suppose 99% of the individuals in the population are not infected. Using our notation this can be written as $P(\overline{D}) = .99$ and it implies that $P(D) = .01$. Suppose the sensitivity of this test is 95% and the selectivity is 90%. In our notation the sensitivity can be expressed as $P(T^+|D) = .95$ and the selectivity as $1 - P(T^+|\overline{D}) = .90$ which implies that $P(T^+|\overline{D}) = .10$. An individual who has just received a positive test result is interested in their chance of having the disease given that the test was positive. Using Bayes' Rule, we find that

$$P(D|T^+) = \frac{(.01)(.95)}{(.01)(.95) + (.99)(.10)} = \frac{.0095}{.0095 + .099} = .0876$$

Therefore, for this hypothetical screening test the person who has tested positive for HIV has less than a 10% chance of actually having the infection! This chance is small because 99% of the people tested do not have the infection but 10% of those get a positive test result. Consequently, there will be many incorrect positive test results.

For other applications of Bayes' Rule it will be convenient to write it using a more general notation. Suppose there are 2 subpopulations indicated by S_1, S_2 and let the probability that an individual belongs to the ith population be $P(S_i)$, and let A be any event. Then the general form of Bayes' Rule can be written as

$$P(S_1 | A) = \frac{P(S_1)P(A|S_1)}{P(S_1)P(A|S_1) + P(S_2)P(A|S_2)} \qquad (2.4)$$

which gives the probability that an individual is from population 1 given that event A has occurred. This form of Bayes' Rule has many applications in the physical and social sciences.

For example, suppose a computer manufacturer obtained memory chips from two suppliers. Supplier 1 supplies 80% of the chips and 0.01% of these are defective. Supplier 2 supplies 20% of the chips and 0.75% of these are defective. If a chip is selected at random from one of the computers and is found to be defective, what is the chance that it came from Supplier 1? In this example, let A be the event that the chip is defective, let S_1 be the event that the chip is supplied by supplier 1, and let S_2 be the event that the chip is supplied by supplier 2. With this notation $P(S_1) = .8$, $P(S_2) = .2$, $P(A|S_1) = .0001$,   and $p(A|S_2) = .0075$. Bayes' rule gives

$$P(S_1 | A) = \frac{P(S_1)P(A|S_1)}{P(S_1)P(A|S_1) + P(S_2)P(A|S_2)}$$

$$= \frac{(.8)(.0001)}{(.8)(.0001) + (.2)(.0075)} = \frac{.00008}{.00008 + .015} = .0053,$$

which implies that $P(S_2 | A) = 1 - P(S_1 | A) = .9947$.   Thus, although Supplier 1 supplied most of the chips, the chip selected at random that was determined to be defective most likely came from Supplier 2.

Exercise Set D

1. Suppose a screening test has been developed for a rare disease that has 99.5% sensitivity and 98% selectivity. If the disease is present in 0.2% of the population, what is the probability that a person who tested positive actually has the disease?

2. Suppose in the previous exercise that the selectivity is increase to 99.5%. What is the probability that a person who tested positive actually has the disease?

3. Suppose that, in a certain state, 55% of the voters are Republicans and 45% of the voters are Democrats. Suppose that 35% of the Republicans are female and 60% of the Democrats are female. What is the chance that a randomly selected voter is a Republican given that the voter is female?

4. Two companies supply car radios to an automotive manufacturer. Company 1 supplies 20% of the radios but about 4% of these are defective. Company 2 supplies 80% of the radios and about 0.5% of these are defective. If a car manufactured by the company has a defective radio, what is the probability that it was supplied by Company 1?

Chapter Review Exercises

1. Fred found a quarter that had been badly damaged. It looked as if someone had put it into a vise and hit it with a hammer. He decided to estimate the chance of flipping a "head" by flipping the coin many times. After 100 flips he obtained 59 heads. After 800 coin flips he obtained 482 "heads".

    a) Give a rough estimate of the chance of flipping a "head" with this coin.

    b) What approach was used to estimate this probability?

2. Suppose a bowl contains:

> 60 red balls
> 30 green balls
> 4 orange balls
> 6 gray balls.

Without looking, a man reaches into a bowl and selects a ball at random.

a) What is the chance of drawing a green ball?

b) What is the chance of drawing a red or an orange ball?

c) What is the chance of drawing a ball that is not red?

3. A fair coin is flipped 7 times.

a) What is the chance of getting 7 heads?

b) What is the chance of getting the sequence HHTTHTH ?

4. An ordinary die is rolled.

a) What is the chance of getting a "3" or a "4"?

b) What is the chance of not getting a "6"?

5. A young couple decide to get married and raise a family. If they have 6 children what is the chance that all 6 will be girls? You may assume that the gender of each child is independent of the gender of all other children in the family and that for each birth the probability that the child will be a girl is $\frac{1}{2}$. What is the chance that at least one child will be a boy?

6. A deck of cards is shuffled and the top ten cards are turned over. If there are no aces in the first ten cards what is the chance that the eleventh card would be an ace?

7. Three cards are dealt off the top of a well-shuffled deck.

    a) If you did not see the first two cards what is the chance that the third card is a Jack?

    b) If you saw the first two cards and determined that the first two cards were not Jacks, what is the chance that the third card is a Jack?

8. A physics student decides to take the final exam even though he knows nothing about the subject. To pass the course he must get a perfect score on the ten question multiple choice final exam that has 4 answers per question. The professor had arranged the correct answers at random and the student decides to guess at each question at random. What is his chance of getting a perfect score?

9. Suppose a new test has been devised to detect bladder cancer. This test has a sensitivity of 90% and a selectivity of 99.9%. If 1% of the population have bladder cancer, what is the chance that an individual who has been tested positive actually has bladder cancer?

10. (Challenging) Recall that a roulette wheel has 18 red slots, 18 black slots, and 2 green slots. If you bet $10 on red and the ball falls into a red slot then you will win $10. If the ball falls into a black or green slot you will lose $10. A man visits a casino to win some money by playing roulette. His strategy is to begin the betting session by betting $10 on red. If he wins he stops that betting session but if he loses he continues playing by betting $20 on red. If he wins he stops the betting session but if he loses he continues playing by betting $40 on red. Regardless of the outcome of the $40 bet he will stop his betting session.

    a) What is the maximum amount he could win in a single betting session?

    b) If he has bad luck what is the maximum amount he could he lose in a single betting session?

c) What is the chance of losing the maximum amount in a single session?

d) Suppose he uses this strategy for 6859 betting sessions. About how many times would he win and lose?

e) Approximately how much would he win or lose using this strategy?

11. (Challenging) A bowl contains 3 red marbles and 2 blue marbles.

a) Two marbles are drawn without replacement from the bowl. What is the chance of drawing at least one red marble?

b) Three marbles are drawn without replacement from the bowl. What is the chance of drawing at least one red marble? Hint: No calculations are necessary.

12. (Challenging) Four people are sitting around a card table playing poker. In poker, a "flush" is a five card hand where all five cards are from the same suite. One player shuffles an ordinary deck of cards and deals five cards to each player. What is the chance that the first player to the right of the dealer is dealt a flush? Note that it could be a heart flush, a diamond flush, a spade flush, or a club flush.

13. Two dice are rolled in a board game. The marker is moved the number of spaces that are equal to the sum of the number of dots shown on both dice. If the marker is three spaces away from a location, what is the chance that, on the next roll of two dice, the marker will be moved three spaces?

Chapter 3. Sampling Distributions

3.1 Populations and Samples

In chapter one we obtained a sample from a population and discussed many ways to describe the sample data. Typically, we take a relatively small sample from a very large population and use the sample mean and sample standard deviation to describe the sample. However, we are usually interested in the population, not in the sample. In this chapter we will learn to use the sample statistics to tell us something about the population.

An example may clarify this important point. Suppose we manufacture light bulbs and need to know approximately how long a typical light bulb will burn before it fails. If we fully test every light bulb that is manufactured by burning it until it fails we can use the sample average to describe the average time to failure. However, if we test every bulb we will have no light bulbs to sell! Clearly we need to take a sample of our production and use the average time to failure that we obtained from the sample as an estimate of the average time to failure for all the light bulbs that we produce. In this example the sample average time to failure is used because it provides an estimate of the population average time to failure.

To describe this situation we will need a few definitions. As we indicated in Chapter 1, a **population** is a set of units about which information is desired and a **sample** is a subset of the population. The relationship between the population and the sample is illustrated in Figure 3.1. We will often take a small sample from a large population and will use the sample mean ($\bar{x}$) to estimate the mean of all units in the population. In our example, the population consists of all light bulbs manufactured and the sample consists of those selected for testing.

Figure 3.1 Populations and Samples

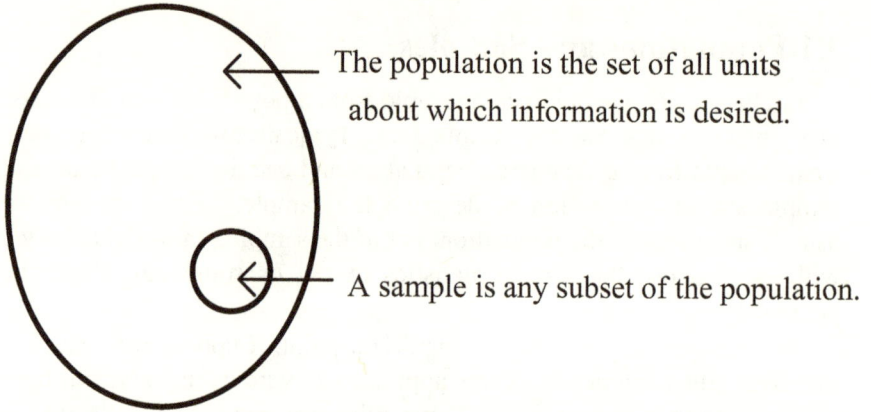

The population is the set of all units about which information is desired.

A sample is any subset of the population.

Political polls can also be used to illustrate the relationship between samples and populations. Suppose we take a sample of 100 likely voters and determine that 53% of the voters will vote for the Republican candidate. We are interested in estimating the population proportion, which is the proportion of all voters in the state who will vote for the Republican candidate. The sample proportion (53%) is used to estimate the population proportion, which is unknown. The population proportion is not known because it is too expensive to interview everyone in the state in order to compute it.

Whenever a sample mean is used to estimate the population mean there is always some chance that the sample mean may underestimate or overestimate the population mean. The methods presented in this chapter allow us to quantify the amount of uncertainty in the estimate.   To quantify the uncertainty we need to understand the relationship between the population and the sample in order to see how close the sample mean is likely to be to the population mean. In the next section we begin to describe the relationship between the population and the sample.

3.2 Random Variables, Population Parameters, and Sample Statistics

In the last chapter we discussed experiments, sample spaces, and events. In many of these experiments we could assign numbers to all events in the sample space. The function that assigns numbers to these events will be called a **random variable**. For example, suppose we roll a die. The sample space consists of all possible outcomes for a single roll and we could define a random variable X to be the number of dots shown on the die.

There are two kinds of random variables. A random variable is said to be **discrete** if its set of possible values is finite or if the set consists of elements that can be listed in an infinite sequence. For an experiment that requires a single die to be rolled, if X is the number of dots shown on a die, then X is a **discrete random variable** since there are only six possible outcomes. For an experiment that requires tossing a coin until a head appears, let X equal the number of tosses. The random variable could assume any number in the sequence of positive integers $\{1, 2, 3, ...\}$ and hence would be discrete.

For some experiments the sample space has an infinite number of outcomes on some interval. A **continuous random variable** is one that can assume an infinite number of values corresponding to all points on some line interval. For example, if we could measure the length of a machine part accurately we would have an infinite number of possible lengths.

The distinction between discrete and continuous random variables may seem a bit confusing but for most variables it is fairly easy to see the difference. Many discrete random variables assume a small number of possible outcomes. Many other random variables, such as height and weight, have so many possible outcomes that we usually consider them to be continuous. This is done because it is easier to work with a continuous random variable that approximates a discrete random variable when the number of outcomes is very large. As a practical matter, if the number of outcomes is less than 6 or 7 we usually consider the random variable to be discrete. If the random variable takes on 20 or more values and these values approximate a continuous random variable, then we generally consider the random variable to be continuous.

In this chapter we will assume that we have obtained a **simple random sample** from a population. Roughly speaking, a simple random sample is obtained by selecting units from a thoroughly mixed population. For example, if we thoroughly shuffle an ordinary deck of cards and then select two cards from the deck, we say we have a simple random sample of size 2. In this example all pairs of cards are equally likely to be drawn. That is, the probability of drawing and Ace of Hearts and a Jack of Diamonds is equal to the chance of drawing any other two cards. A precise definition of a simple random sample is:

A sample of size n is said to be a simple random sample if the selection is done in such a way that all combinations of n units have an equal chance of being selected.

If we need to take a simple random sample of 10 individuals from a population of 100 individuals we could make a card for each individual in the population so that the deck would consist of 100 cards. We could then shuffle the deck thoroughly and select 10 cards from it. The sample of size 10 would be a simple random sample because any other combination of 10 individuals would be equally likely.

Professional polling organizations often use random number tables or computer programs that generate random numbers to select a simple random sample. Once the simple random sample has been obtained, functions of the sample measurements $x_1, x_2, \ldots, x_n$ are used to describe the sample. One such function is the sample mean and another is the sample standard deviation. These are examples of **statistics** which are functions of the sample measurements.

The sample mean is used to describe the location of a sample and the sample standard deviation is used to describe the variability in the sample. We now define corresponding measures for a population having N units. The measure of location for the population is the **population mean** which is defined as

$$\mu = \frac{1}{N} \sum_{i=1}^{N} x_i \; ,$$

where $x_1, x_2, \ldots, x_N$ are the measurements on the units in the population. Note that the formula for the population mean is identical to the formula for the sample mean except the summation extends over the

N units in the population.   The **population standard deviation** is defined as

$$\sigma = \sqrt{\frac{\displaystyle\sum_{i=1}^{N}(x_i - \mu)^2}{N}}$$

and the **population variance** (σ^2) is the square of the standard deviation. Except for the summation extending over the N units in the population, the formula is quite similar to that for the sample standard deviation. The population mean and the population standard deviation are examples of population **parameters.**

For example, consider a population consisting of 10000 balls in a large bowl. Each ball is marked with a number. 5000 of these balls are marked with "0", 3000 of the balls are marked with "10", and 2000 of the balls are marked with "20". The population mean is

$$\mu = \frac{5000(0) + 3000(10) + 2000(20)}{10000} = \frac{70000}{10000} = 7$$

and the standard deviation is

$$\sigma = \sqrt{\frac{5000(0-7)^2 + 3000(10-7)^2 + 2000(20-7)^2}{10000}}$$

$$= \sqrt{\frac{610000}{10000}} = 7.81$$

Now, if we take a sample of size $n = 100$ from the bowl it is quite possible to obtain 52 balls marked "0", 27 balls marked "10", and 21 balls marked "20". The sample average is

$$\bar{x} = \frac{52(0) + 27(10) + 21(20)}{100} = \frac{270 + 420}{100} = \frac{690}{100} = 6.9.$$

In the usual situation we would not know μ so we would use $\bar{x} = 6.9$ to estimate an unknown μ. In this hypothetical situation we can see that

$\bar{x}$ would be close to μ, but in a real situation we would not know how close μ is to $\bar{x}$. The sample standard deviation would be

$$s = \sqrt{\frac{52(0-6.9)^2 + 27(10-6.9)^2 + 21(20-6.9)^2}{100-1}} = \sqrt{\frac{6339}{99}} = 8.00$$

Note that with this sample of $n=100$ the sample standard deviation of $s=8.00$ would be a very good estimate of the population standard deviation. Had we used a smaller sample, the estimates obtained from the sample may have greatly underestimated or overestimated the population values.

Exercise Set A

1. A consumer organization wanted to determine the proportion of light truck owner who are satisfied with the performance of their vehicle. Suppose there are 20,000,000 light trucks on the road and that 500 of these owners were selected to be interviewed.

    a) Identify the population. Does the population consist of trucks or people?

    b) Identify the sample.

    c) What are the values of N and n?

2. In the United States political polling organizations often conduct large surveys within states in the last few months before an election for Governor to determine the support for the candidates. Specify the population that should be sampled in a state.

3. An aircraft company, which manufactured a certain model of airplane for several years, wanted to investigate the planes that were in service to see if a certain part was functioning properly. They took a simple random sample of $n=3$ planes from the population of the $N=200$ planes that were manufactured. The planes were numbered 1 through 200. They selected planes 32, 113, and 147 for

inspection.   A company statistician calculated that the probability of selecting these planes was .00000076 .

a) What is the chance that planes 14, 87, and 152 would have been selected ?

b) What is the chance that planes 8, 9, and 10 would have been selected?

4. An exercise scientist interviewed a sample of 100 adults in a community in order to determine the extend of obesity in the community.   In the interview she obtained information on age, gender, height, and weight.   Which of these variables would be continuous and which would be discrete?

5. A data set has the data on high school seniors who are applying to colleges.   The first few observations are listed below:

Name	Gender	GPA	Admissions test score	Apply for financial aid
Smith, Alice	F	3.21	815	Yes
Jones, John	M	2.87	783	No
Schwartz, Herb	M	2.92	982	No

The "Name" is a discrete identifying variable.   Which of the other four variables are discrete and which are continuous?

6. Suppose a health magazine for men tabulated the results of a questionnaire that was sent in by 2000 readers of the magazine.   The magazine had a circulation of approximately 50,000.

a) Is the sample of 2000 responses a simple random sample?   Why or why not?

b) Do you believe that the results of the responses reflect the views of the readers?   Why or why not?

7. If we roll a die 6000 times we would expect that each face will be shown on approximately 1000 rolls of the die. A frequency histogram of these expected outcomes is

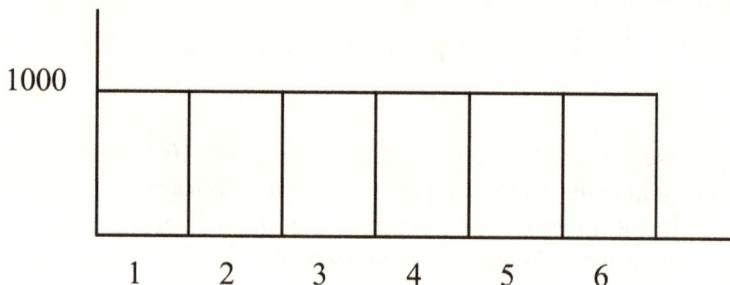

We will consider the 6000 rolls and their expected frequencies sufficient to define a population. Compute the population mean using these $N = 6000$ rolls. Compute the population standard deviation.

8. A die is rolled 10 times and the results are displayed in this frequency histogram:

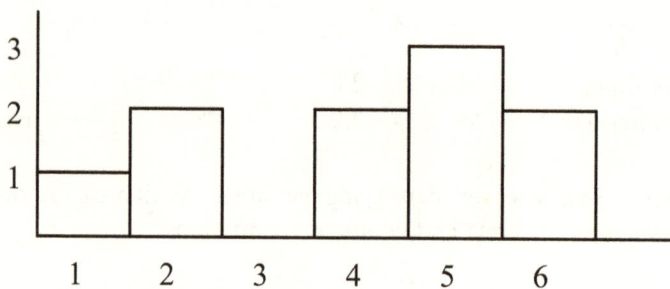

a) Compute the sample mean.

b) Compute the sample standard deviation.

c) Compare your results to the results found in the previous exercise.

3.3 The Sampling Distribution of the Mean and the Central Limit Theorem

The Sampling Distribution of the Mean

One of the most common statistical procedures is to use the sample mean to estimate the population mean. For example, suppose we are interested in estimating the average weight for a population of $N = 2000$ males that are 20 years of age at the University of Timbuktu. If we took a sample of size $n = 10$ from that population we would naturally use the sample mean ($\bar{x}$) as an estimator of the population mean (μ). In this section we will discuss some important ideas that will enable us to estimate how close $\bar{x}$ might be to μ.

Let us assume, for the purpose of this example, that we know the weights for the $N = 2000$ members in the population. A histogram is given in Figure 3.2 along with the population mean ($\mu = 170$) obtained by averaging over the 2000 students in the population. The standard deviation of the population is $\sigma = 42$. Now suppose we take a simple random sample of $n = 10$ males from the population. In the bottom part of Figure 3.2 we display the sample values in a dot diagram, along with the sample mean. Note that the sample mean is reasonably close to the population mean and we would not have been too much in error if we had used the sample mean as our estimate of the population mean. Had we taken a different random sample of size 10 we would have obtained a different value for the sample mean. If we are going to use the sample mean to estimate the population mean, we will need to know how the sample means vary from sample to sample.

Figure 3.2 A histogram of weights from a population of 20 year old males and one sample from that population.

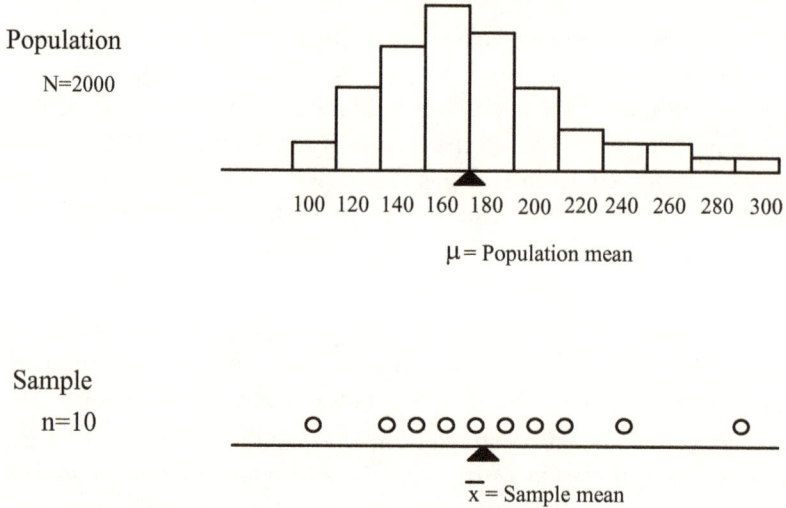

Population
  N=2000

100 120 140 160 180 200 220 240 260 280 300

μ = Population mean

Sample
  n=10

$\overline{x}$ = Sample mean

Now suppose we take 10 samples with each sample consisting of 10 observations. That is, we take a total of 100 observations. Figure 3.3 presents the histogram for the population along with 10 dot diagrams for each of the 10 samples. Note that the sample mean is indicated for each of the 10 samples. There is some variability of the sample means but the means are not nearly as variable as the original observations.

Figure 3.3 A histogram of weights from a population of 20 year old males and ten samples from that population. The mean values are indicated by triangles.

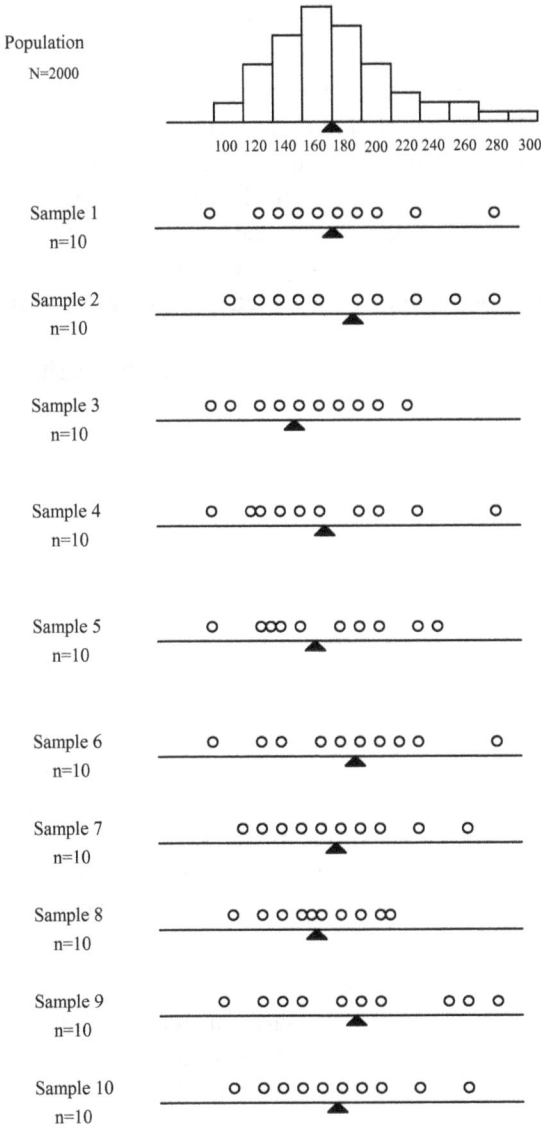

Figure 3.4 gives the histogram of the sample means for the 10 samples of size $n = 10$. This is the **sampling distribution of the mean**. It is important not to confuse the histogram of the sample means with the histogram of the weights in the population that was shown in Figure 3.2. By comparing the location of the population mean in Figure 3.3 and the location of the sample means it can be seen that the sample mean sometimes slightly underestimates the population mean and sometimes slightly overestimates the population mean. However, the average of the sample means, over all samples of size n, equals the population mean μ. We say that the sample mean is an **unbiased estimator** of the mean because it does not consistently understate or overstate the population mean. An estimator that consistently overestimates or consistently underestimates the population mean is called a **biased estimator**.

Figure 3.4 Histogram of the sample mean weights

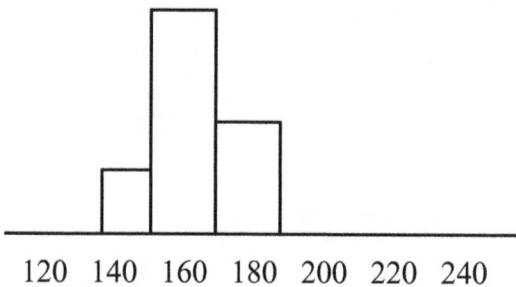

120   140   160   180   200   220   240

It is important to note that the sample means are much less variable than the observations in the population. This can be seen by comparing the sampling distribution of the mean shown in Figure 3.4 to the distribution of the observations in the population shown in Figure 3.2. The variability of the sample means is measured by the standard deviation of the sample means which is called the **standard error of the mean** ($\sigma_{\bar{x}}$).

We could estimate the standard error of the mean if we had repeated samples. In our example, the mean values for our samples are approximately {170, 182, 145, ...} and we could compute the standard deviation of these mean values to provide an estimate of the standard error of the mean.

Of course, in a real example, we would have only one mean value because we would have taken only one sample. So we cannot estimate the standard error of the mean directly. Instead, we must use some results from mathematical statistics to estimate it. We first note that the standard error of the mean should decrease as the sample size increases. This makes sense because, in large samples, the sample mean will be a precise estimate of μ and consequently $\bar{x}$ which will tend to vary little from sample to sample. Statistician have shown that the standard error of

the mean is $\sigma_{\bar{x}} = \dfrac{\sigma}{\sqrt{n}}$, where σ is the population standard deviation.

Thus, when sampling from a population with standard deviation σ, the standard error of the mean decreases as n increases. For samples of size $n = 100$ the standard error of the mean, which is the standard deviation of the mean values, will be one-tenth the standard deviation of the measurements in the population. The main idea can be summarized as follows:

> **The standard error of the mean, which is the standard deviation of the mean values obtained from repeated samples of size n, is given by $\sigma_{\bar{x}} = \dfrac{\sigma}{\sqrt{n}}$.**

For our example using the population of male students at the University of Timbuktu, the standard deviation of the weights is $\sigma = 42$. Thus, the standard error of the mean, is $\sigma_{\bar{x}} = \dfrac{\sigma}{\sqrt{n}} = \dfrac{42}{\sqrt{10}} = 13.28$. Since the standard error of the mean is the standard deviation of the sample mean in repeated samples, there is much less variability in the sample mean than there is in the original population.

We now know two important facts that describe the relationship between the population and the sample. The first fact is that the sample mean ($\bar{x}$) is an unbiased estimator of the population mean (μ). The second fact is that the standard deviation of the sample means, which we call the standard error of the mean, can be computed as $\sigma_{\bar{x}} = \dfrac{\sigma}{\sqrt{n}}$. Thus, if we know the population mean and standard deviation we can easily compute the   location and   variability of the sampling distribution. However, in most applications we also need to know the appropriate distribution of the sample means. For large samples the distribution is determined by the central limit theorem.

The Central Limit Theorem

For large samples the shape of the sampling distribution is easy to determine. The **central limit theorem** states that the sampling distribution of the sample means will be approximately normal if the sample size is sufficiently large. This is true even if the relative frequency histogram of the population is highly skewed or has longer or shorter tails than the normal curve. The central limit theorem can be illustrated with the histogram of weights from a population of 20 year old males shown in Figure 3.5. This distribution is not normal; it is skewed to the right. Yet, if we take repeated samples of 36 students from that population, the central limit theorem tells us that the sampling distribution of the sample means will be approximately normal.

The central limit theorem is illustrated in Figure 3.5 which gives the sample histogram of weights for the mean values obtained from 100 samples having $n = 36$ observations in each sample. Note that the sampling distribution, unlike the original population, is approximately normal. Since the population mean was $\mu = 170$, the mean of these sample means should be near 170. Since the population standard deviation   was   42   the   standard   error   of   the   mean   is $\sigma_{\bar{x}} = \dfrac{\sigma}{\sqrt{n}} = \dfrac{42}{\sqrt{36}} = \dfrac{42}{6} = 7$. Consequently, the standard deviation of the 100 mean values should approximate 7. As we shall see, knowledge of the mean, standard deviation, and the distribution of the mean value will allow us to estimate the chance of $\bar{x}$ begin close to μ.

Figure 3.5 A histogram of weights from a population of 20 year old males and the sampling distribution of the sample mean based on 100 samples of *n*=36 observations.

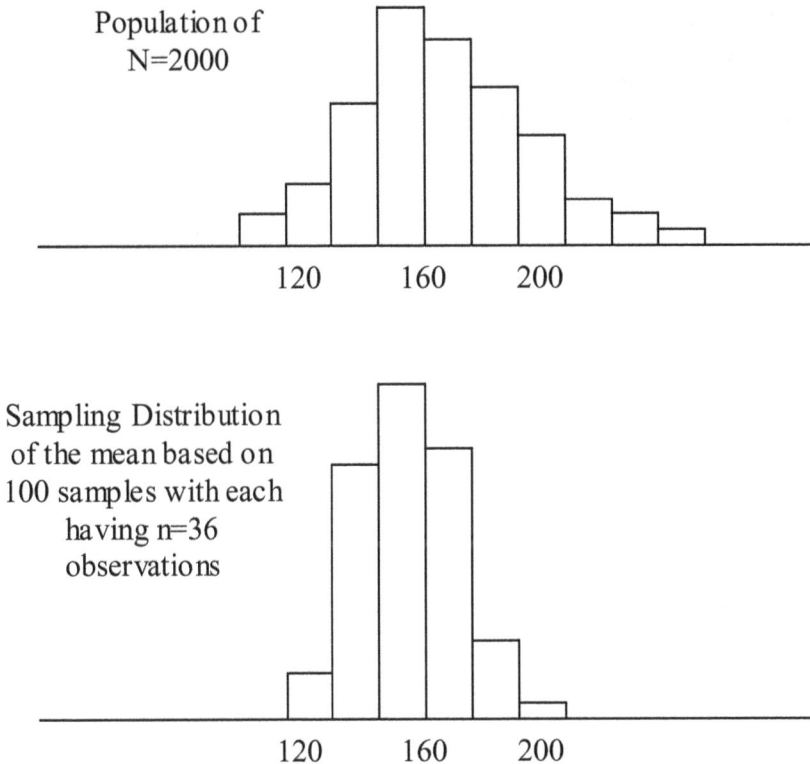

Population of
N=2000

120   160   200

Sampling Distribution
of the mean based on
100 samples with each
having n=36
observations

120   160   200

The central limit theorem states that for large samples the sampling distribution of the sample means will be approximately normal but it does not state just how large the sample must be to insure that the sampling distribution approximates the normal distribution. However, statisticians have shown that samples of $n \geq 30$ are usually sufficient if the population distribution is symmetric or slightly skewed. If the population distribution is extremely skewed the sample size may need to be 100 or more for the sampling distribution to approximate the normal. In this text we will not work with extremely skewed or bizarre distributions so the sampling distribution of the sample mean will closely approximate the normal distribution in this text whenever $n \geq 30$.

The main idea of the central limit theorem is illustrated in Figure 3.6. While the distribution of the population is quite skewed, note that the distribution of the sample means for $n=4$ is less skewed than the population distribution. Also note that for $n=36$, the sampling distribution is very close to normal. That is, as the sample size increases the overall shape of the distribution of sample means tends to more closely approximate the normal distribution.

**Figure 3.6 Sampling distribution of the mean for samples of
size *n*=4 and *n*=36.**

Population

skewed

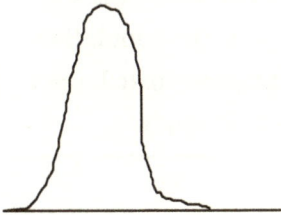

Sampling Distribution of the Mean

n=4

slightly skewed

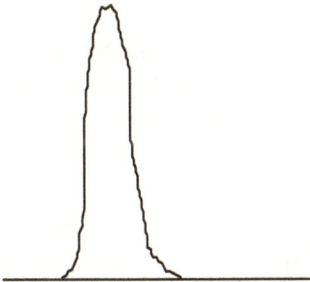

n=36
approximately

normal

Although the central limit theorem states that, for samples sufficiently large, the distribution of sample means will be approximately normal, it says nothing about the small sample properties of the sample means. We will now investigate the small sample distributions of the sample means for samples having fewer than 30 observations.

Suppose we have a very small sample of size $n = 4$ from a population. Since we have such a small sample we really don't have a clear idea about the shape of the distribution of the population. If the population is normal the distribution of the sample means will be exactly normal and if the population closely approximates the normal then the sample means will very closely approximate the normal. However, the mean values based on small samples from highly skewed distributions are not normally distributed but tend to be slightly skewed. Therefore it is not possible to compute probabilities based on the normal distribution when small samples are taken from highly skewed populations.

We can summarize what we know about the distribution of sample means as follows:

> **If the sample size exceeds 30 and the population distribution is not extremely skewed, the sampling distribution of the means will closely approximate the normal distribution. If the population distribution is approximately normal, the sampling distribution will be normal even if the sample size is small.**

Using The Central Limit Theorem

In the first chapter we noted that, for normal distributions, 95% of the observations were within 2 standard deviations of the mean. Since, for large samples, the distribution of the sample mean is approximately normal, we can predict where an individual sample mean might fall. For our example concerning a population of 20 year old males we know that the population distribution is approximately normal with a mean of $\mu = 170$ and a standard deviation of $\sigma = 42$. Thus, with $n = 36$, the sampling distribution of the mean will be normal and the standard error of the mean will be $\sigma_{\bar{x}} = \dfrac{\sigma}{\sqrt{n}} = \dfrac{42}{\sqrt{36}} = \dfrac{42}{6} = 7$. Since the standard error of the mean is the standard deviation of the sample means, we know, with repeated sampling, that 95% of these mean values will be within 2 standard errors of the mean from the population mean. That is, they will be in the interval $[170 - 2(7), 170 + 2(7)] = [156,184]$. Note that we used the standard error of the mean in this calculation because we are interested in the mean. The basic rule is that, when using formulas to compute intervals for a random variable, you must use the mean and standard deviation for that random variable. In this case we were interested in computing an interval for the sample mean for samples of

$n = 36$, so it was necessary to use the standard deviation of the means from samples having size $n = 36$, which is the standard error of the mean.

Exercise Set B

1. Suppose that the distribution of the height of adult females is 64 inches and that the population standard deviation is 2.5 inches and that the distribution is approximately normal in shape. Which of the following histograms could be the sampling distribution of the sample mean for samples of size n=25.

a)   b)   c)

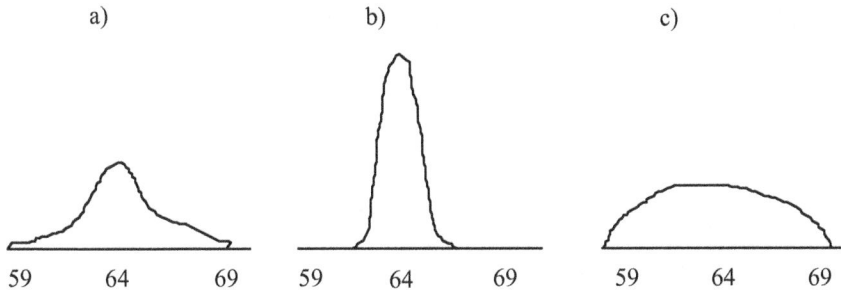

| 59 | 64 | 69 | 59 | 64 | 69 | 59 | 64 | 69 |

2. Suppose the population distribution of the number of hours per weeks that the average high school student spends on the World Wide Web is :

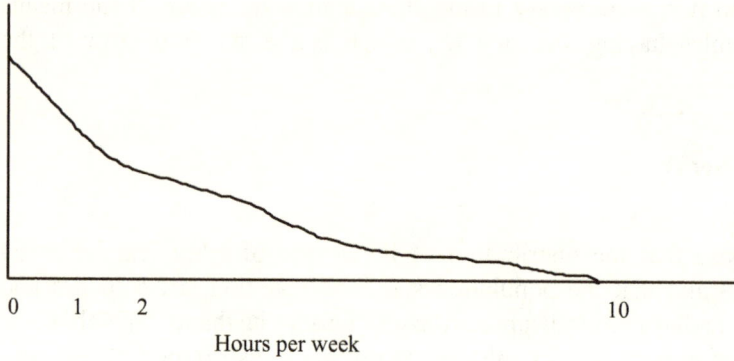

Hours per week

One of the following distributions is the sampling distribution of the sample mean for a sample size of $n = 4$ and one is for a sample size if $n = 25$. Match the sample size to the distributions.

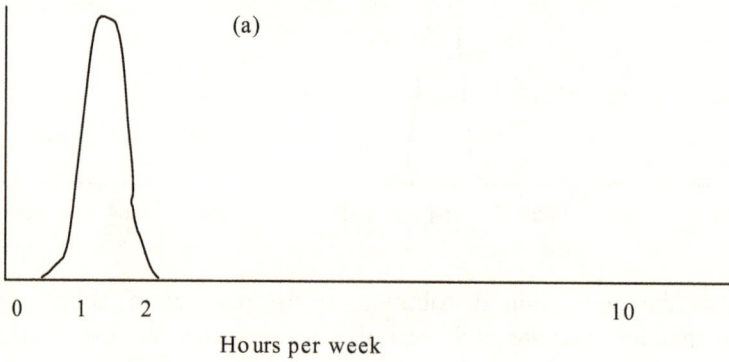

(a)

Hours per week

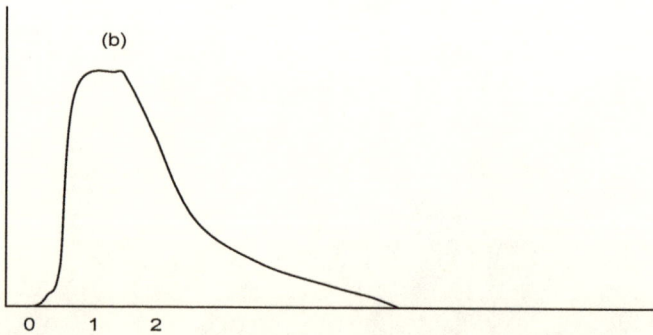

(b)

3. Suppose that 36 students in a statistics class each draw, with replacement, four tickets from the box below. Each student computes a mean value based on these 4 draws. If the population standard deviation of the tickets in the box is 1.41, what is the standard error of the mean? Will the standard deviation of the mean values obtained by students be close to 1.41 or will it be close to the standard error of the mean?

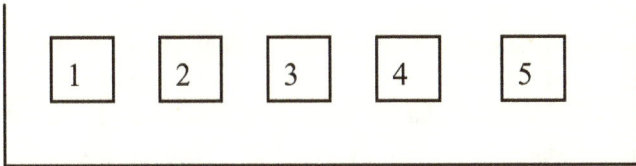

4. Suppose that in a state or country the average family income is $40,000 per year and the standard deviation is $30,000 per year. What is the standard error of the mean for samples of size $n = 100$ selected at random from this population?

5. A college offers an intensive 2 week course in statistical methods in the summer. After teaching the course a few years an instructor determines that the population distribution of the final exam scores looks something like this:

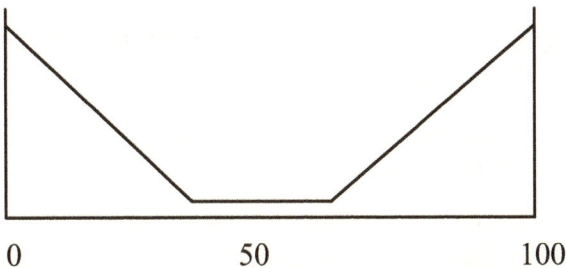

The distribution shows that many students do very well and that many students do very poorly.

a) Will the sampling distribution of the mean for samples of size $n = 2$ be approximately normal? Why or why not?

b) Will the sampling distribution of the mean for samples of size $n = 50$ be approximately normal? Why or why not?

6. Suppose a social scientist knows a great deal about the television viewing habits of adults in the U.S. Suppose the number of viewing hours per week by adults has a population mean of $\mu = 20$ and a population standard deviation of $\sigma = 15$. Answer the following questions based on these population parameters and the histogram of TV viewing hours.

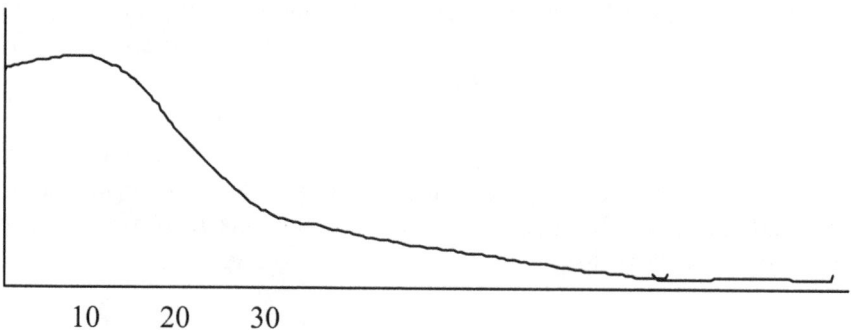

```
     10    20    30
```

a) For a sample of size $n = 225$, what is the standard error of the mean $(\sigma_{\bar{x}})$?

b) Is the sampling distribution of the sample mean normally distributed for samples of size $n = 255$?

c) Approximately what percentage of TV viewers watch between $(20 - 2\,\sigma_{\bar{x}})$ and $(20 + 2\,\sigma_{\bar{x}})$ hours per week?

5%   50%   68%   95%

c) If you take many samples, each of size $n = 225$, what percentage of these means will fall between $(20\text{-}2\,\sigma_{\bar{x}})$ and $(20\text{+}2\,\sigma_{\bar{x}})$ hours per week?

5%   50%   68%   95%

7. Suppose we have obtained the population distribution of the time to obtain pain relief after taking two aspirin tables. The relative frequency histogram is shown below.

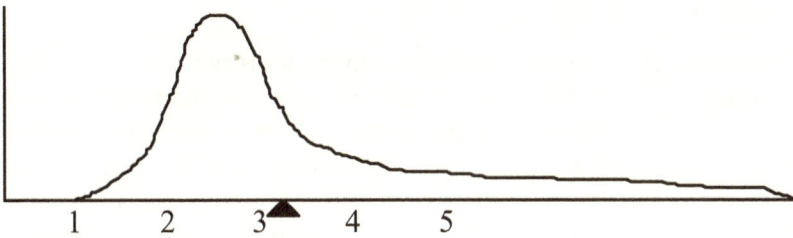

The population parameters are :
$$\mu = 3.2 \qquad\qquad \sigma = 2.0$$
Now suppose we take 200 samples, each of size $n = 50$, from this population.

a) Is the distribution of time to pain relief approximately normal?

b) Would the distribution of the sample means be approximately normal?

c) The average of 200 sample means should be approximately equal to _____.

d) The standard deviation of the 200 sample means should be approximately equal to ___.

e) Approximately 95% of the 200 sample means should fall between ___ and ___.

8. Suppose we try to design a survey to estimate the number of cigarettes smoked per day by adult females between the ages of 20 and 30. Before we decide on a sample size, we want to calculate the standard error for several sample sizes. If $\sigma = 8$, what would be the standard error for studies of size $n = 25$, 64, 100, and 400. Make a plot with $\sigma_{\bar{x}}$ on the y-axis and n on the x-axis and connect the points. What does this graph indicate about the relationship between n and $\sigma_{\bar{x}}$?

3.4 The Normal Distribution

The normal distribution is the most important distribution in statistics. Its importance derives from the fact that the sample mean, which is often used to estimate the population mean, is approximately normally distributed if the sample size is large. The normality of the sample mean is guaranteed by the central limit theorem for large samples. The central limit theorem also guarantees that the sample proportions are approximately normally distributed for large samples.

The normal distribution can also be used as an approximation to some distributions of real data. For example, the heights of adult men and the heights of adult women are approximately normally distributed. Although most real world distributions are skewed to the right, many roughly approximate the normal. Also, as we shall see in Chapter 7, the normal distribution can be used to calculate the chance of obtaining a certain number of heads in a coin flipping experiment. Because there are many applications of the normal distribution the student needs to become very familiar with the properties of this distribution.

The Standard Normal Distribution

We will often use random variables that are normally distributed, have a mean of zero, and have a standard deviation of one. The normally distributed variables having $\mu = 0$ and $\sigma = 1$ are called **standard normal variables** and the normal distribution having $\mu = 0$ and $\sigma = 1$ is called the **standard normal distribution**. If we had thousands of observations from a standard normal distribution and constructed a histogram using a large number of intervals we would obtain a histogram something like that shown in Figure 3.7 . The histogram approximates the smooth bell-shaped curve given in Figure 3.7, which was first used by DeMoivre in 1733 to simplify certain calculations and which is usually called the normal curve. The density function of the standard normal curve, which has $\mu = 0$ and $\sigma = 1$, can be written as

$$f(x) = \frac{e^{-x^2/2}}{\sqrt{2\pi}}$$

where $e \cong 2.71828$ and $\pi \cong 3.14159$. As we shall see, there is no need to calculate $f(x)$ because we generally need to calculate the areas under the normal curve rather than $f(x)$ itself.

Figure 3.7 Histogram for normally distributed data and the standard normal curve.

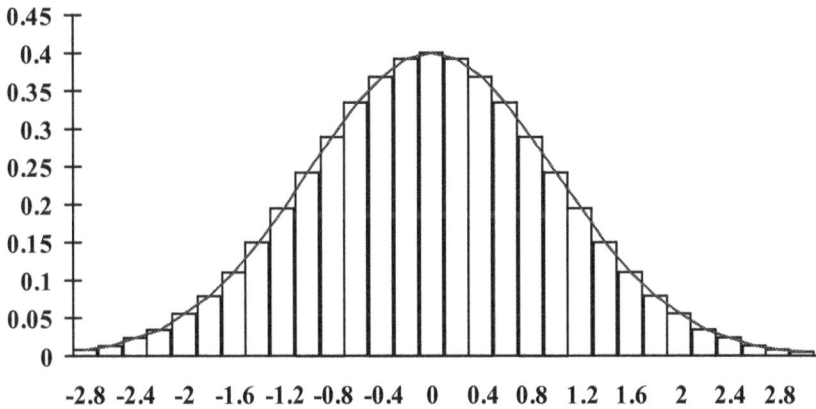

In working with histograms that approximate the normal distribution we often need to estimate the proportion of observations that fall in an interval. Using the relative frequency histogram we can compute this proportion by summing the heights of the vertical bars in that interval. This technique works well if there are a few vertical bars, but becomes cumbersome if many bars are involved.

A more convenient way to calculate the proportion of observations in an interval is to rescale the histogram so that the relative frequency is represented by the area of the vertical bar rather than by its height. Since the area of a vertical bar equals the height of the bar times the width of the bar, we can force the area to equal the proportion by setting the height of the bar equal to the relative frequency divided by the width. By doing the same recoding for every vertical bar we construct a **density histogram**.

The construction of a density histogram can be illustrated with the following example. Suppose we know that 10 of 50 students score between 60 and 70 on a test. On a frequency histogram we would indicate the frequency by the height of the vertical bar. To construct a density histogram, we need to represent the proportion, which is $\frac{10}{50} = .2$, by the area in the vertical bar. From geometry we know that, for rectangles,

$$width \times height = area$$

or that

$$height = \frac{area}{width}.$$

In this example, we want the area to equal the proportion, which is .2, and the width of the interval between 60 and 70 equals 10. Thus, the height should be

$$height = \frac{area}{width} = \frac{.2}{10} = .02$$

for that interval. We would use the same formula to compute the heights of all intervals. By constructing the density histogram using these heights the areas will sum to 1.0 . If the widths are the same for all intervals it is easy to see that the density histogram will have the same overall appearance on the frequency histogram. So why do we bother making a density histogram?

The advantage of using the density histogram is that proportions can be represented by the areas in the vertical bars. Since the normal curve is an approximation to the density histogram, proportions for density histograms can be approximated by the areas under the normal curve, which have been tabulated in Table 1 of the Appendix. Note that the values in the Table give the area to the right of z for any $z \geq 0$. It is traditional to use Z for the normally distributed random variable that has $\mu = 0$ and $\sigma = 1$. The key features of the standard normal distribution can be summarized as follows:

The standard normal distribution:

- **Has a mean of zero and a standard deviation of one.**

- **Is symmetric about zero.**

- **Has a total area under the curve of one.**

Let $P(Z > z)$ be the probability that the random variable Z exceeds a specified value z. Table 1 is set up to read the area to the right of any specified value. Thus, the probability that Z exceeds 1.0, which is represented by the shaded area in Figure 3.8, can be read directly from the table as $P(Z > 1) = .1587$. The area to the left of any value can be computed easily since the total area must equal one. Consequently, since $P(Z < 1) + P(Z > 1) = 1$, we can compute

$$P(Z < 1) = 1 - P(Z > 1) = 1 - .1587 = .8413$$

We can easily compute $P(Z < -z)$ from the table because the standard normal curve is symmetric about zero. Therefore, $P(Z < -z) = P(Z > z)$ for all $z > 0$. For example,

$$P(Z < -2.1) = P(Z > 2.1) = .0179$$

It is important to remember that for the normal distribution the probability associated with a single point is zero. That is, if the distribution of Z is continuous, $P(Z = z) = 0$ for any value of z. The probability is zero because, for continuous distributions, the random variable can equal any one of an infinite number of values in the interval. If we use histograms that approximate the normal distribution we will sometimes be concerned with the probability associated with a single point, but for the standard normal random variable Z we know $P(Z \geq z) = P(Z > z)$ for all values of z.

Figure 3.8 Sketch used to calculate P(Z>1) for the standard normal distribution.

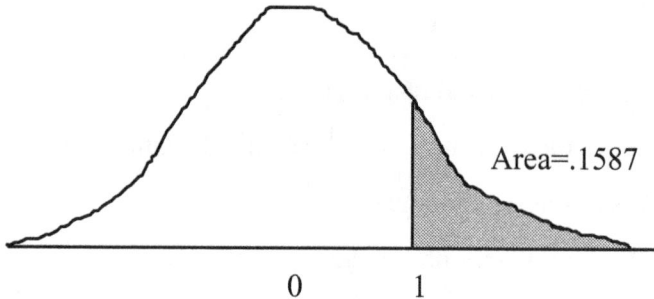

Area=.1587

0   1

With a little ingenuity we can compute the probability associated with any interval. In Chapter 1 we stated that approximately 95% of the observations are within 2 standard deviations of the mean. Using the rough sketch of the area in Figure 3.9 and the values from Table 1 we can compute the more accurate value

$$P(-2 < Z < 2) = P(Z > -2) - P(Z > 2) = .9773 - .0228 = .9545$$

Calculations of this sort can be understood more easily if you make a small sketch of the standard normal distribution and shade in the appropriate area as we have done in Figure 3.9.

Figure 3.9 Sketch used to calculate P(-2<Z<2) for the standard normal distribution.

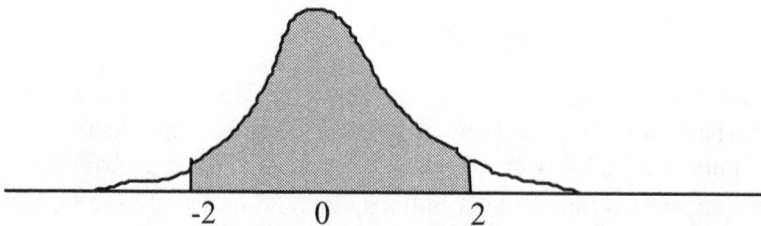

-2   0   2

Table 1 is usually used to compute the right tail area, which is the probability that a standard normal random variable exceeds a certain value. Sometimes we have a probability, which we will denote by α, and we want to determine a value such that α is the probability in the right tail. For a right tail area of $\alpha = 0.05$, we can see from Table 1 that $P(Z > 1.645) = 0.05$. We will say that 1.645 is the **critical value** of the standard normal distribution for $\alpha = 0.05$ and we will represent this critical value by $z_{.05}$. For any probability α that we select, the critical value is found by looking at Table 1 for the value z_α that has α in the right tail. We observe that for $\alpha = 0.05$ the critical value is the 95th percentile of the standard normal. In later chapters we will often specify the probability $\alpha = 0.025$ which gives a critical value of $z_{.025} = 1.96$. This critical value is the 97.5th percentile of the standard normal. In general, the critical value z_α is the 100(1-α) percentile of the standard normal distribution.

Standardizing Variables

If the random variables are normally distributed, we would like to work with standard normal random variables because we can compute probabilities using Table 1. Yet most random variables have a non-zero mean and a variance that differs from one. To work with these variables, we will often transform, or **standardize**, a random variable to obtain a random variable that has a mean of zero and a standard deviation of one. The transformed variable is called a **standardized random variable.** If the original variable was normally distributed, the standardized random variable will also be normally distributed and Table 1 can be used to compute probabilities.

There are two steps in the standardization process. In the first step the mean is subtracted from each observation so that the values $\{x_1 - \bar{x}, x_2 - \bar{x}, \ldots, x_n - \bar{x}\}$ have a mean of zero. The first step is illustrated by an example shown in Figure 3.10. In this sample of n=8 observation, the original observations $\{x_1, x_2, \ldots, x_8\}$ have $\bar{x} = 3$ and $s = 2$. By subtracting the mean from each observation we obtain the **centered observations** $\{x_1 - \bar{x}, x_2 - \bar{x}, \ldots, x_8 - \bar{x}\}$ which have a mean of zero and are shown in the second dot diagram of Figure 3.10. The standard deviation of these observations equals those of the original observations because subtracting a constant from all observations does not change the variability in the data. In the second step, the centered observations are transformed to obtain a standard deviation of one. This is accomplished by dividing each observation by the standard deviation to obtain the standardized observations

$$\left\{ \frac{x_1 - \overline{x}}{s}, \frac{x_2 - \overline{x}}{s}, \ldots, \frac{x_8 - \overline{x}}{s} \right\},$$

which are displayed in the third dot diagram of Figure 3.10. These standardized variables have a mean of zero and a standard deviation of one. If the original observations were normally distributed the standardized variables will have a standard normal distribution. Consequently, probabilities concerning the transformed observations can easily be computed using Table 1. Generally, instead of the two step approach, we combine the two steps by using the formula

$$z_i = \frac{x_i - \overline{x}}{s} \text{ for } i = 1, \ldots n.$$

If the original variables were normally distributed then the standardized variables will also be normally distributed. However, any set of observations can be standardized, even variables that are not normally distributed, but the set of standardized variables may not be normally distributed. In this text the letter z_i will be reserved for standardized observations.

Figure 3.10 A dot diagram of 8 observations, their values
after subtracting the mean, and their standardized values.

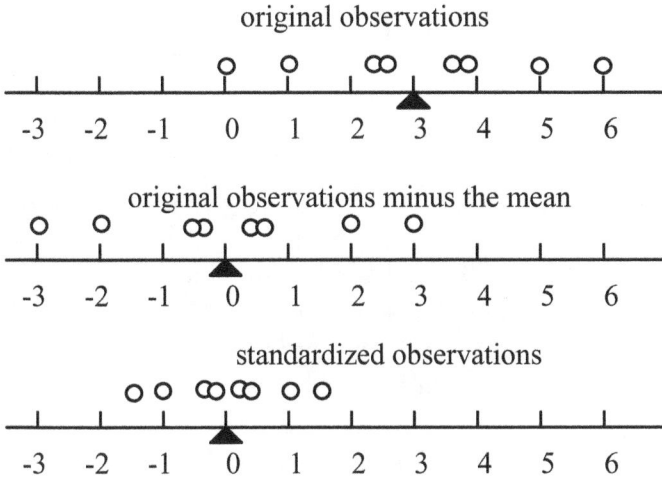

original observations

original observations minus the mean

standardized observations

The Normal Distribution

We often need to work with normal variables that are not standard
normal variables. For example, consider the histogram of heights from a
population of adult females shown in Figure 3.11 . The distribution has a
normal shape with $\mu = 64$ inches and $\sigma = 2.5$ inches. Suppose we want
to compute the chance that a randomly selected female is taller than 70
inches.   Since the only table for the normal distribution is for the
standard normal distribution, we need to standardize the values in the
problem.   Therefore, instead of computing $P(X > 70)$ directly, we
standardize the values on both sides of the inequality to obtain

$$P(X > 70) = P\left(\frac{X - \mu}{\sigma} > \frac{70 - \mu}{\sigma}\right).$$

If we let $Z = \dfrac{X - \mu}{\sigma}$, then Z will be normally distributed with a mean of
zero and a standard deviation of one. Since the transformed variable Z is
distributed as a standard normal distribution, probabilities concerning Z

can be computed using Table 1. We substitute $\mu = 64$ and $\sigma = 2.5$ to obtain

$$P(X > 70) = P\left(Z > \frac{70 - 64}{2.5}\right) = P(Z > 2.4) = .0082.$$

When we compute $z = \frac{x - \mu}{\sigma}$ we say we are expressing the observed value x in **standard units**. The z value indicates the number of standard deviations that x falls above or below the mean value. In our example $z = 2.4$ indicates than $x = 70$ is 2.4 standard deviations above the mean value of $\mu = 64$. This allowed us to compute $P(Z > 2.4)$ directly from Table 1. Note that we indicated the height in inches on the figure and indicated the corresponding standard units below the height.

Figure 3.11 Calculating P(X>70) for a normal distribution with a mean of 64 inches and a standard deviation of 2.5 inches.

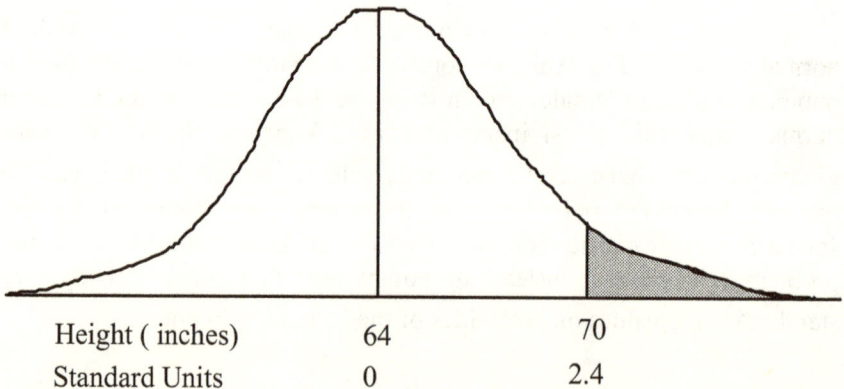

| Height (inches) | 64 | 70 |
| Standard Units | 0 | 2.4 |

It should be noted that there are many valid approaches to solving probability problems for normal distribution. However, these kinds of problems can usually be solved by:

- Drawing a normal curve and indicating the area that corresponds to the probability that is required.
- Standardizing by subtracting the mean and dividing by the standard deviation.
- Using Table 1 to compute the desired probabilities.

For example, suppose we wanted to determine the chance that a randomly selected adult female will have a height between 59 inches and 69 inches. The probability can be represented by the shaded area in Figure 3.12. If we standardize the variables and values we obtain

$$P(59 < X < 69) \;=\; P\left(\frac{59-\mu}{\sigma} < \frac{X-\mu}{\sigma} < \frac{69-\mu}{\sigma}\right)$$

$$= \; P\left(\frac{59-64}{2.5} < Z < \frac{69-64}{2.5}\right)$$

$$= \; P(-2 < Z < 2)$$

Now we observe that the shaded area cannot be obtained directly from Table 1, but can be obtained by subtracting the area above 2 from the area above -2. Thus,

$$P(59 < X < 69) = P(-2 < Z < 2)$$

$$= P(Z > -2) - P(Z > 2) = .9772 - .0228 = .9544$$

As can be seen, Table 1 can be used to calculate any probability for any normal distribution, provided you are able to identify the probabilities that need to be added or subtracted.

Figure 3.12 Calculating P(59<X<69) for a normal distribution with a mean of 64 inches and a standard deviation of 2.5 inches.

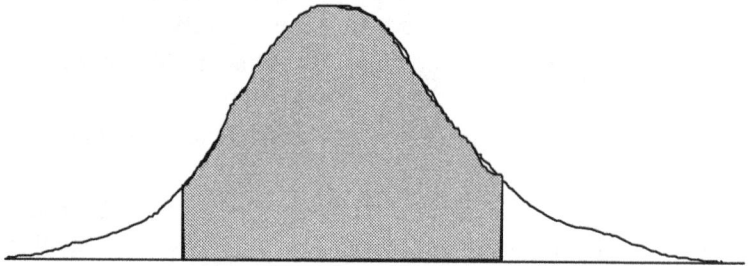

Height (inches)	59	64	69
Standard Units	-2	0	-2

This same approach can be used to calculate probabilities for any variable that has a normal distribution. It is commonly used with the sample mean which we know is approximately normally distributed if the sample size is large. For example, suppose we took a sample of size $n = 36$ and wanted to determine if the sample mean would fall between 63.5 inches and 64.5 inches. Since we are working with the sample mean we will need to use the standard deviation of the sample mean, which is the standard error of the mean $(\sigma_{\bar{x}})$. Since $\sigma = 2.5$ the standard error is $\sigma_{\bar{x}} = \dfrac{\sigma}{\sqrt{n}} = \dfrac{2.5}{\sqrt{36}} = .417$ and we can compute the probability as

$$
\begin{aligned}
P(63.5 < \bar{X} < 64.5) \ &= \ P\left(\frac{63.5 - \mu}{\sigma_{\bar{x}}} < \frac{\bar{X} - \mu}{\sigma_{\bar{x}}} < \frac{64.5 - \mu}{\sigma_{\bar{x}}}\right) \\
&= \ P\left(\frac{63.5 - 64}{.417} < \frac{\bar{X} - \mu}{\sigma_{\bar{x}}} < \frac{64.5 - 64}{.417}\right) \\
&= \ P(-1.2 < Z < 1.2) = .8849 - .1151 = .7698
\end{aligned}
$$

Calculations of this kind are often used in statistical work because $\bar{x}$ will be approximately normally distributed if the sample size is large enough. This calculation assumes that we know σ, but we will see in the next section how to avoid making this assumption.

Exercise Set C

1. Assume that Z is a standard normal random variable with a mean of zero and a standard deviation of one. Compute

    a) $P(Z > 1)$

    b) $P(Z > -1)$

    c) $P(Z < -1)$

    d) $P(-1 < Z < 1)$

2. Suppose that X is normally distributed with $\mu = 100$ and $\sigma = 15$.

    a) Compute $P(X > 100)$.

    b) Is $P(X > 102) = P(X < 98)$?

    c) Is $P(X > 102) = P(X < 102)$?

    d) Compute $P(X > 115)$.

    e) Compute $P(X < 70)$.

    f) Compute $P(70 < X < 115)$.

3. A college uses an admissions test that has a mean of 500 points and a standard deviation of 100 points.

    a) What is the chance that a randomly selected student will have a score greater than 400?

    b) If the college admits students who have a score of at least 325 points, what percentage of those taking the test would meet the admissions criterion?

4. (Challenging) A coffee vending machine dispenses coffee into 10 ounce cups. The machine can be adjusted so that the average amount

of coffee dispensed, over many cups, can be set to any value between 5 and 15 ounces. If the standard deviation of the coffee that was dispensed was $\sigma = 1$ ounce, what mean values should the machine be set to so that 95% of the time the coffee does not overfill the cups?

5. Suppose we roll two dice and count the number of dots on both dice. If we rolled the dice a large number of times we would find that $\mu = 7$ and $\sigma = 2.415$.   If we roll the pair of dice 25 times what is the approximate chance that the average of the 25 rolls is between 6 and 8? You may use the normal approximation to the sampling distribution of the mean.

6. A body-builder who eats 3 bananas every day decided that he needs to get 1500 mg. of potassium every day. Suppose the distribution of potassium is normal with a mean of 600 mg. and a standard deviation of 100 mg.

    a) What is the average amount he needs in the three bananas to obtain 1500 mg. per day?

    b) What is the standard error of the mean for $n=3$ bananas?

    c) What is the chance that the total potassium intake of the three bananas exceeds 1500 mg. per day?

7. (Challenging) A tire company claims that their tires last 60,000 miles. A consumer organization claims that the tires last only 55,000 miles. Both agree that there is quite a bit of variability in tire durability and that $\sigma = 10000$. The consumer organization takes a sample of 25 tires from different cars and computes an average life of $\bar{x} = 54,000$ miles. The tire company claims that the observed mean is a "fluke" and that if another sample was taken the average might be over 60,000 miles.

    a) If the population mean really is 60,000 miles, what is the chance of obtaining a sample mean of 54,000 or less?

    b) If the population mean really is 55,000 miles what is the chance of obtaining a sample mean of 54,000 miles or less?

    c) Who do you believe and why?

3.5 The Sampling Distribution of the Mean with Unknown Variance

In the last section we were able to compute probabilities for the normal distribution because we knew μ and σ. However, in most real-world situations the values of μ and σ are not known and hence we cannot compute the standardized value $z = \dfrac{x - \mu}{\sigma}$. In this section we will discuss the sampling distribution of the mean when we know μ but do not know σ.

There are some situations where we believe we know the value of μ but do not know the value of σ. For example, a researcher may have no reason to believe that a low salt diet would increase or decrease weight. That is, if the diet has no influence on weight gain, the weight gain should have a mean of zero. In a typical research situation we would take a random sample of people on this diet and compute their sample mean weight gain. In this situation we can assume that $\mu = 0$ but cannot standardize the data because σ is unknown.

One approach is to standardize using the sample standard deviation instead of the population standard deviation in the formula for the standard error of the mean. That is, we will use $s_{\bar{x}} = \dfrac{s}{\sqrt{n}}$ as an estimate of the standard error of the mean instead of $\sigma_{\bar{x}} = \dfrac{\sigma}{\sqrt{n}}$, because σ is unknown. To compute probabilities associated with a mean value when μ is assumed known we standardize the mean values using $s_{\bar{x}}$ instead of $\sigma_{\bar{x}}$ to obtain the standardized variable T

$$T = \frac{\bar{x} - \mu}{s_{\bar{x}}} = \frac{\bar{x} - \mu}{s/\sqrt{n}}.$$

Because the sample standard deviation was used to standardize $\bar{x}$, the distribution of T does not equal the distribution of $Z = \left(\dfrac{\bar{x} - \mu}{\sigma_{\bar{x}}} \right)$.

However, for very large samples it can be shown that $s_{\bar{x}}$ is an excellent estimate of $\sigma_{\bar{x}}$ and that the distribution of Z closely approximates the distribution of T. For small samples the T statistic is more variable than the Z statistic, because s sometimes seriously underestimates σ which can cause T to be quite large. The standard deviation will be underestimated with small samples when the observations happen to be clumped together.

The distribution of T was investigated by William Sealy Gosset, who worked as a brewer in the Guinness brewery in Dublin. He often used small samples to estimate the barley yields from farms in the area, and in 1908 he published a paper which demonstrated that the random variable T followed a distribution that was later called the "Student's" t distribution. In his early writings Gosset used the pseudonym "Student" to publish his results and that name became attached to the distribution. In this text we will refer to the distribution simply as the t distribution.

The t distribution is actually not one distribution but is a family of distributions. For a sample of size n we will need a certain kind of t distribution which is determined by the sample size. If n is large we will use a t distribution that closely approximates a standard normal distribution. If n is small we will use a t distribution that has much longer tails than the standard normal. In order to determine the appropriate t distribution, we first need to calculate the **degrees of freedom,** which equals $n-1$. Gosset was able to demonstrate that the standardized variable

$$T = \frac{\bar{x} - \mu}{s_{\bar{x}}} = \frac{\bar{x} - \mu}{s/\sqrt{n}}$$

is distributed as a t distribution with $v = n-1$ degrees of freedom, where v is the Greek lowercase nu. The standardized variable is based on a hypothesized μ and an estimated standard error.

Figure 3.13 shows t distributions for $v = 3$ and $v = 35$ degrees of freedom, which correspond to samples of $n=4$ and $n=36$ respectively. Note that a random variable that follows the t distribution with 3 degrees of freedom is much more variable than a standard normal random variable. It can be shown that a random variable that has a t distribution with v degrees of freedom has a standard deviation of $\sqrt{v/(v-2)}$. Thus, if $v = 3$ the random variable has a standard deviation of 1.732. Conversely, the t distribution with $v = 35$ has a standard deviation of 1.030 and it closely approximates the standard normal distribution. For $v > 30$ there is little practical difference between the normal and the t distribution and it is possible to use the standard normal distribution instead of the t distribution for most applications.

Figure 3.13 The normal distribution and two t distributions

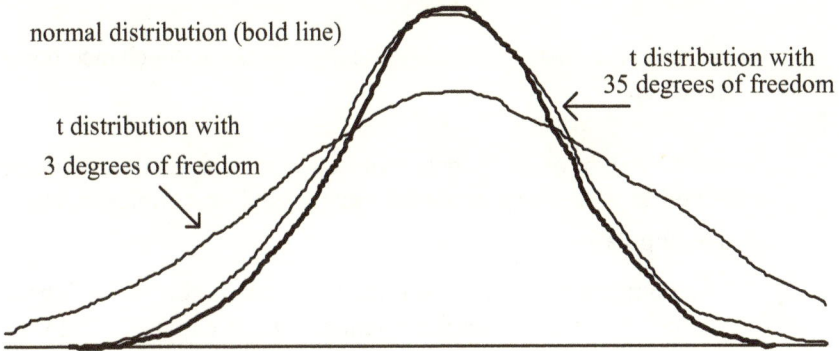

normal distribution (bold line)

t distribution with 35 degrees of freedom

t distribution with 3 degrees of freedom

Probabilities for t statistics are given in Table 2, which is set up in a different manner than that used for the standard normal distribution. In Table 2 the degrees of freedom are listed in the left column and the α levels are listed on the top of the columns. If we need a critical value for a t statistic at $\alpha = 0.05$ with $v = 3$ degrees of freedom we look in the row corresponding to v and the column corresponding to α. Thus, from the third row under $\alpha = 0.03$ we find the critical value of 2.353. This means that in samples of size $n = 4$ from a normal distribution with mean $\mu = 0$, there is a 5% chance that the value of t will exceed 2.353. In subsequent chapters we will use $t_{\alpha, v}$ to indicate the critical value of the t distribution for α with v degrees of freedom. With this notation $t_{.05, 2} = 2.353$.

Exercise Set D

1. Suppose we select a sample of size $n = 10$ from a normal population having $\mu = 80$. If we compute $\bar{x}$ and s from the sample and compute

$$t = \frac{\bar{x} - 80}{s/\sqrt{10}},$$

   what is the chance that t will exceed 2.25? What is the chance that t will exceed 3.25?

2. Suppose the height of adult males follows a normal distribution with $\mu = 69$ and $\sigma = 3$. Suppose we take many random samples of size 5 from that population.

   a) If we compute the sample standard deviations for each of these samples, will the standard deviation always equal 3? Why or why not?

   b) If we compute $(\bar{x} - 69)$ for each sample what is the approximate average value of $(\bar{x} - 69)$ over all samples?

   c) What is the standard error of the mean for samples of size 5?

   d) If we compute $z = \left(\dfrac{\bar{x} - 69}{3/\sqrt{5}} \right)$ for each sample, what is the standard deviation of these z values?

   e) If we compute $t = \left(\dfrac{\bar{x} - 69}{s/\sqrt{5}} \right)$ for each sample, with s computed each time from the sample, what is the standard deviation of these t values.

3. Suppose the height of adult males follows a normal distribution with $\mu = 69$ and $\sigma = 3$. Also, suppose we take repeated samples of size

$n = 5$ from   this   population   and   compute $z = \left(\dfrac{\bar{x} - 69}{3/\sqrt{5}} \right)$ and

$t = \left(\dfrac{\bar{x} - 69}{s/\sqrt{5}} \right)$ for each sample.

a) What are the standard deviations of z and t?

b) Does the sampling distribution of t approximate that of z? Why or why not?

4. Suppose that test scores on a reading test are normally distributed with $\mu = 80$. If we take a random sample of 10 scores what is the approximate chance that $\left(\dfrac{\bar{x} - 80}{s/\sqrt{10}} \right)$ will exceed 2.4 ?

5. Suppose a college admissions test has a mean score of $\mu = 500$. If we take a random sample of $n = 14$ test scores, what is the chance that $\left(\dfrac{\bar{x} - 500}{s/\sqrt{n}} \right)$ will be between -1.35 and 1.35 ?

6. Assume that the distribution of the heights of adult males is normal with $\mu = 69$ inches. Suppose we take a sample of $n = 5$ males at random.

a) What is the chance that $\left(\dfrac{\bar{x} - 69}{s/\sqrt{5}} \right)$ exceeds 0? No computations should be necessary.

b) What is the chance that $\left(\dfrac{\bar{x} - 69}{s/\sqrt{5}} \right)$ exceeds 1.533?

c) What is the chance that $\left(\dfrac{\bar{x} - 69}{s/\sqrt{5}} \right)$ is less than 1.533?

7. Suppose that a mathematics achievement test had a mean of $\mu = 100$. Two random samples, each of size $n = 8$, were taken. The results are shown below:

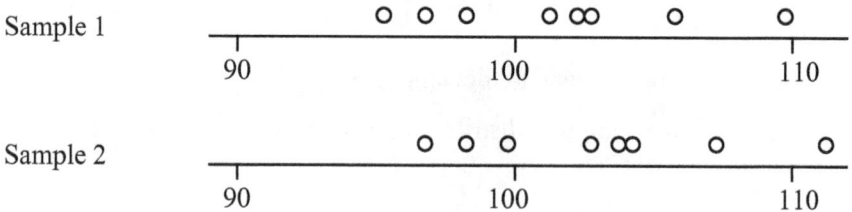

Sample 1

Sample 2

Which sample should produce the largest t statistic? No calculations should be necessary.

3.6 The Sampling Distribution for Proportions

In this chapter we have only, so far, considered continuous distributions. However, many random variables are not continuous and many of these discrete random variables have only two possible values. For example, a question on a survey may require either a "yes" or a "no" response.   In this situation we are interested in the proportion of responses that are positive. Or we might classify manufactured parts as "acceptable" or "defective".   We would then be interested in the proportion of parts that are defective.

Discrete random variables that take on only two values are called **binary random variables**. It is possible to analyze binary random variables by using the formulas and methods that have already been presented in this chapter; we need only to set up an appropriate coding system for the binary responses. If we define

$$X_i = \begin{cases} 1 & \text{if the ith response is positive} \\ \\ 0 & \text{if the ith response is negative} \end{cases}$$

then the sum $\sum_{i=1}^{n} X_i$ equals the number of positive responses. Using this notation, the proportion of positive responses ($\hat{p}$) is the sample mean

$$\hat{p} = \frac{\sum_{i=1}^{n} X_i}{n} \; ,$$

and the population proportion (p) is the population mean

$$p = \frac{\sum_{i=1}^{N} X_i}{N} \; .$$

It can be shown that, by using this coding and by simplifying the formula for the population standard deviation, that the population standard deviation of the X values is $\sigma = \sqrt{p(1-p)}$. In most applications we are interested in the standard error of the sample proportion which is

$$\sigma_{\hat{p}} = \sigma_{\bar{x}} = \frac{\sigma}{\sqrt{n}} = \frac{\sqrt{p(1-p)}}{\sqrt{n}} \; .$$

It can also be shown that the sample proportion is an unbiased estimator of the population proportions. Thus, the sampling distribution of the sample proportion has mean p and standard deviation $\dfrac{\sqrt{p(1-p)}}{\sqrt{n}}$.

For small samples the total number of positive responses follows the **binomial distribution,** which will be discussed in Chapter 8. For large samples the sampling distribution of the proportion approximates the normal distribution. This normal approximation can be used whenever the number of positive responses and the number of negative responses equals or exceeds 5 . Thus, the normal approximation is valid for a survey having $n = 100$ and $p = 0.4$ because there were 40 positive and 60 negative responses. However, it cannot be used for a survey having $n = 1000$ and $p = 0.002$ since there are only 2 positive responses.

Suppose we take repeated samples of size $n = 20$ from a very large population. If 40% of the people in the population give positive responses the sampling distribution will have a mean of $\mu_{\hat{p}} = .4$ and the standard error of the proportion will be

$$\sigma_{\hat{p}} = \frac{\sqrt{p(1-p)}}{\sqrt{n}} = \frac{\sqrt{(.4)(.6)}}{\sqrt{20}} = .11$$

The exact distribution, which is roughly normal, is shown in Figure 3.14 and we can use a normal distribution to compute probabilities.

Figure 3.14 Sampling distribution of the proportion for samples of size n=20 for p=.4

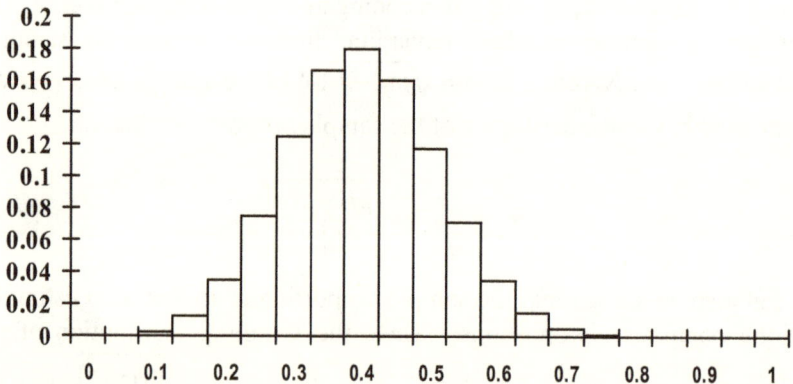

Many political polls rely on samples with $n > 500$ in order to estimate the degree of support for a particular issue or for a candidate. Since the percentage of voters who favor one candidate is usually between 10% and 90%, the number of responses for each candidate will usually exceed 5. Therefore, the normal distribution can usually be used as a good approximation to the sampling distribution. The standard error of the proportion must be small whenever n is large because $\sigma_{\hat{p}} = \dfrac{\sqrt{p(1-p)}}{\sqrt{n}}$

and because it can be shown that $\sqrt{p(1-p)}$ never exceeds 0.5. In repeated sampling, if the sampling distribution approximates a normal, then about 95% of the $\hat{p}$'s should be within $2\left(\dfrac{\sqrt{p(1-p)}}{\sqrt{n}}\right)$ of the population proportion p.

The normal distribution can be used to estimate the chance that the number of successes will fall between any two values. For example, suppose we take a sample of $n=100$ observations from a population that has a probability of a success of $p=.4$ and we want to estimate the chance that the number of successes exceeds 50. The standard error of the proportion will be $\sigma_{\hat{p}} = \dfrac{\sqrt{p(1-p)}}{\sqrt{n}} \leq \dfrac{\sqrt{(.4)(.6)}}{\sqrt{100}} = .04899$. We can compute the chance that the number of successes exceeds 50 by computing the chance that the sample proportion exceeds .5 . Therefore,

$$P(\text{number of successes exceeds } 50) = P(\hat{p} > .5)$$

$$= P\left(\dfrac{\hat{p}-p}{\sigma_{\hat{p}}} > \dfrac{.5-p}{\sigma_{\hat{p}}}\right)$$

Now, for a large sample, $\hat{p}$ is approximately normally distributed with mean p and a standard error of $\sigma_{\hat{p}}$, so that $Z = \dfrac{\hat{p}-p}{\sigma_{\hat{p}}}$ is approximately a standard normal variable and we can look up the probabilities for Z in Table 1. Thus,

$$P(\hat{p} > .5) \approx P\left(Z > \dfrac{.5-.4}{.04899}\right) = P(Z > 2.041) \approx .021.$$

The same procedure can be used to roughly estimate $P(a < \hat{p} < b)$ for any two proportions a and b.

A small modification to this procedure can be used to provide a more accurate approximation. Figure 3.15 gives the sampling distribution for the proportion along with the normal approximation, for a sample of size $n = 100$ having $p = 0.4$. Note that if we are interested in the chance that the number of successes exceeds 50, the exact answer is given by the shaded area which is closely approximated by finding $P(X \geq 50.5)$ from the normal approximation. Therefore a more accurate approximation is

$$P(\text{number of successes exceeds 50}) \ = (\hat{p} > .505) = P\left(Z > \frac{.505 - .4}{.04899} \right)$$

$$= P(Z > 2.143) = .016.$$

This estimate probability is slightly smaller and is more accurate than the probability computed without this adjustment.

Figure 3.14 Sampling distribution for the proportion for
$n=100$ and $p=.4$

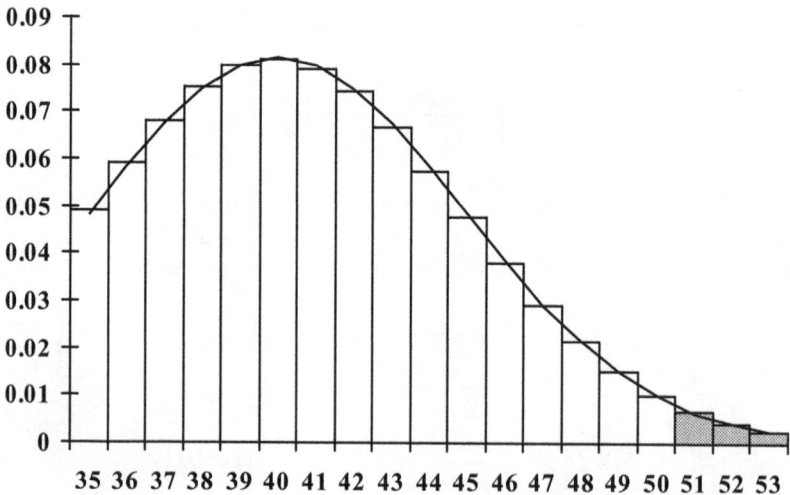

This adjustment is called the **continuity correction**. To understand this better, suppose we want to estimate the chance of obtaining exactly 50 successes. We could use the area under the normal curve between

49.5 and 50.5 to approximate P(X=50). Similarly, we would use the area between 52.5 and 53.5 to approximate p(X=53). This more accurate method of approximating probabilities can be used with any normal approximation to the sampling distribution of the proportion. The best way to determine if .5 should be added or subtracted from the number of successes is to draw a diagram and shade in the desired area.   For example, if we wanted to determine the chance of obtaining a number of successes greater than or equal to 30 and less than or equal to 50, we would first compute

$$P(30 \le X \le 50) = P(29.5 \le X \le 50.5)$$

$$= P\left(\frac{29.5}{100} \le \frac{X}{100} \le \frac{50.5}{100}\right) = P(.295 \le \hat{p} \le .505).$$

Then, by standardizing and using the standard normal distribution, we find, with $p = 0.4$ and $\sigma_{\hat{p}} = .04899$, that

$$P(30 \le X \le 50) = P\left(\frac{.295 - .4}{.04899} \le \frac{\hat{p} - p}{\sigma_{\hat{p}}} \le \frac{.505 - .4}{.04899}\right)$$

$$= P(-2.143 \le Z \le 2.143) = .9679.$$

Keep in mind that the normal distribution is an approximation to the true distribution for these discrete probabilities. Hence the probabilities computed using the normal distribution are approximations.   The continuity correction, which is used to obtain a more accurate estimate of a probability, may not be needed if n is large or if a rough estimate of the probabilities is sufficient.

Exercise Set E

1. A fair coin is flipped 100 times. A head is considered a "success" and a tail is considered a "failure". We are interested in the proportion of heads.

    a) What is p?

    b) Compute the standard error of the proportion.

c) What is the approximate chance of obtaining 40 to 60 heads, inclusive? That is, what is the chance that $.4 \leq \hat{p} \leq .6$?

2. Suppose that, in a certain city, 1% of the households do not have a refrigerator.

   a) Will the sampling distribution of the proportion approximate the normal distribution if $n = 100$? Why or why not?

   b) Will the sampling distribution of the proportion approximate the normal distribution if $n = 2000$? Why or why not?

3. Suppose you roll a fair die many times looking for a "6".

   a) What is the chance of rolling a "6" on a single roll?

   b) If you roll a die 10 times and want to compute the probability of getting more than 3 "6"s, are you justified in using the normal approximation?

   c) If you roll a die 600 times, what is the approximate chance of getting more than 120 "6"s?

4. A political organization believes that 60% of the voters agree with their position on a certain issue. Suppose that only 45% of the voters actually agree with their position. If they take a small sample of $n = 50$ voters, what is the chance that they will conclude (incorrectly) that the majority of voters favor their position.

5. Suppose that 10% of the undergraduates attending a large university are taking prescription medication. If a random sample of 100 undergraduates are interviewed, what is the approximate chance that 20% of those in the sample are taking prescription medication?

6. A large university claims that only 5% of their undergraduates smoke cigarettes. However, a sociologist at the university believes that 20% of the undergraduates smoke. The sociologist interviews 400 students and finds that 22% of those interviewed smoke.

   a) If the population proportion is $p = 0.05$, estimate $P(\hat{p} \geq .22)$?

b) If the population proportion is $p = 0.20$, estimate $P(\hat{p} \geq .22)$?

c) Do you believe that $p = 0.05$ or that $p = 0.20$? Why?

7. There are 18 red slots, 18 black slots, and 2 green slots on a roulette wheel. The wheel is spun and a ball is allowed to fall into one of the slots. A gambler who bets one dollar on red has a $p = \frac{18}{38} = .4737$ chance of winning one dollar. Suppose the gambler's sole objective is to leave the casino with more money than he took into it.

a) If he bets 100 times what is the chance that he will win more often than he will lose?

b) If he bets 10000 times what is the chance that he will win more often than he will lose?

8. Some airlines "overbook" airplanes by taking more reservations than seats because they expect that some people with reservations will not show up. Suppose an airline, which uses planes that have a capacity of 200, takes 210 reservation for each plane. If the airlines know that 92% of those who make reservations actually show up, what is the approximate chance that more than 200 will show up to take a certain flight?

Chapter Review Exercises

1. A health services organization at a large university wanted to determine the proportion of undergraduates who smoke. They took a random sample of $n=100$ undergraduates and determined that 22 smoked.

a) The response variable is the smoking status of an individual. Is this variable continuous or discrete?

b) What is the sample proportion?

2. A nutritionist wanted to obtain the height, weight, and age of high school wrestlers in the state of Wisconsin. Specify the population that should be sampled and indicate which variables are continuous and which are discrete?

3. A scientist conducted an experiment using rats and collected the age, weight, blood pressure, and gender of the rats. Which variables are discrete and which are continuous?

4. Suppose the heights of female Air Force personnel are normally distributed with a mean of 64.7 inches and population standard deviation of 2.8 inches. A new fighter plane is developed that will accommodate pilots between 62.5 and 73.5 inches in height. What proportion of females in the Air Force will fit into this plane?

5. Suppose the annual cost of maintaining an automobile has a distribution that is shown below. The population mean is $800 per year with a population standard deviation of $900 per year.

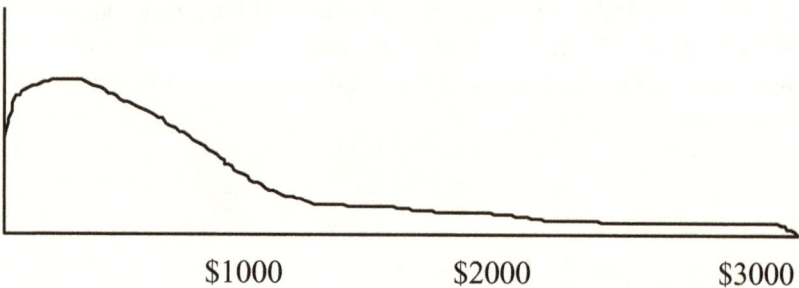

$1000 $2000 $3000

a) If we take many samples of size $n = 81$ and compute the sample mean for each sample, what will the standard deviation of these means approximate?

b) Will 95% of the observations in a sample fall in the interval $(800 - 2\sigma_{\bar{x}}, 800 + 2\sigma_{\bar{x}})$? Why or why not.

6. Assume that automobiles are driven 12,300 miles per year and that the population standard deviation is 5,100 miles per year. Suppose that repeated samples of size $n = 50$ are taken from this population.

   a) Will the sample means be approximately normally distributed?

   b) Compute the standard error of the mean.

   c) In repeated samples of size $n = 50$, approximately 95% of the sample means should be between _____ and _____.

7. The heights of female soccer players are approximately normally distributed. If we take many small samples, each of size $n=4$, from this population, will the distribution of the sample means be approximately normal? Why or why not?

8. Assume that the alcohol consumption per week on a certain campus is known to have a population mean of $\mu = 6$ drinks per weeks with a population standard deviation of $\sigma = 5$ drinks per week.

   a) Do you believe that the distribution of the alcohol consumption in the population is normal? Why or why not?

   b) A graduate student takes a survey of $n = 100$ students and obtains a sample mean of $\bar{x} = 3$ drinks per week. What is the chance of getting a sample mean of 3 or less?

9. A sociologist took a survey of $n = 1000$ people who have recently received a B.S. or a B.A. degree from an accredited college or university. They wanted to determine the proportion of undergraduates who completed college within 4 years. Assume that, in the population of all undergraduates, the proportion of those who complete college within 4 years is $p = 0.65$.

   a) What is the chance that the sample proportion will equal or exceed .70?

   b) What is the chance that the sample proportion will be less than or equal to .60?

   c) What is the chance that the sample proportion will be within ±.05 of the true proportion?

10. Suppose that 15 people enter an elevator that has a capacity of 3000 pounds. If the population has a mean weight of 150 pounds and a standard deviation of 30 pounds, what is the chance that these 15

people will exceed the maximum capacity of the elevator? You may assume that these 15 people entering the elevator constitute a random sample from the population and that the weights in the population are approximately normally distributed. [Hint: If the people on the elevator had an average weight of 200 pounds the elevator would be operating at its capacity.]

11. As part of a quality control method at a brewery, beer cans are sampled to determine if the filling machine is filling the cans to the specified level. The brewer believes that a properly functioning filling machine has a mean of $\mu = 12.0$ ounces with $\sigma = 0.3$ and that the distribution of the volume is approximately normal. A quality control inspector takes a sample of 16 cans and finds $\bar{x} = 11.5$. If the filling machine if functioning properly, what is the chance of obtaining a sample mean of 11.5 ounces or less?

Chapter 4. Confidence Intervals Based on a Single Sample

Introduction

In the first three chapters we developed the tools necessary to estimate population means and population proportions. We found that there will always be some uncertainty whenever we use data from a sample to estimate a population mean or proportion. In this chapter we will learn the best ways to estimate a population mean and will learn to compute confidence intervals around our sample estimate. We will also learn how to compute a confidence interval for a population proportion. Here are some examples of the problems we will address:

- We have collected data on SO_2 emissions from 16 coal burning power plants. Should we use a mean, a median, or some other estimator to describe this data?
- We have obtained weight loss data from 15 men who are enrolled in a weight loss program and have used the sample mean to estimate the population mean. How accurately can we estimate the average weight loss that a large group of men would experience if they had used the same program?
- An opinion poll shows that a politician is supported by a certain percentage of people who were interviewed. How can we describe the accuracy of our estimate?

Our objective in this chapter is to find the best ways of estimating a population mean or proportion. We begin our discussion by considering the choice of estimator. That is, after we have collected our data, should we use a mean, median, or some other estimator? We will take up this important question in section 4.1. In the remaining sections we will develop confidence intervals for means and proportions.

4.1 Choosing the correct estimator

In chapter one we described several estimators that could be used to estimate the population mean, including the mean, median, and trimmed mean. Although the mean is easy to compute, we found that it was sensitive to the presence of a few outliers.   We also described the median, which was not sensitive to outliers, but was computed using only the middle observations. The trimmed mean avoided many of these problems.   It was defined as the average of the observations that remained after trimming, or discarding, a proportion of the largest and smallest values. A commonly used trimming proportion is .10, so that the 10% trimmed mean is computed by averaging the observations remaining after trimming the smallest 10% and largest 10% of the observations.

The choice of the most appropriate estimator depends largely on the objective of the research. If the objective is to estimate the population mean, the sample mean is the recommended estimator because it is an unbiased estimator of the population mean. If the objective is to estimate the population median, the sample median is the recommended estimator because, in large sample, it should be reasonably close to the population median. However, the sample median is not a good estimator of the population mean because it can be seriously biased.   That is, if the sample median is used to estimate the population mean and the distribution is skewed, as shown in Figure 4.1, the sample median will consistently underestimate the population mean.

Figure 4.1 The relationship between the median and the mean when the distribution is skewed.

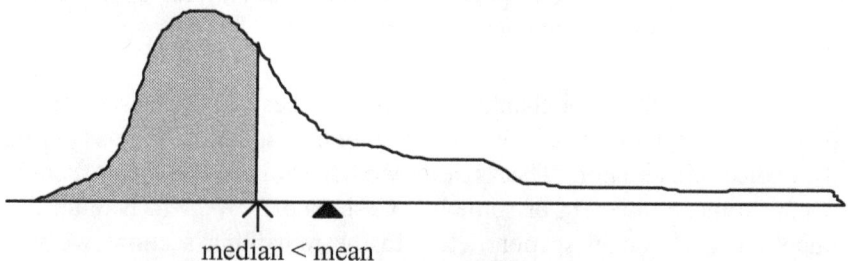

median < mean

It is not difficult to choose an estimator if the sole objective is to obtain an unbiased estimate of a population total. In these applications if we take a sample from a population of N units, we use the sample mean ($\bar{x}$) as an estimator because $N\bar{x}$ will be an unbiased estimator of the population total. For example, if we wanted to obtain a good estimate of the total value of all farms in a state, we would take a random sample of farms and calculate the sample mean. The total value could be estimated by multiplying the average farm value by the total number of farms in the state. If we take a sample of $n = 100$ farms from the $N = 20000$ farms in the state, and the mean value of the farms in the sample was $\bar{x} = \$350,000$ then the estimate of the total value of all farms would be $N\bar{x} = 20000 \times \$350,000 = \$7,000,000,000$. We would not have been able to obtain an unbiased estimate of the total if we had used the median as an estimator because the median tends to underestimate the population mean.

If the objective is to estimate the population median, then the sample median is often recommended because the population median can be estimated by the sample median with little bias. For example, a sociologist who is interested in estimating the median family income for a state would use the sample median. The sample mean would not be used because it would tend to overestimate the population median and would be more sensitive to outliers than the median.

Occasionally, the researcher may know that the population is symmetric. If it really is symmetric, as shown in Figure 4.2, the population mean will equal the population median. For these situations either the mean, median, or trimmed mean could be used. It can be shown that the mean is the most precise estimator if the population distribution is normal. However, if the population distribution has longer tails than the normal the median may be more precise than the mean.

Although the choice of the best estimator depends on the exact shape of the population distribution, for many symmetric distributions the 10% trimmed mean is a good choice because it is nearly as precise as the mean and it is not sensitive to the presence of a few outliers. Consequently, if the distribution of the observations in the population is symmetric, a good choice is the 10% trimmed mean because there is little loss in precision by trimming a few observations.

**Figure 4.2 The mean equals the median when the
distribution is symmetric.**

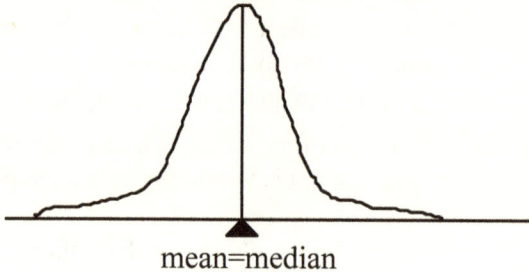

mean=median

Consequently, before choosing an appropriate estimator, it is necessary to clearly understand the objective of the study. It is also helpful to
know if the distribution is skewed or symmetric. With these points in
mind the estimator can be selected by following these recommendations:

- The sample mean is recommended if the objective is to estimate the population mean or the population total.

- The sample median is recommended if the objective is to estimate the population median.

- If the distribution is believed to be symmetric, then either the
  mean, trimmed mean, or the median can be used. The 10%
  trimmed mean is often preferable because it is more precise
  than the median and it is more robust than the mean.

Controversies in Statistics

Should we use the mean or the median?

Some people feel that the mean should always be used to summarize data; other are of the opinion that the median should be used because it better represents the "typical" value. Who is right?

Usually, researchers choose whichever estimator is most appropriate for the job. Some may feel that the mean is the only estimator that should be considered because it is the only unbiased estimator of the population mean, even if the observations come from a skewed distribution. However, if the objective is to estimate the population median, because it is considered the "typical" value, then it is perfectly appropriate to use the median. The choice of an estimator depends mainly on the objective of the researcher. If the objective is to estimate the population total or a population average then a sample mean would ordinarily be used. If the objective is to estimate a typical value, the median may be more appropriate.

Exercise Set A

1. As part of a larger study, ten subjects were used in an experiment to determine if triglycerides could be reduced using an oat-bran diet for 21 days. (Based on Anderson, J. W., et al., "Hypocholesterolemic effects of oat-bran or bean intake for hypercholesterolemic men," *The American Journal of Clinical Nutrition* 40, (1984) pp. 1146-1155.) Before going on these diets the triglycerides were recorded (in mg/dl) as follows:

163, 170, 180, 121, 148, 173, 217, 398, 617, 700

a) Compute the mean, median, and the 10% trimmed mean.

b) In your opinion, which estimate best reflects the "typical" value?

2. In an experiment a researcher needs to estimate the average weight of rats given an experimental treatment. Suppose the distribution of the weights is known to be symmetric and suppose the seven rats have weights (in grams) of:

330, 339, 340, 364, 365, 369, 371

a) Draw a dot plot for the data.

b) Compute the mean and the median.

c) In your opinion, which of these estimates would be the best estimate of the population mean? Explain.

3. Suppose an agricultural economist needs to estimate the total amount of corn produced in the state of Iowa. She takes a sample of 100 one-acre plots and measures the amount of corn in each plot. Which estimator will allow her to produce an unbiased estimate of the total amount of corn produced in the state? She is considering using either the mean or the median. You should not assume that the population distribution is symmetric.

4. An educational researcher investigated the amount of time spent by undergraduates studying outside of class. Suppose we knew that the distribution of study time (in hours per week) looked something like:

Proportion
of
Students

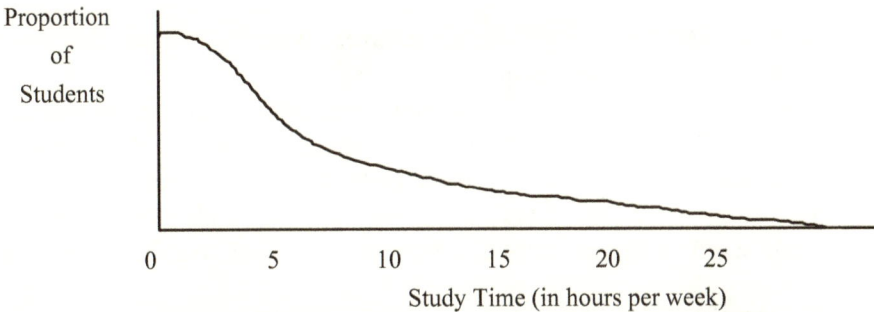

Study Time (in hours per week)

a) If the researcher took a large sample from the population and computed the mean and the median of the observed sample values, would you expect the median to be larger than the mean? Explain.

b) If the researcher wanted an unbiased estimate of the population mean, would she use the sample mean or sample median as an estimator?

4.2 Large Sample Confidence Intervals for the Population Mean

Large Sample Confidence Intervals

In the last section we discussed several estimators of the population mean. In this section we will assume that we have taken a sample from the population and have used the sample mean to estimate the population mean. Since we used an estimate from a sample, we know that the population mean could be somewhat above or below that estimate. We would like to be able to quantify how far we might be from the population mean by constructing an interval that should contain the population mean. An interval that contains the population mean, for a large proportion of repeated samples, is called a **confidence interval**. That is, we imagine that if we took many samples and constructed intervals based on the data from each of these samples, then most of these intervals would contain the population mean. One of the most popular confidence intervals is the 95% confidence interval. A 95% confidence interval indicates that, if 100 confidence intervals were constructed based on the data from 100 samples of the same size, about 95 of these intervals would cover the population mean.

We will now construct a 95% confidence interval for a large sample having $n \geq 30$. Since we have a large sample we know that the sampling distribution of $\bar{x}$ will be approximately normal, no matter how the observations were distributed. For large samples we also know that the sample standard deviation (s) can be used to estimate (σ), so that we can use $\left(\dfrac{s}{\sqrt{n}}\right)$ as an estimate of the standard error $\left(\dfrac{\sigma}{\sqrt{n}}\right)$.

In the last chapter we noted that, in repeated sampling, approximately 95% of the values of $\left(\dfrac{\bar{x}-\mu}{s/\sqrt{n}}\right)$ will fall between $-z_{.025} = -1.96$ and $z_{.025} = 1.96$. The proportion of time that $\left(\dfrac{\bar{x}-\mu}{s/\sqrt{n}}\right)$ falls above 1.96 or below -1.96 are indicated by the shaded regions in Figure 4.3.

Figure 4.3 The Sampling Distribution of $\left(\dfrac{\bar{x}-\mu}{s/\sqrt{n}}\right)$.

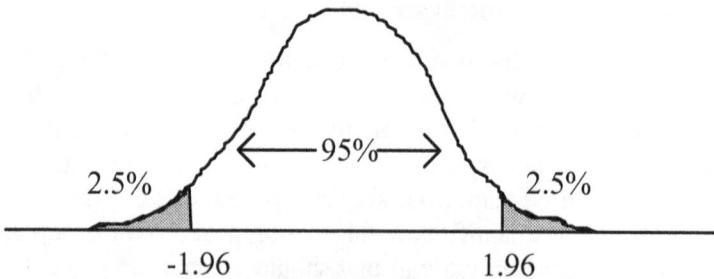

2.5%   ←—95%—→   2.5%

-1.96   1.96

We can use the fact that $\left(\dfrac{\bar{x}-\mu}{s/\sqrt{n}}\right)$ is approximately normally distributed to construct an approximate 95% confidence interval for the population mean. Since approximately 95% of the values of $\left(\dfrac{\bar{x}-\mu}{s/\sqrt{n}}\right)$ will fall between -1.96 and 1.96, we can state that

$$P\left[-1.96 \leq \frac{\bar{x} - \mu}{s/\sqrt{n}} \leq 1.96\right] \approx .95$$

We now rearrange the inequalities to obtain a confidence interval. The inequalities can be written as

$$-1.96\frac{s}{\sqrt{n}} \leq \bar{x} - \mu \leq 1.96\frac{s}{\sqrt{n}},$$

and, by further algebraic manipulation we obtain

$$-\bar{x} - 1.96\frac{s}{\sqrt{n}} \leq -\mu \leq -\bar{x} + 1.96\frac{s}{\sqrt{n}}.$$

If we multiply these inequalities by -1 we will change the direction of the inequalities. After rearrangement we obtain the probability statement

$$P\left[\bar{x} - 1.96\frac{s}{\sqrt{n}} \leq \mu \leq \bar{x} + 1.96\frac{s}{\sqrt{n}}\right] \approx .95$$

In this probability statement μ is an unknown constant. The probability refers to the chance that the random interval

$$\left[\bar{x} - 1.96\frac{s}{\sqrt{n}}, \bar{x} + 1.96\frac{s}{\sqrt{n}}\right]$$

covers the population mean μ. If we take 100 independent samples, each of size n, we would obtain 100 different confidence intervals. Approximately 95 of these intervals would cover μ. Note that the confidence interval is for μ, not for $\bar{x}$.

We can construct a 90% confidence interval in the same way, except that we would use $z_{.05} = 1.645$ instead of $z_{.025} = 1.96$. The 90% confidence interval for μ is

$$\left[\bar{x} - 1.645 \frac{s}{\sqrt{n}}, \bar{x} + 1.645 \frac{s}{\sqrt{n}} \right].$$

It can be seen that the 90% confidence interval is narrower than the 95% confidence interval. The interpretation of the 90% confidence interval is that, approximately 90% of these confidence intervals will cover μ.

Occasionally we need a level of confidence other than 90% or 95%. If we let α be the proportion in the tails of the distribution then we would use $z_{\alpha/2}$ in the derivation. The large sample $100(1-\alpha)\%$ confidence interval for μ, which can be derived in the same manner as the 95% confidence interval, is

$$\left[\bar{x} - z_{\alpha/2} \frac{s}{\sqrt{n}}, \bar{x} + z_{\alpha/2} \frac{s}{\sqrt{n}} \right].$$

The values of α that are most commonly used are .05 and .10, which correspond to 95% and 90% confidence intervals, but any value can be used. Note that a 50% confidence interval will be narrow but it will often not cover μ.

Example 4.1

The following example will illustrate the use of a confidence interval. Researchers measured the serum transferrin receptor concentrations, which is a measure of iron deficiency, in $n = 176$ women in their third trimester of pregnancy. In these women the mean receptor concentration was $\bar{x} = 5.96$ mg/L with $s = 2.37$ mg/L. (Based on Carriaga, M. T., et al. "Serum Transferrin Receptor for the Detection of Iron Deficiency in Pregnancy, *American Journal of Clinical Nutrition,* 54 (1991) pp. 1077-1081.). A 95% confidence interval for the population mean receptor concentration would be

$$\left[\bar{x}-1.96\frac{s}{\sqrt{n}},\bar{x}+1.96\frac{s}{\sqrt{n}}\right]=\left[5.96-1.96\frac{2.37}{\sqrt{176}},5.96+1.96\frac{2.37}{\sqrt{176}}\right]$$

$$=\left[5.96-.35,\ 5.96+.35\right]=\left[5.61,6.31\right]$$

Because the sample size is quite large, the confidence interval is quite narrow. Note that the unknown population mean (μ) either is in the interval or is not in the interval. After we take the sample we still don't know if the interval covers μ. We do know, however, that if we took 100 samples of pregnant women, each of size $n=176$, and we computed an interval for each sample, then about 95 of those intervals would cover the population mean μ.

If we wanted a 99% confidence interval, instead of the 95% confidence interval, we would use $z_{.005}=2.576$ in the general formula to obtain

$$\left[5.96-2.576\frac{2.37}{\sqrt{176}},5.96+2.576\frac{2.37}{\sqrt{176}}\right]$$

$$=\left[5.96-.46,\ 5.96+.46\right]=\left[5.50,6.42\right]$$

Note that the 99% confidence interval is wider than the 95% confidence interval. This makes sense because, in repeated samples, wider intervals will cover μ more often than narrow intervals. ■

We noted, based on the one sample of $n=176$ pregnant women, that the 95% confidence interval for μ is [5.61, 6.31]. Unfortunately, we cannot be certain that this interval will include μ. We stated that when we compute a 95% confidence interval for μ we compute a random interval that may or may not cover μ but, if we take many sample of the same size from this population, 95% of these random intervals would include μ. To clarify this idea consider Figure 4.4 which gives 10 intervals corresponding to 10 samples, each of size $n=176$, from the same population. In an actual situation we would not know μ, but suppose for this explanation that we know that $\mu=5.88$. Note that most of the intervals do include μ. In a real situation we would not take repeated samples and would not know μ so we would never know if the confidence interval for the one sample that we used included μ. We do know that if we use 95% confidence intervals for many of our estimates, that over a long period of time we will be correct about 95% of the time.

Figure 4.4 Confidence intervals for the population mean using 10 samples each of size $n=176$

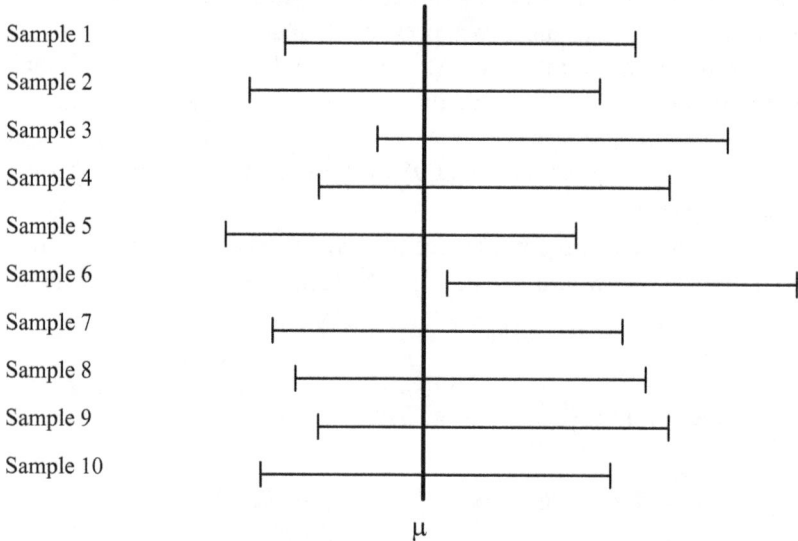

Calculating the Sample Size

We observed, from the formula for the $100(1-\alpha)$% confidence inter-val, that the width of the interval depends on the standard deviation and the sample size. If we want a narrower confidence interval we can decrease the width of the interval by increasing the sample size. Thus, if we have an estimate of the standard deviation, we can calculate the sample size that would be required to produce a confidence interval having a specified **bound**, which equals half the width of the confidence interval. To calculate the sample size we begin by expressing the bound as

$$\text{Bound} = z_{\alpha/2}\frac{s}{\sqrt{n}}.$$

By algebraic manipulation we can solve for n to obtain

$$n = \left(z_{\alpha/2} \frac{\sigma}{\text{Bound}} \right)^2.$$

Since we don't know σ and have not yet taken a sample we usually don't have a precise estimate of σ. The most common procedure is to roughly estimate σ from a prior study. For example, suppose we are planning to take a sample to estimate the mean height of oak trees in a forest. If we are satisfied with a bound of 10 feet for a 95% confidence interval and believe, based on a prior study, that the standard deviation is approximately 30 feet, then the required sample size would be

$$n = \left(z_{\alpha/2} \frac{\sigma}{\text{Bound}} \right)^2 = \left(1.96 \frac{30}{10} \right)^2 = 5.88^2 = 35 \text{ trees.}$$

This small sample should be sufficient to achieve a bound of 10 feet. If we wanted a narrower bound of 5 feet then we would need a sample of

$$n = \left(z_{\alpha/2} \frac{\sigma}{\text{Bound}} \right)^2 = \left(1.96 \frac{30}{5} \right)^2 = 11.76^2 = 140 \text{ trees.}$$

As you can see, if you decrease the bound that you want, you will need to greatly increase the sample size. Also, an accurate sample size calculation depends on an accurate estimate of σ. Since there is usually some uncertainty about σ, there will usually be some uncertainty about the sample size. Consequently, the estimated sample size should be viewed as a rough estimate of the actual sample size that would be required.

Finite Population Correction

In many areas of scientific research the population is very large, containing thousands or millions of objects. For these situations we can assume that taking an observation from the population will not greatly change the characteristics of the remaining observations. However, if the population itself is not large, the removal of a single observation may change the remaining population. Therefore, when sampling without replacement from a small population, we need to pay attention to the size of the population when we compute the confidence interval. It can be shown when a sample of size n is drawn from a population of size N, that the standard error of the mean is

$$\sqrt{\frac{N-n}{N-1}}\frac{\sigma}{\sqrt{n}}.$$

The factor $\sqrt{\dfrac{N-n}{N-1}}$ is called the **finite population correction** factor and has the effect of reducing the standard error. Table 4.1 gives the finite population correction factor for several values of N and n. Note that the finite population correction factors are close to 1.0 except when the sample size (n) is a large proportion of the population size (N).   In general, the finite population correction can be ignored if the sample is less than 10% of the population. This is usually the case in political polling because the sample sizes are usually less than 3000 and the population sizes are usually greater than 100,000. The finite population correction can also be ignored in large health surveys of the U. S. population because the population size is very large and the sample size is relatively small.

Table 4.1   The finite population correction for several sample sizes (n) and for several population sizes (N).

N	n	$\sqrt{\dfrac{N-n}{N-1}}$
1000	10	.9955
1000	100	.9492
1000	500	.7075
1,000,000	100	.9999
1,000,000	1,000	.9995
1,000,000	10,000	.9900

If the sample size exceeds 10% of the population size, the finite population correction should be included in the calculation and the estimated standard error becomes $\sqrt{\dfrac{N-n}{N-1}}\dfrac{s}{\sqrt{n}}$, which should be used in place of $\dfrac{s}{\sqrt{n}}$ in the formulas for confidence intervals. The following example will illustrate the use of the finite population correction.

Example 4.2

For example, suppose we have a sample of $n = 67$ coal burning power plants from a population of $N = 413$ coal burning power plants. These power plants had an average SO$_2$ emission of $\bar{x} = .241$ pounds per million Btu and a standard deviation of $s = .044$ pounds per million Btu. Since the size exceeds 10% of the population size we use the finite population correction to estimate the standard error as

$$\sqrt{\frac{N-n}{N-1}}\frac{s}{\sqrt{n}} = \sqrt{\frac{413-67}{413-1}}\frac{.044}{\sqrt{67}} = (.9164)(.00537) = .00492$$

We note that by using the finite population correction we have reduced the standard error from .00537 to .00492. A 95% confidence interval for the population mean would be

$$\left[.241 \pm 1.96(.00492)\right] = \left[.241 \pm .00964\right] = \left[.231, .251\right]$$

This is a slightly narrower confidence interval than we would have obtained had we ignored the finite population correction. ■

Exercise Set B

1. A professor of sociology asked each student in his class of 40 graduate students to conduct a sample survey of 50 families. Each family was interviewed to determine how much gasoline they consumed in all of their automobiles and trucks in the last week. Each student was then asked to compute a 90% confidence interval based on the data that they collected. There were 40 sample surveys that

produced 40 confidence intervals.   How many of these intervals should cover the population mean? Explain.

2. As part of a larger study, researchers obtained dietary data on $n = 97$ pregnant adults. These women reported an average caloric intake of $\bar{x} = 2134$ Kcal per day with a standard deviation of $s = 498$ Kcal per day.   (Based on Giddens et al. "Pregnant Adolescent and Adult Women have Similarly Low Intakes of Selected Nutrients," *Journal of the American Dietetic Association,* 100 (2000) pp. 1334-1340.)

   a) Compute a 95% confidence interval for the population mean.

   b) Compute a 99% confidence interval for the population mean.

   c) Which confidence interval is wider?

3. A shoe manufacturer wanted to know how many miles per week their customers actually ran in their running shoes. They took a random sample of 100 customers and computed a mean of $\bar{x} = 4.5$ miles and standard deviation of $s = 6.4$ miles.

   a) Compute a 90% confidence interval for the population mean.

   b) Do you believe the distribution of mileage in the population is symmetric or skewed? Explain.

   c) For samples of size $n = 100$, is the sampling distribution of the mean symmetric or skewed? Explain.

4. A county health department was concerned about the health effects of radium in the drinking water. The department decided to obtain information about the drinking water consumption for the $N = 10000$ children in the county.   A random sample of 100 children was taken and it was determined that the sample mean was $\bar{x} = .8$ liters per day with a sample standard deviation of $s = 1.2$ liters per day.

   a) Compute a 95% confidence interval for the sample mean.

b) What size sample would be needed to obtain a confidence interval that has half the width of the one you computed for part (a)? [Hint: $\frac{s}{\sqrt{n}}$ would need to be smaller.]

c) If the department took a sample of $n=1000$ students, would they be certain that the confidence interval actually covers the population mean. Why or why not?

5. The Epidemiologic Catchment Area program was an effort to collect psychiatric data from the general population over 18 years of age. In one city 3004 people were interviewed out of a population of approximately one million people. Do you believe that the finite population correction factor is required for the analysis of this data. Why or why not?

6. The body mass index (BMI) equals the weight (in Kilograms) divided by the square of the height (in meters). Suppose you took a simple random sample of $n = 30$ college students from a small college that has a population of 150 students and found that the sample mean was $\bar{x} = 24.5$ kg/m^2 with a standard deviation of $s = 6.2$ kg/m^2. Suppose you want to estimate the mean BMI for this small college.

a) Calculate the finite population correction factor for this sample.

b) Calculate a 95% confidence interval for the population mean BMI.

7. Suppose you want to determine the amount of drinking water that children drink per day. You want to have a very precise estimate of the population mean, so you specify that the bound on the 95% confidence interval should be .1 liters per day. If the standard deviation is estimated to be around 1.4 liters per day, what sample size would be required to produce a 95% confidence interval with a bound of .1 liters per day?

8. A sample of $n = 67$ coal burning power plants was obtained from a population of $N = 413$ coal burning power plants to determine the average SO_2 emissions. These power plants had an average SO_2 emission of $\bar{x} = .241$ pounds per million Btu and a standard deviation of $s = .044$ pounds per million Btu.

a) Compute a 90% confidence interval for the population mean. Include the finite population correction factor.

b) In order for the confidence interval to be valid, was it necessary to assume that the distribution of SO_2 is approximately normal?

c) In order for the confidence interval to be valid, was it necessary to assume that the sampling distribution of the mean is approximately normal?

4.3 Small Sample Confidence Intervals for the Mean

The confidence interval that we described in the last section is appropriate for large samples that have at least 30 observations. With large samples having $n \geq 30$ the central limit theorem guarantees that the sampling distribution of the mean will be approximately normal. This fact allowed us to compute the confidence intervals.   However, if the sample size is less than 30 there are several problems. First, data from a small sample may provide a poor estimate of the standard deviation. Second, if we have taken a small sample from a skewed or long-tailed distribution the sampling distribution of the mean may not approximate a normal distribution. The approach that we will take is to assume that the observations are from a normal population, and make the modifications that are necessary to adjust for the fact that the standard deviation is not known.

From the last chapter we know that the statistic $\left(\dfrac{\bar{x} - \mu}{s/\sqrt{n}} \right)$ has a t distribution with $v = n - 1$ degrees of freedom, provided the observations are obtained from a normal distribution. Consequently, to derive a $100(1-\alpha)\%$ confidence interval for μ we begin with the probability statement

$$P\left[-t_{\alpha/2,n-1} \le \frac{\bar{x}-\mu}{s/\sqrt{n}} \le t_{\alpha/2,n-1}\right] = 1-\alpha.$$

By performing the same algebraic manipulations that we used in section 4.2 we find the $100(1-\alpha)\%$ confidence interval for the population mean μ is

$$\left[\bar{x} - t_{\alpha/2,n-1}\frac{s}{\sqrt{n}}, \bar{x} + t_{\alpha/2,n-1}\frac{s}{\sqrt{n}}\right].$$

Note that this interval looks much like the large sample confidence interval except that the critical values are obtained from a t distribution rather than from a normal distribution. However, it must be emphasized that the small sample test assumes that the observations come from a normal distribution. This assumption was unnecessary with the large sample confidence interval.

Example 4.3

Consider the following example of a confidence interval from a small sample. A cigarette company wanted to check the nicotine content of a certain brand of cigarettes. The company took a random sample of $n = 13$ cigarettes and obtained an average nicotine content of $\bar{x} = 12.2$ mg with a standard deviation of $s = 2.7$ mg. If the assumption is made that the nicotine content is normally distributed, the researcher can compute a 95% confidence interval as:

$$\left[\bar{x} - t_{\alpha/2,n-1}\frac{s}{\sqrt{n}}, \bar{x} + t_{\alpha/2,n-1}\frac{s}{\sqrt{n}}\right]$$

$$= \left[12.2 - 2.18\frac{2.7}{\sqrt{13}}, 12.2 + 2.18\frac{2.7}{\sqrt{13}}\right]$$

$$= [10.56, 13.83].$$

The validity of this confidence interval depends on the validity of the normality assumption. If the population distribution of the nicotine content is skewed then, in repeated samples of size $n = 13$, the proportion of confidence intervals that cover the population mean may not equal 95%. Note that by using $t_{.025,12} = 2.18$ for the small sample instead of the large sample value of $z_{.025} = 1.96$, the small sample confidence interval is wider than the large sample confidence interval. This increased width is necessary because, with only 13 observations, the standard deviation is not known accurately. ■

Although the large sample formula cannot be used for small samples, the small sample formula can be used for large samples. It can be seen in Table 2 that, for $n \geq 30$, the value of $t_{\alpha/2,n-1}$ approximates $z_{\alpha/2}$, so that the width of the small sample confidence interval approximates the width of the large sample confidence interval. Thus, it is convenient to use the t table for construction confidence intervals for small and large sample because $z_{\alpha/2} = t_{\alpha/2,\infty}$.

The major difference between small samples and large sample confidence intervals is that the validity of the small sample interval rests heavily on the assumption of normality. With large samples the central limit theorem guarantees the normal distribution of the sample mean, so there is no need to be concerned about the shape of the distribution when large samples are taken from populations having skewed or long-tailed distribution. If the sample is small and the population distribution is not normal, there is no simple formula that can be used to construct a confidence interval.

Exercise Set C

1. As part of a larger study, ten subjects were used in an experiment to determine the effect of an oat-bran diet on men who had hypercholesterolemia. Their triglycerides averaged 289 mg/dl before they were put on a oat-bran diet. After 21 days on the oat-bran diet their average triglycerides dropped to an average of 235 mg/dl with a standard deviation of approximately 142 mg/dl. (Based on Anderson, J. W., et al., "Hypocholesterolemic effects of oat-bran or bean intake for hypercholesterolemic men," *The American Journal of Clinical Nutrition* 40, (1984) pp. 1146-1155.)

    a) Compute a 90% confidence interval on the triglyceride level after 21 days on the diet.

b) What assumptions are necessary to ensure the validity of the confidence interval?

2. A researcher attempted to determine the relationship between vitamin D and the calcium content in bones. Calcium content was believe to be normally distributed in the population. Eight rats were fed a diet deficient in vitamin D and their calcium content was determined to be

$$7.2, \quad 8.4, \quad 7.5, \quad 6.9, \quad 8.3, \quad 7.9, \quad 7.1, \quad 7.3$$

a) Compute a 95% confidence interval for the population mean.

b) Was it necessary to assume that the population distribution was normal?

3. A graduate student needed to do a quick sample of families in order to estimate the average family income of all families in the state who had children in day care centers. He took a random sample of 7 families and obtained $\bar{x} = \$36,000$ with $s = \$12,500$.

a) Do you believe that family income is normally distributed? Explain.

b) In your opinion, is it possible to validly state a 95% confidence interval for the population average family income? Explain.

4. An agricultural researcher wanted to determine the average weight of hogs that were fed a certain feed. He selected 10 hogs at random and found a sample mean of $\bar{x} = 615$ pounds and a sample standard deviation of s=52 pounds. Suppose the weights were believed to be normally distributed. Compute a 95% confidence interval for the population mean weight.

5. As part of a larger study to determine the relationship between platelet calcium and blood pressure, researchers measured the platelet calcium in $n = 9$ male patients 25 to 57 years of age with borderline hypertension. These subjects had a mean platelet calcium of $\bar{x} = 127$ nM with a standard deviation of $s = 10$ nM. (Based on Erne,

P. et al. "Correlation of Platelet Calcium with Blood Pressure," *The New England Journal of Medicine,* 310 (1984) pp. 1084-1087.) Compute a 90% confidence interval for the platelet concentration in the population of all male patients 25 to 57 years of age with borderline hypertension.

6. Expectant mothers often discover that the actual delivery date is not close to the expected delivery date. Researchers obtained the error in the delivery date, which is defined as the actual date minus the expected date, for $n = 18$ expectant mothers. They found an average error of $\bar{x} = 6.3$ days with a standard deviation of $s = 11.6$ days. You may assume that the errors are normally distributed.

    a) Compute the 95% confidence interval for the average error.

    b) Are you certain that this confidence interval will cover the sample average? Explain.

    c) Are you certain that this confidence interval will cover the population average? Explain.

7. Suppose the weights of adult rats are normally distributed. Two researchers take independent simple random samples from a large population of rats and compute the mean, standard deviation, and a 95% confidence interval.

    a) Researcher A takes a sample of $n = 7$ animals and obtains a mean of 510 grams and a standard deviation of 120 grams. Compute the 95% confidence interval.

    b) Researcher B takes a sample of $n = 28$ animals and obtains a mean of 520 grams and a standard deviation of 120 grams. Compute the 95% confidence interval.

    c) In the confidence intervals computed above, which had the smaller value for $t_{\alpha/2, n-1}$?

    d) Compute the widths of the 95% confidence intervals that were found in (a) and (b). Give two reasons why Researcher B has a smaller confidence interval than Researcher A.

4.4 Large Sample Confidence Intervals for the Population Proportion

Calculating a Confidence Interval

In section 4.2 we presented large sample confidence intervals for the population mean. In this section we are concerned with estimating the population proportion, which we will denote by p, by using the sample proportion, which we will denote by $\hat{p}$. For example, suppose we are interested in determining the proportion of adults who have an intolerance for milk products. If we take a sample of $n = 200$ individuals and find that 50 of these are intolerant, the sample proportion of milk intolerant individuals is $\hat{p} = \dfrac{50}{200} = .25$. This sample proportion is used to estimate the proportion of milk intolerant individuals in the larger population. In this section we will learn how to construct confidence intervals for the population proportion. This will give us some sense of the accuracy of the estimate.

We will use

$$\hat{p} = \frac{\text{sum of positive responses}}{n}$$

as our estimate of the population proportion p. It can be shown that the standard error of the proportion ($\sigma_{\hat{p}}$) can be written as

$$\sigma_{\hat{p}} = \frac{\sqrt{p(1-p)}}{\sqrt{n}}.$$

We do not know p, and hence cannot compute $\sigma_{\hat{p}}$, but we can obtain a good estimate of it by using $\hat{p}$ in place of p, provided we have a sample large enough so that $\hat{p}$ is reasonably close to p.

In our example we found $\hat{p} = .25$ based on a sample of $n = 200$ individuals. Consequently the estimate of $\sigma_{\hat{p}}$ is

$$\frac{\sqrt{\hat{p}(1-\hat{p})}}{\sqrt{n}} = \frac{\sqrt{.25(1-.25)}}{\sqrt{200}} = \frac{.433}{14.14} = .0306$$

Because our estimate of p has an estimated standard error of 3.06%, it follows that we do not have a very precise estimate of p.

To compute a large sample confidence interval we use the same development as was used in the last section to obtain the $100(1-\alpha)\%$ confidence interval for the population proportion

$$\left[\hat{p} - z_{\alpha/2} \frac{\sqrt{\hat{p}(1-\hat{p})}}{\sqrt{n}}, \hat{p} + z_{\alpha/2} \frac{\sqrt{\hat{p}(1-\hat{p})}}{\sqrt{n}} \right].$$

This formula is based on the normal approximation to the sampling distribution of $\hat{p}$. If we wanted to compute the 95% confidence interval for the population proportion of milk intolerant individuals, we would use $z_{\alpha/2} = 1.96$ in the confidence interval formula to obtain

$$\left[.25 - 1.96 \frac{\sqrt{.25(1.-75)}}{\sqrt{200}}, .25 + 1.96 \frac{\sqrt{.25(1.-75)}}{\sqrt{200}} \right]$$

$$= \left[.25 - 1.96(.0306), .25 + 1.96(.0306) \right]$$

$$= \left[.25 - .060, .25 + .060 \right] = \left[.19, .31 \right]$$

Thus, our 95% confidence interval is from 19% to 31%.

One very common application of these confidence intervals is in the analysis of political polls.   These studies often have sample sizes between $n = 500$ and $n = 3000$ and they usually use 95% confidence intervals. It is traditional in these studies to state the **margin of error**, which is $1.96 \frac{\sqrt{\hat{p}(1-\hat{p})}}{\sqrt{n}}$, to express the accuracy of the results. The following example illustrates the use of these procedures in political polling.

Example 4.4

Suppose a polling organization takes a sample of $n=900$ voters and finds that 60% of the voters in the sample favor one candidate. Using

$\hat{p} = .6$ we can compute the 95% confidence interval for the population proportion as

$$\left[\hat{p} - 1.96 \frac{\sqrt{\hat{p}(1-\hat{p})}}{\sqrt{n}}, \ \hat{p} + 1.96 \frac{\sqrt{\hat{p}(1-\hat{p})}}{\sqrt{n}} \right]$$

$$= \left[.6 - 1.96 \frac{\sqrt{.(6)(.4)}}{\sqrt{900}}, \ 6 + 1.96 \frac{\sqrt{.(6)(.4)}}{\sqrt{900}} \right]$$

$$= \left[.6 - .032, \ .6 + .032 \right] = \left[.568, \ .632 \right].$$

Therefore the 95% confidence interval for the population proportion is the interval between 56.8% and 63.2%. The sample estimate of the support for the candidate is 60%, with a margin of error of 3.2%.   ∎

This confidence interval of 56.8% to 63.2% for the population proportion is an approximation based on the normal approximation to the sampling distribution of the proportion. For the normal approximation to be valid we need to have at least 5 positive responses and at least 5 negative responses. This restriction does not generally limit its usefulness because in most applications the samples sizes are large. However, if we were to take a sample of $n=1000$ individuals to determine what proportion of those individuals had a certain disease, we might find that 2 have the disease and 998 do not. For this sample we cannot use the formula for the confidence limits because the smaller of the positive and negative counts does not exceed 5. (For a confidence interval formula that is more accurate with small samples see exercise 10 in this section.)

We will use the following approach to large sample confidence intervals. If the number of positive responses and the number of negative responses exceed 5, the large sample $100(1-\alpha)\%$ confidence interval for the population proportion is

$$\left[\hat{p} - z_{\alpha/2} \frac{\sqrt{\hat{p}(1-\hat{p})}}{\sqrt{n}}, \hat{p} + z_{\alpha/2} \frac{\sqrt{\hat{p}(1-\hat{p})}}{\sqrt{n}} \right],$$

where $\hat{p}$ is the sample proportion.

Calculating the Sample Size

We note, from the formula for the $100(1-\alpha)\%$ confidence interval for the population proportion, that the bound can be written as

$$Bound = z_{\alpha/2} \frac{\sqrt{\hat{p}(1-\hat{p})}}{\sqrt{n}},$$

where $\hat{p}$ is a prior estimate of p. If we have not yet taken a sample, but have some rough estimate $\hat{p}$ of p, and wish to obtain a certain bound, we can calculate the sample size by solving the equation for n to obtain

$$n = \left(z_{\alpha/2} \frac{\sqrt{\hat{p}(1-\hat{p})}}{Bound} \right)^2.$$

Unlike the sample size calculation for the population mean, we can often obtain a reasonably accurate estimate of the sample size because $\sqrt{\hat{p}(1-\hat{p})}$ is usually a reasonable estimate of the standard deviation.

Example 4.5

For example, suppose we want to estimate the proportion of voters who would vote for a candidate, and we want the bound of a 95% confidence interval to be .03. If we use a rough estimate that $\hat{p} = .4$ then

$$n = \left(z_{\alpha/2} \frac{\sqrt{\hat{p}(1-\hat{p})}}{Bound} \right)^2 = \left(1.96 \frac{\sqrt{(.4)(.6)}}{.03} \right)^2 = \left(1.96 \frac{.49}{.03} \right)^2 = 1025 \text{ voters.}$$

Note that if we had used an estimate of $\hat{p} = .5$ instead of $\hat{p} = .4$ then the sample size would be

$$n = \left(1.96 \frac{\sqrt{(.5)(.5)}}{.03} \right)^2 = \left(1.96 \frac{.50}{.03} \right)^2 = 1068 \text{ voters..} \quad \blacksquare$$

Thus, if p really was .4, our estimate of n would have been close to the correct value if we had guessed any p between .4 and .5 . In political poling it is a good idea to use $\hat{p} = .5$ because it will always give the largest value of $\sqrt{\hat{p}(1-\hat{p})}$, which will guarantee that the actual bound will not be wider than the bound specified with $\hat{p} = .5$. Thus, the choice of $\hat{p} = .5$, which gives $\sqrt{\hat{p}(1-\hat{p})} = \sqrt{.5(.5)} = .5$ can be seen as a slightly conservative value for any proportions between .3 and .7. For values of p outside of this range it is preferable to obtain a rough estimate of p before estimating a sample size.

Exercise Set D

1. Are women less likely to be left handed?   To help answer this question a researcher obtained information on the handedness of 709 females and 400 males. The results of the survey showed that 42 of the females were left-handed while 40 of the males were left-handed. (Based on Oldfield, R. C. "The Assessment and Analysis of Handedness: The Edinburgh Inventory," *Neuropsychologia* , 9, (1971) pp.97-113.)

   a) Compute a 95% confidence interval for the proportion of females who are left-handed.

   b) Compute a 95% confidence interval for the proportion of males who are left-handed.

2. Researchers at the University of Padua and at the University of Pittsburgh obtained biopsy specimens from 556 patients with myopathy. They found that the levels of α-sarcoglycan were decreased on 54 of the patients. (Based on Duggan, et al. "Mutations in the Sarcoglycan Genes in Patients with Myopathy," *New England Journal of Medicine*, 336 (1997) pp. 618-624.)

   a) What is your best estimate of the incidence of decreased α-sarcoglycan levels?

   b) Compute a 95% confidence interval for the proportion of myopathy patients who have decreased α-sarcoglycan.

3. Suppose a certain state, which now has a drinking age of 21, is considering changing it to 19. A poll of 300 registered voters was conducted to measure the support for legislation to change the drinking age. It was found that 85 voters supported the legislation and 215 voters opposed the legislation. If possible, compute a 95% confidence interval for the proportion in the state who support this legislation.

4. A researcher is interested in the proportion of adults who are on a diet at any given time. She assigns two graduate assistants to construct two surveys.

   a) The first graduate assistant takes a simple random sample of $n=100$ adults and finds that 34 are on a diet at the time that they were interviewed. Compute a 95% confidence interval.

   b) If the second graduate assistant takes a simple random sample of $n=400$ adults from the same population as the first graduate student, would she expect her 95% confidence interval to be narrower or wider than the confidence interval computed by the first graduate assistant? Approximately how much narrower or wider would it be?

5. A public health inspector took a simple random sample of 49 cooked hamburgers from 49 fast food restaurants in a large city. Each hamburger was tested to see if it was cooked to 155° F and it was found that eight of these hamburgers were determined to be undercooked. If possible, compute a 95% confidence interval for the proportion of all hamburgers that were undercooked. If it is not possible to construct such an interval, explain why it is not possible.

6. A sociologist wanted to estimate the proportion of people who have had difficulty paying their credit card bills. She believes that the proportion is near .15, but wanted to do a survey that would have a bound of .04 for a 95% confidence interval. What sample size would be required?

7. An appliance manufacturer wanted to know what proportion of households in a certain country have indoor plumbing. Based on a

random sample of 300 households it was determined that 297 households in the sample did have indoor plumbing. If possible, construct
a 95% confidence interval for the population proportion. If it is not
possible to construct a 95% confidence interval explain why it is not
possible.

8. A political candidate wanted to determine the support he had in a
large voting district. Instead of taking a random sample, he relied on
the results from a poll done by a television station which asked their
viewers to call the station to express their opinion. If 125 of 200
callers supported his candidacy, should he be confident that he will
win?

9. A political scientist asked controversial questions, which usually
produced proportions of positive responses between .3 and .7 . To
make the confidence interval calculations easier, she decided to
make a table of $\sqrt{\hat{p}(1-\hat{p})}$ for several values of $\hat{p}$.

a) Compute $\sqrt{\hat{p}(1-\hat{p})}$ for several values to fill in the table:

$\hat{p}$	$\sqrt{\hat{p}(1-\hat{p})}$
.3	
.4	
.5	
.6	
.7	

b) Does $\sqrt{\hat{p}(1-\hat{p})}$ vary much between $\hat{p}=.3$ and $\hat{p}=.7$? Why
or why not?

c) If she did not mind making the intervals slightly wider than they
should be, could she set $\sqrt{\hat{p}(1-\hat{p})}$ to a constant? Explain.

10. For small samples it has been shown that a better approximation to the exact confidence limits can be obtained by using the following formula for a $100(1-\alpha)\%$ confidence interval for the population proportion:

$$\left[\frac{\hat{p} + \dfrac{z_{\alpha/2}^2}{2n} \pm z_{\alpha/2}\sqrt{\dfrac{\hat{p}(1-\hat{p})}{n} + \dfrac{z_{\alpha/2}^2}{4n^2}}}{1 + (z_{\alpha/2}^2)/n} \right]$$

(See Agresti, A. and Coull, B. "Approximate is Better Than Exact for Interval Estimation of a Binomial Proportion," *The American Statistician*, 52 (1998), pp.119-126) Use this formula to compute the 95% confidence interval for the data given in exercise 5. Compare this confidence interval to that obtained by using the large sample formula for the confidence interval.

Chapter Review Exercises

1. Do students spend too much time surfing the World Wide Web? If a researcher wanted to estimate the median amount of time, per week, that students spend surfing the web, should she use the sample median or the sample mean as an estimator?

2. A government agency wanted to estimate the total amount of money spent eating food outside the home. After they took a sample of $n = 87$ individuals they computed the mean and the median of the data in the sample, while noting that only a few individuals in the sample spent a great deal of money eating outside the home. Should the government agency use the mean or the median to estimate the total amount of money spent outside the home? Explain.

3. Employees of an industry were screened for the presence of hypertension. In an experiment designed to reduce blood pressure in those employees who has hypertension, $n = 95$ subjects were assigned to a biofeedback group, which was given training in relaxation and stress management. After eight months the biofeedback

group had an average reduction in systolic blood pressure of $\bar{x} = 15.3$ mm Hg with a standard deviation of $s = 1.55$ mm Hg. (Based on Patel, C., Marmot, M. G., and Terry, D. J., "Controlled trial of biofeedback-aided behavioural methods in reducing mild hypertension," *British Medical Journal,* 282(1981) pp. 2005-2008.)

a) Compute a 99% confidence interval for the average reduction in systolic blood pressure.

b) In your opinion, does it appear that the blood pressure would be reduced if this biofeedback training were given to a very large population of individuals who had hypertension. Explain.

4. In order to determine the proportion of females between 15 and 25 years of age who would be diagnosed as having anorexia nervosa, a psychiatrist interviewed 600 females in that age range. She found that 3 met the criteria for anorexia nervosa. You may assume that the females who were interviewed were selected at random from a large population. Will a large sample confidence interval for the proportion be valid? Why or why not?

5. A public health official wanted to obtain a precise estimate of the proportion of women between the ages of 18 and 24 who smoke cigarettes. He wanted the 95% confidence interval to have a bound of .005 and believed that the proportion of smokers was around .2 .

a) What sample size would be required to achieve that bound?

b) If the bound was increased to .03, what sample size would be required?

c) Why is there a large difference in sample size between the answers to (a) and (b)?

6. A consumer organization wanted to estimate how much money the average car owner spent on repairs in the last year. A random sample of $n=100$ owners produced a sample mean of $1250 per year and a sample standard deviation of $630 per year. The distribution of the repair expenses was roughly normal.

a) Compute the 95% confidence interval for the population mean.

b) If the distribution of the sample observations was roughly normal, approximately what percentage **of the sample values** fell in the 95% confidence interval? [Hint: 95% is not the correct answer.]

16%   45%   95%   98%   99%

7. A social scientist took a random sample of $n=50$ observations and computed the 90%, the 95%, and the 99% confidence intervals for the population mean. These intervals are shown below. Identify the three confidence intervals.

i) |———————————————————————————|

ii) |————————————|

iii) |——————————————————————|

8. A nutritionist working at a wellness program at a large university wanted to determine the average caloric intake of male professors. The nutritionist selected 40 professors at random from the population of $N=900$ male professors and found that they consumed an average of $\bar{x}=2650$ calories per day with a standard deviation of $s=1210$ calories per day.

a) Construct a 95% confidence interval for the population mean without using the finite population correction factor.

b) Construct a 95% confidence interval for the population mean using the finite population correction factor.

c) Does the finite population correction make a big difference in the width of the confidence interval? Why or why not?

9. In a study that involved 81 women who had a myocardial infarction, 43 had smoked at least 20 cigarettes per day. Compute a 95% confi-

dence interval on the percentage of myocardial infarction patients who had smoked at least 20 cigarettes per day. (Based on Oliver, M. F., "Ischaemic Heart Disease in Young Women," *British Medical Journal*, 2 (1974) pp.253-259.)

10. Data on height and weight was obtained from the premenopausal women who took part in the third National Health and Nutrition Examination Survey (NHANES III). The researchers reported the average percent of body fat in a subset of $n = 435$ premenopausal women who met the inclusion criteria was 30.13%, with a standard error of 10.92%. (Based on Heim, D. L., Holcomb, C. A., and Loughin, T. M. "Exercise mitigates the association of abdominal obesity with high-density lipoprotein cholesterol in premenopausal women: Results from the third National Health and Nutrition Examination Survey," *Journal of the American Dietetic Association, 100* (2000), pp. 1347-1353.)

a) Is the percent of body fat a continuous random variable or a binary random variable?

b) Is it appropriate to use the formula $\dfrac{\sqrt{\hat{p}(1-\hat{p})}}{\sqrt{n}}$ to estimate the standard error? Is this how they computed the standard error? Explain.

c) Compute a 95% confidence interval for the mean percent body fat in the population of premenopausal women.

11. Suppose the lifetime of a certain brand of automobile batteries is normally distributed. If we take a sample of $n=9$ batteries and find the average lifetime to be $\bar{x} = 58$ months with a standard deviation of $s = 9$ months, what is the 95% confidence interval for the population mean?

12. Suppose you wanted to estimate the lifetime of a certain brand of automobile battery, but have not yet drawn a sample. You want a bound of 2 months for a 95% confidence interval for the population average lifetime, and believe the standard deviation is approximately 9 months. What sample size would be required?

13. A quality control inspector at a food processing plant wanted to determine if the cheese packages that were labeled "16 ounces" really weighed 16 ounces. He selected $n=25$ cheese packages at random from a warehouse containing a very large number of these packages and found that the sample mean was $\bar{x} = 15.1$ ounces and the standard deviation was $s = 1.2$ ounces.

   a) Compute a 95% confidence interval for the population mean. You may assume that the distribution of weights is normal.

   b) Does the confidence interval include 16 ounces?

   c) Do you believe that the packages are labeled properly?

14. A health care administrator was interested in estimating the average length of stay in a hospital. After she selected 50 admissions records at random she determined that the sample mean was 5.2 days per admission and the sample standard deviation was 6.8 days per admission.

   a) Compute a 90% confidence interval for the population mean.

   b) Can there be any negative values in this data set? Do you believe that the distribution of lengths of stay is approximately normal? Explain.

   c) If it necessary that the data be normally distributed for the confidence interval in (a) to be valid? Explain.

Chapter 5. Tests of Hypotheses Based on a
Single Sample

Introduction

In the last chapter we saw how we could construct a confidence interval for a population mean or a population proportion. In this chapter we take a different approach. We will test theories based on data obtained from one sample that was selected from one population. The methods used to test theories are called **tests of hypotheses** or **significance tests**. Two examples of these tests are:

1) A researcher believes that a drug can reduce blood pressure in certain patients. She takes a sample of 20 patients and records the drop in blood pressure in these patients. The data is then used to help the researcher determine if the drug was effective in reducing blood pressure.

2) An educator wants to see if a one-week intensive reading program can increase the reading comprehension of high school students. Sixty students are enrolled in the program, which requires measuring the reading comprehension of the students before and after the reading program. The reading scores can be used, along with the statistical procedures described in this chapter, to determine if the reading program really was beneficial.

Two methods for testing these theories will be presented in this chapter. The t-test, which is the more popular test, will be presented in section 5.3. The signed rank test, which has some advantages over the t-test in some situations, will be presented in section 5.4. The advantages and disadvantages of these two testing procedures will be discussed in section 5.5.

5.1 Basic Concepts of Tests of Hypotheses

Often, in basic research, statistical tests are more useful than confidence intervals. For example, a geneticist may want to use the data from a sample to determine if a new variety of corn has a greater yield than the old variety of corn. The geneticist may not be too interested in a confidence interval, preferring instead to concentrate on the question of whether the new variety offers any improvement over the old variety. A hypothesis testing procedure provides the information necessary to determine if the new variety really does have a greater yield that the old variety.

In most statistical testing situations there is one theory that states that the new procedure is not any better than the old procedure, and another theory that states that the new procedure is better than the old procedure. The theory that the new procedure is not any better than the old procedure is called the **null hypothesis**. The other theory, which states that the new procedure is beneficial, is called the **alternative hypothesis**. Researchers usually want to reject the null hypothesis in order to demonstrate that the alternative hypothesis is true. As we shall see, researchers are not always able to reject the null hypothesis; the null hypothesis may not be false or there may be insufficient evidence to demonstrate that the null hypothesis is false.

An example may help to clarify the role that these two hypotheses play in statistical inference. Suppose a biochemist wants to know if a new drug lowers blood pressure. In this example the alternative hypothesis would be that the drug really does lower blood pressure, since that is what the researcher wants to demonstrate. The null hypothesis is that the drug is not effective in lowering blood pressure. The null hypothesis is usually a statement that the treatment is not effective or that there is no difference between a new treatment and a standard treatment. Since the usual research objective is to demonstrate that the alternative hypothesis is true, it is sometimes called the **research hypothesis**.

There are several steps in any statistical test. The first step is to set up the null and alternative hypotheses. The next step is to design the experiment, collect the data, and compute a test statistic from the data. The last step is to decide whether you should reject, or not reject, the null hypothesis. You should always keep in mind that the overall objective is to reject the null hypothesis.

We will first use an example to give an overview of the general testing procedure; the detailed procedure will be described in the next two sections. Suppose a government agency wanted to determine if a decongestant had the adverse effect of raising the diastolic blood pressure above the average diastolic blood pressure of 80 mm Hg. Since they wanted to detect an increase in blood pressure, the alternative hypothesis was that the diastolic blood pressure of those taking the drug was greater than 80 mm Hg. The null hypothesis was that the decongestant had no effect on the diastolic blood pressure or that the diastolic blood pressure of those taking the drug equaled 80 mm Hg. Now suppose that we obtained a random sample of $n = 100$ individuals who are taking this decongestant and we find that the average diastolic blood pressure from that sample is $\bar{x} = 83$ mm Hg. Is this average sufficiently high to reject the null hypothesis?

To answer this question we need to determine the variability in the sampling distribution of the mean. Suppose the population standard deviation is $\sigma = 10$ mm Hg. We know from chapter 3 that the standard error of the mean would be $\sigma_{\bar{x}} = \dfrac{\sigma}{\sqrt{n}} = \dfrac{10}{\sqrt{100}} = 1$. We also know that, for a large sample, the sampling distribution of the mean will be approximately normal. Thus, the sampling distribution of the mean, which is shown in Figure 5.1, indicates that, under repeated sampling from a population having a mean of $\mu = 80$ mm Hg, we would rarely obtain a sample mean of $\bar{x} = 83$ mm Hg or greater. Since our observed mean is 3 standard errors above the population mean, we find, from Table 1 with $z = 3$, that the chance of obtaining a sampling average of $\bar{x} = 83$ mm Hg or greater is approximately .0013 .

Figure 5.1 The sampling distribution of the mean with
$$\mu = 80 \text{ mm Hg.}$$

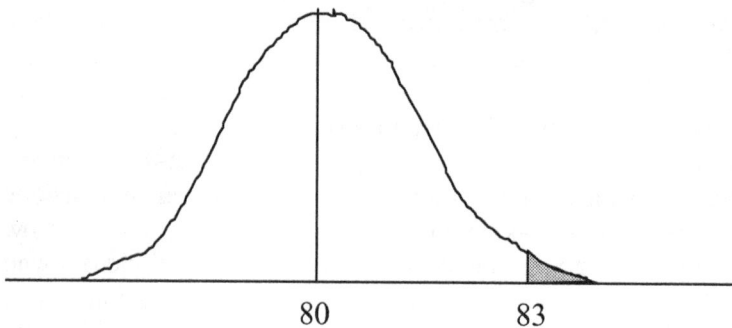

80   83

We now have the facts we need to make our decision. If the null hypothesis really is true then we have been quite lucky to have obtained such a large average in our sample. The other possibility is that the decongestant really does raise the diastolic blood pressure. Most researchers, if they had obtained a sample of $\bar{x} = 83$ mm Hg, would reject the null hypothesis and conclude that the decongestant does raise blood pressure.

It is traditional to denote the null hypothesis by H_0 and the alternative hypothesis by H_a. Using this notation we can state the null hypothesis as:

$$H_0 : \mu = 80 \text{ mm Hg,}$$

and the alternative hypothesis as:

$$H_a : \mu > 80 \text{ mm Hg.}$$

This is called a **one-tailed** test because the alternative hypothesis only considered the possibility that the decongestant would raise the blood pressure. What would happen if the decongestant lowered the blood pressure? If the researchers felt that the decongestant would lower blood pressure then the one-tailed test would specify the alternative as $H_a : \mu < 80$ mm Hg, and we would reject the null hypothesis if we obtained a sample mean far below 80 mm Hg. If the researchers felt that the decongestant could decrease or increase the diastolic blood pressure then the **two-tailed** alternative would be specified as $H_a : \mu \neq 80$ mm Hg, and we would reject the null hypothesis if we obtained a sample mean far from 80 mm Hg in either direction.

Exercise Set A

1. Suppose it is widely believed that high-school students, on average, get eight hours of sleep per night. If you believe that high-school students actually get less than eight hours of sleep per night, what should you state as the null and alternative hypotheses?

2. Cheese of often sold in one pound packages. A consumer advocate believes that the packages of cheese averaged less than 16 ounces, and wants to show that the average weight is less than 16 ounces. State the null and alternative hypotheses.

3. Suppose that a pain relief medicine has already been approved by the government for general use and that it is widely believed that it will have no effect on systolic blood pressure. Suppose you are doing some research with this drug and believe that the drug may increase systolic blood pressure. If you were to do a statistical test, what would you specify as the null and alternative hypothesis? State the hypotheses in terms of the change in the systolic blood pressure.

4. An organization that opposes gambling believes that many college students spend too much money gambling. The gambling industry claims that the average college student spends about $20 per year gambling. The anti-gambling organization believes that the average is much higher and wants to do a test of significance to demonstrate that fact. State the null and alternative hypothesis.

5. An automobile manufacturer wanted to determine if a new high-pressure tire would increase the gas mileage of their subcompact car. They used a sample of $n = 15$ identical cars in an experiment that had each car use the standard tires for 300 miles and then had each car use the new high-pressure tires for another 300 miles. The differences in gas mileage were recorded for the $n = 15$ cars.

State the null and alternative hypotheses that would be appropriate to detect an improvement in gas mileage. State the hypotheses in terms of the change in the gas mileage.

5.2 More on Tests of Hypotheses

The last section contained an overview of the testing process. We described the null and alternative hypotheses and we worked through one simple example. In our decongestant example we rejected the null hypotheses even though we were not absolutely certain that it was false. Every time we perform a test of significance there is some chance that we could be making an error. We would make an error if we reject the null hypothesis when the null hypothesis is really true, or if we fail to reject a null hypothesis that really is false.

Because errors could be made in the testing procedure, we need to describe all possible outcomes of the testing process. If we error in rejecting a null hypothesis when the null hypothesis is really true it is called a **Type I error**, and the chance of this happening is called the **level of significance** of the test. On the other hand, if we do not reject the null hypothesis when the null hypothesis is really false then it is called a **Type II error**. For a given sample size there is no way to completely control the chances of making these two types of errors. It is traditional in statistics to control the level of significance, which we will designate by α, by setting it equal to a small amount, say $\alpha = .05$. The probability of a type II error is then determined by the number of observations and by the level of significance. Table 4.2 shows two possible states of nature and the two decisions that could be made for each of these states of nature.

Table 5.1 Decision Table for a Test of Significance

	Decision	
True State of Nature	Do not reject H_0	Reject H_0
H_0 is true	Correct Decision	Type I error
H_0 is false	Type II error	Correct Decision

In the next section we will see how we can design a test to have a specified level of significance. For now, we note that if we obtain a sample mean that departs too far from what we would expect to find if the null hypothesis were true, then we would reject the null hypothesis. In the example in the last section we decided to reject the null hypothesis because the sample mean was 3 standard errors above the population mean that was specified by the null hypotheses. The number of standard errors above or below the population mean specified by the null hypothesis is the **test statistic**. If we decide to reject the null hypothesis for values of a test statistic that exceed a certain value, then that value is called the **critical value** of the test. If a test statistic falls in the interval above the critical value then we would reject the null hypothesis. That interval is called the **critical region.** It is shown as the shaded region to the right of the critical value in Figure 5.2.

For example, if we had decided, before collecting the data, that we would reject the null hypothesis if the sample mean was more than 2 standard errors above the population mean (μ_0) that is specified by the null hypothesis $H_0 : \mu = \mu_0$, then the critical value would be 2. The level of significance, which can be obtained from Table 1, is $\alpha = .0228$ for a critical value of $z = 2$.

We often want to control the level of significance at $\alpha = .05$, which corresponds to a critical value of $z = 1.645$. In our example, if the researcher would reject the null hypothesis when the sample mean was more than 1.645 standard errors above μ_0, then the researcher would have a 5% chance of making a Type I error. That is, if we set $\alpha = .05$

we will reject the null hypothesis if $z > 1.645$.   The relationship between the critical value of a test and the level of significance is shown in Figure 5.2.

Figure 5.2 The sampling distribution of the test statistic under the null hypothesis.

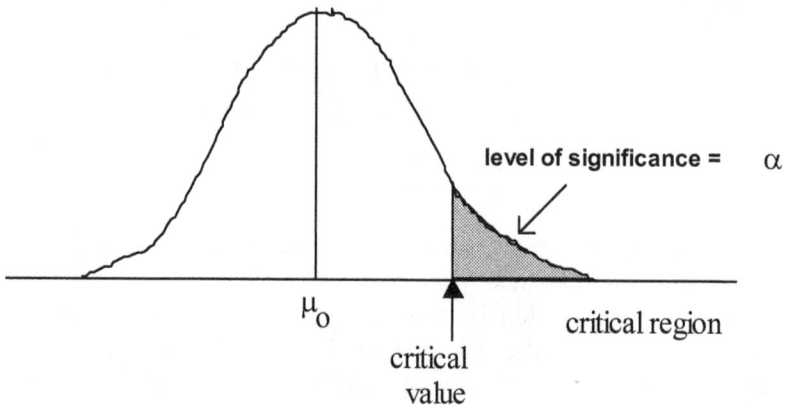

Once we fix the critical value for our test we can determine the probability of rejecting the null hypothesis when the null hypothesis is false. But a null hypothesis can be false in many ways. In our example, the null hypothesis states that the mean diastolic blood pressure equals 80 mm Hg exactly. If the alternative hypothesis is true, the population mean may be 81 mm Hg or it may be 84 mm Hg. The alternative hypothesis is vague; it only states that the population mean exceeds 80 mm Hg. However, the probability of rejecting the null hypothesis, which is called the **power** of the test, depends on the exact value of the population mean μ, which is usually unknown. Since the population mean is unknown we usually state the power for several values of the population mean or make a chart showing the power as a function of the population proportion.

The P-value Approach

Many people use critical values, based on a predetermined level of significance, to determine if a null hypothesis should be rejected. Another approach to significance testing does not require the calculation of critical values. With this approach we determine the chance of obtaining a test statistic as large or larger than the one we actually observed when the null hypothesis is true. This chance is called the **p-value**. If the p-value is less than the significance level we reject the null hypothesis. That is, we calculate, under the null hypothesis, the chance of obtaining a test statistic as extreme or more extreme than the one we obtained, and if that chance is smaller than α we reject the null hypothesis. In our example concerning the effect of decongestant on blood pressure, the sample mean was 3 standard errors above the mean. With $z = 3$ we obtain, from Table 1, a p-value of $p = .0013$, which would cause us to reject the null hypothesis for $\alpha = .05$.

Before computers were commonly available the p-value approach was seldom used because it was very difficult to compute p-values without the aid of a computer. In more recent times the p-value approach has become more popular because p-values can easily be computed and because the p-value itself gives the chance of obtaining a test statistic that is as extreme or more extreme than the value we observed.   In this text we will use the p-value approach rather than simply stating that the test was, or was not, rejected at a certain level of significance. We also encourage the publication of p-values because the p-value gives the reader additional information to judge the significance of the results. If a researcher, using the critical value approach with $\alpha = .05$, rejects the null hypothesis, then the reader does not know if the p-value was .045 or was .0002 . If we found that $p = .045$ there would be a sizable chance that the result could have been due to chance.   However, if we found that $p = .0002$ there is very little chance, under the null hypothesis, that the result could have been due to chance.

Two-tailed Tests

In our example, we decided to reject the null hypothesis if the sample mean diastolic blood pressure was large. But what would we do if we obtained a sample mean much smaller than 80 mm Hg? The answer depends on how we stated our hypotheses. If the alternative hypothesis was $H_a : \mu > 80$, then we would not reject the null hypothesis. However, if the alternative hypothesis was that $H_a : \mu \neq 80$ then we might reject the null hypothesis.   This is an example of a **two-tailed test** that can reject the null hypothesis for large or small values of the test statistic.

The p-values associated with one-tailed and two-tailed tests are displayed in Figure 5.3. For the one-tailed test of $H_0 : \mu = \mu_0$ against $H_a : \mu > \mu_0$ we reject the null hypothesis only for large values of the test statistic. If we obtain a test statistic z, the p-value is represented by the shaded area to the right of z in Fig 5.3(a). For the one-tailed test of $H_0 : \mu = \mu_0$ against $H_a : \mu < \mu_0$ the p-value is represented by the shaded area to the left of z in Fig. 5.3(b). For the two-tailed test of $H_0 : \mu = \mu_0$ against $H_a : \mu \neq \mu_0$ the test statistic is symmetrically distributed about zero, as shown in Figure 5.3(c), and the p-value is represented by the shaded areas to the left of $-|z|$ and the right of $|z|$. For these tests the null hypothesis is rejected if the absolute value of the test statistic exceeds the critical value.

Figure 5.3 p-values for one-tailed and two-tailed tests corresponding to a test statistic z.

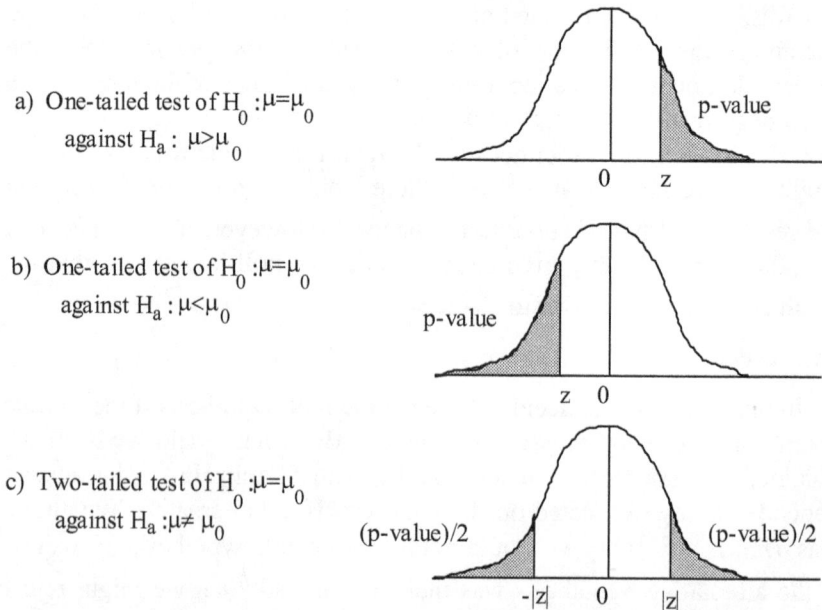

a) One-tailed test of $H_0 : \mu = \mu_0$
   against $H_a : \mu > \mu_0$

b) One-tailed test of $H_0 : \mu = \mu_0$
   against $H_a : \mu < \mu_0$

c) Two-tailed test of $H_0 : \mu = \mu_0$
   against $H_a : \mu \neq \mu_0$

There is sometimes controversy about whether a one-tailed or a two-tailed test should be used. If the researcher believes that the test statistic will be positive and is interested in detecting a positive effect of a treatment, then a one-tailed test would be appropriate. A serious problem with one-tailed tests is that other scientists, who will read the results after they are published, may feel that a two-tailed test would be more appropriate because the treatment effect could be positive or negative. Since a two-tailed p-value is twice that of a one-tailed p-value, the questionable use of a one-tailed test may give the impression that the researcher is trying to artificially lower the p-value. For example, suppose a researcher reports a one-tailed p-value of $p = .035$, which leads the researcher to reject the null hypothesis with a level of signifi-cance of $\alpha = .05$. If the treatment effect could be positive or negative the two-tailed p-value would be $p = .07$, which would prevent the researcher from rejecting the null hypothesis. It then becomes crucial to be able to justify the use of the one-tailed test. For this reason many researchers use two-tailed tests unless they have a very good reason for using a one-tailed test.

Exercise Set B

1. A university administrator, who knew that the average GPA of all students at a large university was 2.73, wanted to determine if stu-dents who lived off campus had a higher or lower average GPA. Because the administrator wanted to be quite certain of the results, she decided to use a level of significance of $\alpha = .01$.

   a) In your opinion, would a one-tailed or a two-tailed test be most appropriate?

   b) Suppose the administrator collected the data, computed the test statistic, and found the p-value to be $p = .072$. State the con-clusion.

2. Suppose two researchers wanted to test the null hypothesis $H_0 : \mu = \mu_0$ against $H_a : \mu > \mu_0$ by collecting data from a sample of size $n=50$. Suppose researcher A obtained a test statistic z_A that was greater than the test statistic z_B that was obtained by Researcher B. Which researcher obtained the smaller p-value?

3. An electrical component manufacturer produced "20 Amp" fuses, which were designed to blow when the current exceeds 20 amperes. To insure that the fuses meet the specifications, the manufacturer selected $n = 34$ fuses for inspection to determine if it is reasonable to conclude that the average current needed to blow the fuses is 20 amperes. They want to detect any increase in current required to blow the fuses, using a level of significance of $\alpha = .05$.

    a) State the null and alternative hypotheses. Is this a one-tailed test or a two-tailed test?

    b) Suppose the p-value, based on the data and the null hypothesis, is $p = .023$. Would they reject the null hypothesis? State your conclusion.

    c) Suppose they want to make a chart showing the power of this test for various mean currents for all fuses in the population. Which one of the following charts might best describe the power as a function of the mean current? [Hint: What is the power of the test for a mean current of 20 amperes?]

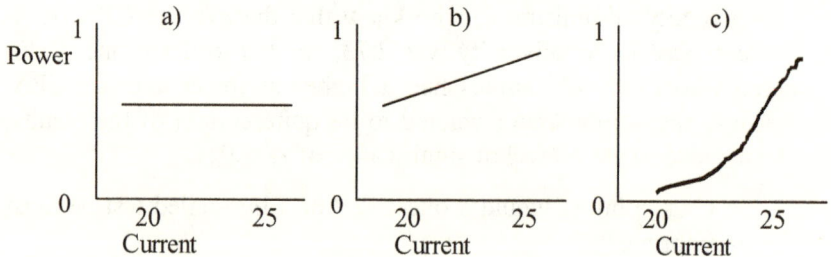

4. Suppose we are testing $H_0 : \mu = \mu_0$ against the alternative $H_a : \mu > \mu_0$ and have determined a critical value based on a level of significance (see Figure 5.2). If we decrease the level of significance, do we decrease or increase the critical value? Explain.

5. A fair coin is tossed 20 times. It can be shown that the probability of obtaining exactly 17 heads is .001087. The probability of obtaining exactly 18 heads is .000181, the probability of obtaining exactly 19 heads is .000019, and the probability of obtaining exactly 20 heads is .000001.

a) Make a vertical bar chart showing the probabilities of obtaining 17, 18, 19, and 20 heads.

b) If you are testing the null hypothesis that the coin is fair and you obtain 18 heads out of 20 tosses, what is the p-value?

c) If the level of significance is $\alpha = 0.05$ should you reject H_0?

5.3 The One-sample t-test for Paired Comparisons

A t-test for normal populations

The t-test is the most popular test for testing the null hypothesis $H_0 : \mu = \mu_0$ against the alternative $H_a : \mu > \mu_0$, where μ_0 is a specified value of the population mean. Our general approach will be to calculate the test statistic, estimate the p-value, and then reject the null hypothesis if the p-value is less than α. We will assume in this section that $X_1,...,X_n$ are independent observations from a normal distribution with mean μ. In Chapter 3 we noted that, if H_0 is true, the standardized variable

$$T = \frac{\bar{x} - \mu}{s/\sqrt{n}}$$

has a t distribution with $v = n-1$ degrees of freedom. Therefore, if the null hypothesis is true the distribution of t, in repeated sampling of size n, should look like the distribution shown in Figure 5.4. Note that the probability of obtaining a t statistic greater than $t_{\alpha,n-1}$ is α. If we obtain a test statistic of

$$t = \frac{\bar{x} - \mu_o}{s/\sqrt{n}}$$

and find that it exceeds the critical value $t_{\alpha,n-1}$, then we know that the p-value must be less than α. That is, if $t > t_{\alpha,n-1}$ we reject the null hypothesis because the p-value is less than α. The critical values for the t distribution are listed in Table 2 in the Appendix.

Figure 5.4 Sampling distribution of t for samples of size n

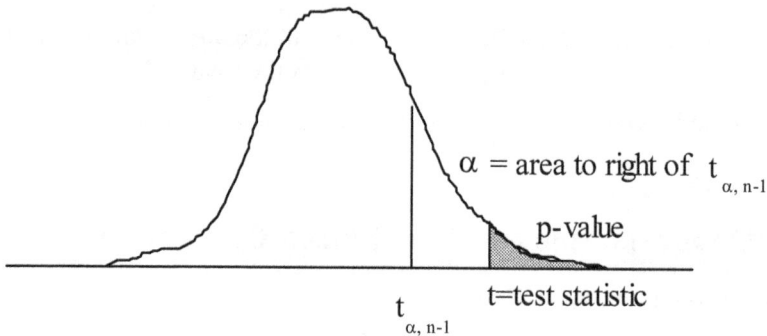

α = area to right of $t_{\alpha, n-1}$

p-value

$t_{\alpha, n-1}$

t=test statistic

We would also reject the null hypothesis if we computed the p-value directly and found that the p-value was less than α. The p-values can be roughly estimated from the critical values of the t distribution found in Table 2. If computer software is used to compute the test statistic the p-value will usually be included in the computer output.

The steps in a one-sample t-test are as follows:

1. State the null and alternative hypotheses.

2. Compute the test statistic $t = \dfrac{\bar{x} - \mu_o}{s/\sqrt{n}}$.

3. Estimate the p-value using the t distribution with $v = n - 1$ degrees of freedom.

4. Reject the null hypothesis if $p < \alpha$.

5. State the conclusion.

Example 5.1

For example, suppose a researcher is interested in determining if a new reading method is more effective than an old method, which had a mean of 70. If the distribution of scores using the new reading method is normal, then the researcher can use the t-test to determine if the mean of the reading scores is greater than 70. For this example the alternative hypothesis is that the mean of the reading scores is greater than 70. The null hypothesis, which the researcher hopes to reject, is that the mean is equal to 70. That is, $H_0 : \mu = 70$ and $H_a : \mu > 70$.

Now suppose the researcher measures the reading ability of 25 children, who were selected at random, and finds that the sample mean is $\bar{x} = 76$ and the sample standard deviation is $s = 12$. The test statistic would be

$$t = \frac{\bar{x} - \mu_o}{s/\sqrt{n}} = \frac{76 - 70}{12/\sqrt{25}} = \frac{6}{2.4} = 2.5$$

The p-value is roughly estimated from Table 2 in the Appendix, with $v = n - 1 = 24$ degrees of freedom, to be near $p = .01$. If we use a level of significance of $\alpha = .05$ we would reject the null hypothesis and conclude that the new reading method is more effective than the old method. ∎

One very common research design uses **paired comparisons** to increase the power of the test. In these designs the information is obtained from each member of the pairs, and the difference between these observations is analyzed with the one-sample t-test. In this next example a genetics company is interested in determining if a new variety of corn produces greater yields than the old variety.

Example 5.2

Suppose ten plots were selected at random and each of these plots was divided into two subplots, with one planted in the new variety and one planted in the old variety. The yield per acre for the two varieties in the 10 plots are:

Plot	Old Variety	New Variety	Difference
1	122	125	3
2	121	126	5
3	117	119	2
4	127	132	5
5	117	115	-2
6	120	121	1
7	110	118	8
8	128	125	-3
9	107	119	12
10	124	126	2

We are interested in the difference in yields between the two varieties. Since the fertility of the soils can vary from plot to plot, it makes sense to compare the varieties within the plots. For each plot the best way to compare the new variety to the old variety is to calculate the difference in yields between the old and new varieties. Let μ denote the population mean difference between the old and new varieties over all plots. The null hypothesis is that there is no difference in yields, which implies that $\mu = 0$, and the alternative hypothesis is that the new variety has a greater yield than the old variety, which implies that $\mu > 0$. Therefore, we wish to test $H_0 : \mu = 0$ against the alternative $H_a : \mu > 0$. The test statistic will be based on the mean of the differences, which is $\bar{x} = 3.3$, and the standard deviation of the differences, which is $s = 4.47$. The test statistic is

$$t = \frac{\bar{x} - \mu_o}{s/\sqrt{n}} = \frac{3.3 - 0}{4.47/\sqrt{10}} = \frac{3.3}{1.414} = 2.33,$$

and the approximate p-value, with $v = 9$ degrees of freedom, can be found in Table 2 to be between $p = .025$ and $p = .01$. Therefore, we reject the null hypothesis using $\alpha = .05$ and conclude that the new variety has a greater yield than the old variety. ∎

Figure 5.5 gives the Excel computer output, which shows the t-test statistics to be 2.33, when rounded. The computer output, which came from a data analysis tool in Excel, also gives the p-value for the one-tailed test on the next line as $p = .02227$. This result agrees with the bounds that we found from Table 2. Note that the output also gives the hypothesized mean difference of $\mu_0 = 0$, which can be specified by the user, and the two-tailed p-value.

Figure 5.5 Output from the t-test data analysis tool in Excel with paired data.

	Variable 1	Variable 2
Mean	122.6	119.3
Variance	25.6	46.23333333
Observations	10	10
Pearson Correlation	0.753162088	
Hypothesized Mean Difference	0	
df	9	
t Stat	2.332804467	
P(T<=t) one-tail	0.022271049	
t Critical one-tail	1.833113856	
P(T<=t) two-tail	0.044542097	
t Critical two-tail	2.262158887	

A Large Sample Test

The t-test can also be used with large samples. We noted in Chapter 3 that the t-distribution closely approximates the normal distribution if the degrees of freedom exceeds 30. Therefore, if $n > 30$ the p-value obtained from the t-distribution will approximate that obtained from the normal distribution and it makes little practical difference which one is used. Consequently, for large samples, it is traditional to compute the test statistic

$$z = \frac{\bar{x} - \mu_0}{s/\sqrt{n}},$$

which is then used with a table of the standard normal distribution to obtain the p-value.

The big difference between the large sample and small sample test is that the large sample test requires fewer assumptions. With small samples the observations were assumed to be normally distributed so that the test statistic would follow a t-distribution. With large samples we can use the central limit theorem to demonstrate that the distribution of the test statistic z approximates a normal distribution, even if the observations are from a skewed or long-tailed distribution.

Example 5.3

For example, suppose a consumer organization decided to challenge the claim made by a pharmaceutical company that their pain reliever gives relief within 5 minutes, on average. The consumer organization, which believes that the average time to pain relief exceeds 5 minutes, will not pursue the matter further unless there is convincing evidence that the average time to pain relief really does exceed 5 minutes. They design an experiment to test $H_0 : \mu = 5$ against the alternative $H_a : \mu > 5$, using $\alpha = .05$. The $n = 47$ subjects that were selected at random for the study had pain for an average of $\bar{x} = 5.7$ minutes, with a standard deviation of $s = 5.5$ minutes. The large sample test statistic is

$$ z = \frac{\bar{x} - \mu_o}{s/\sqrt{n}} = \frac{5.7 - 5.0}{5.5/\sqrt{47}} = \frac{.7}{.80} = .87 $$

Although the time to pain relief may be skewed, the distribution of z should be approximately normal because the sample size is large. Therefore, they can use the standard normal distribution in Table 1 to find the approximate p-value of $p = .192$, which would indicate that they should not reject the null hypothesis if they had used a significance level of α=.05. They would conclude that there is insufficient evidence to show that the average time to pain relief exceeds 5 minutes. ∎

Note that it is not correct to conclude that the null hypothesis is true. They have failed to reject the null hypothesis but this should not be interpreted to mean that they know that the pain reliever works in exactly 5 minutes. The time to pain relief may be somewhat larger than 5 minutes, but they do not have sufficient evidence to demonstrate that it must be greater than 5 minutes. This result may be disappointing to the consumer organization; in most situations researchers want to reject the null hypothesis so they can make a clear statement about the results of the experiment.

Exercise Set C

1. As part of a larger study to determine the effect of the menstrual cycle on food intake, researchers measured the food intake of 8 women before and after their menstrual period. The data given below is the average caloric intake over the 10 days before and after a menstrual period. (Based on "The effect of the menstrual cycle on patterns of food intake," *The American Journal of Clinical Nutrition*, 34 (1981) pp. 1811-1815)

Average Caloric Intake

Subject	Premenstrual	Postmenstrual
1	2378	1706
2	1393	958
3	1519	1194
4	2414	1682
5	2008	1652
6	2092	1260
7	1710	1239
8	1967	1758

a) Compute the differences (Postmenstrual-Premenstrual). How many of the women reported a decrease in food intake?

b) Compute the mean and standard deviation of the differences.

c) Suppose the researcher would like to detect an increase or a decrease in food intake using $\alpha = .05$. Set up the appropriate null and alternative hypotheses.

d) Compute the appropriate test statistic and find the approximate p-value. You may assume that the differences are normally distributed.

e) State your conclusion.

2. Some health organizations have used dental offices to identify people who have high blood pressure. However, it is possible that dental offices are not appropriate locations for these screening tests because some people may experience an increase in blood pressure simply because they are visiting a dentist. To determine the effects of the dental setting on blood pressure measurements, researchers compared the systolic blood pressure of $n = 74$ adults taken in the dental setting to their blood pressure in a medical setting. They found that the average of the differences (Dental blood pressure - Medical blood pressure) was 4.47 mm Hg with a standard deviation of 8.77 mm Hg. (Based on "The effect of the dental setting on blood pressure measurement," *American Journal of Public Health*, vol. 73, no. 10, pp. 1210-1212.)

a) Will it make a difference, with this data, if a large sample or a small sample t-test is used? Explain.

b) Compute the appropriate test statistic to test $H_0 : \mu = 0$ versus $H_a : \mu > 0$, and roughly estimate the p-value.

c) State your conclusion.

3. The dots plots show the data obtained by two researchers who each took samples of $n = 10$ to test $H_0 : \mu = 0$ against the alternative $H_a : \mu > 0$. If the one-sample t-test was used which researcher would obtain the largest test statistic? No computations should be necessary.

Researcher A

0
5

Researcher B

0
5

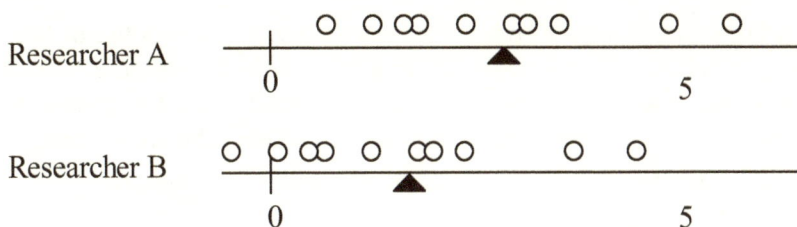

4. Two researchers each take samples of $n=10$ to test $H_0:\mu=0$ against the alternative $H_a:\mu>0$. If the one-sample t-test was used which researcher would obtain the largest test statistic? No computations should be necessary.

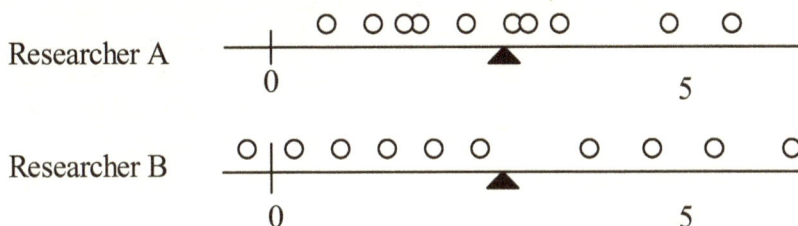

Researcher A

0
5

Researcher B

0
5

5. An electric equipment manufacturer produces circuit breakers that are designed to cut off the power if the current exceeds 20 amperes. The quality control inspector suspects that the circuit breakers need a current of more than 20 amperes to break the circuit. He decides to perform a test with $H_0:\mu=20$ amperes versus $H_a:\mu>20$ amperes by taking a sample of $n=100$ circuit breakers. He obtained a mean current of $\bar{x}=21.6$ with a sample standard deviation of $s=4.2$.

a) Compute the appropriate test statistic.

b) What is the approximate p-value?

c) Should he reject the null hypothesis using $\alpha=.05$?

d) What is the conclusion?

6. A brewery attempts to put 12 ounces of beer in each bottle of beer they produce. However, their filling machine may put more beer or less beer into the bottles. The quality control inspector decides to test H_0: $\mu = 12$ ounces versus H_a: $\mu \neq 12$ ounces. She takes a sample of 16 cans and finds that $\bar{x} = 11.8$ ounces and $s = .57$ ounces.

    a) Compute the appropriate test statistic.

    b) What is the approximate p-value?

    c) Should you reject H_0 : $\mu = 12$ using $\alpha = .05$?

    d) What is your conclusion?

7. A pharmaceutical company wants to market a drug that they believe will lower the temperature of individuals who have a fever. In order to obtain permission to market the drug, the company must demonstrate that the drug is effective in lowering temperature. They conducted a trial of the drug in 25 volunteers who had an initial temperature between 100 F° and 102 F° . The average decrease in temperature (Before - After) was 1.3 F°. The standard deviation of the difference was 1.6 F° .

    a) This is an example of a _____ _____ test.

    b) State the null and alternative hypotheses.

    c) Compute the appropriate test statistic.

    d) What is the p-value?

    e) Does the drug appear to be effective? Why or why not?

8. A petroleum company developed an energy saving motor oil that the company believes will improve fuel economy. To find out if it really does save energy they performed a test of significance. The null hypothesis was that there is no difference in fuel economy between the standard oil and the energy saving oil; the alternative hypothesis was that the gas mileage is greater with the energy saving oil. In order to validate this claim, they selected 8 new cars at random and

observed the gas mileage on each car using the standard oil and using the energy saving oil. The results were:

Fuel Consumption (miles per gallon)
Car

	1	2	3	4	5	6	7	86	mean	st.dev.
Standard Oil	24.1	26.2	19.1	31.0	27.2	25.6	22.2	27.6	25.375	3.620
Energy Saving	25.3	25.9	19.6	31.4	27.1	26.6	23.3	28.2	25.925	3.470
Difference	1.2	-.3	.5	.4	-.1	1.0	1.1	.6	.550	.548

a) Which row of data should be analyzed in the table in order to test the null hypothesis?

b) What assumptions are necessary to proceed with the analysis?

c) Compute the appropriate test statistic and roughly estimate the p-value.

d) Using a significance level of $\alpha = .05$, what is your conclusion?

9. A researcher was interested in determining if undergraduates consumed more coffee when they were stressed. In order to obtain some information on this topic she measured the coffee consumption, in cups per day, of 25 undergraduates during the first week of the semester and during the last week of the semester. The undergraduates that were measured in the last week of the semester were the same undergraduates that were interviewed in the first week. The data for these 25 undergraduates are as follows:

Coffee consumption (cups per day)

	Mean	Standard Deviation
First Week	1.7	2.1
Last Week	2.3	2.6
Difference (last-first)	.6	1.3

a) State the null and alternative hypotheses.

b) Compute the test statistic and approximate the p-value.

c) Would the null hypothesis be rejected with a level of signifi-
    cance of $\alpha = .05$?

d) Would the null hypothesis be rejected with a level of signifi-
    cance of $\alpha = .01$?

e) Using $\alpha = .01$, state the conclusion.

5.4 The Signed Rank Test

The t-test that we discussed in the last section is the recommended
test when the observations are normally distributed because it can be
shown that it is the most powerful test that can be devised for normally
distributed data. That is, the t-test has the best chance of detecting a
departure from the null hypothesis with normally distributed data.
However, in actual practice researchers often do not know the distribution of the observations. This can be a serious problem because the t-test
may not be the most powerful test if the distribution is not normal. One
problem is that $\bar{x}$ and s, which are used in the formula for the test
statistic of the t-test, are sensitive to outliers. The presence of outliers
will inflate s, which may reduce the magnitude of the test statistic and
weaken the power of the test.

The **signed rank test** avoids many of these problems. The distribution of the signed rank test statistic requires that the observations follow
a symmetric distribution, but does not require that the observations
follow a normal distribution.   Since the normal distribution is just one
kind of symmetric distribution, the symmetric assumption of the signed
rank test is less restrictive than the normal assumption of the t-test. Also,
the signed rank test is often more powerful than the t-test if the observations are from a non-normal distribution.

The **signed rank test statistic (S)**, which is sometimes called the
Wilcoxon signed rank test statistic, can be used to test the null hypothesis
$H_0 : \mu = 0$ against the one-tailed alternative $H_a : \mu > 0$. Let $\{X_1, ..., X_n\}$
be a random sample of n observations from a population having a
symmetric distribution. We begin by taking absolute values of the

observations to obtain $\{|X_1|,...,|X_n|\}$, and then we rank these absolute values to obtain the ranks of the absolute values $\{R_1^+,...,R_n^+\}$. With this notation, R_i^+ is the rank of the absolute value of the ith observation among the absolute values of all observations. The signed rank test statistic(S) is the sum of the signed ranks over the positive observations. That is, $S = \sum R_i^+$ where the summation extends over all observations having $X_i > 0$. The p-values, which are a function of S, are given in Table 3 of the Appendix for small samples having fewer than 10 observations. An example will help illustrate the calculation of S.

Example 5.4

Suppose a petroleum company wanted to determine if a new formula would get greater gas mileage than the older formula. Suppose 8 automobiles were used to test the gasoline mileage of the formulas and we want to test the null hypothesis that the new formula has the same gas mileage as the old formula. The alternative is that the new formula gives greater mileage than the old formula. Now suppose that each automobile is driven for 500 miles on each brand of gasoline to obtain the following data:

Automobile	Formula Old	New	Difference (New-Old)	$\lvert X_i \rvert$	R_i^+
1	24.2	26.5	2.3	2.3	7
2	25.1	25.8	.7	.7	4
3	26.2	25.7	-.5	.5	3
4	28.1	29.3	1.2	1.2	5
5	26.7	26.6	-.1	.1	1
6	22.5	24.1	1.6	1.6	6
7	25.2	27.9	2.7	2.7	8
8	25.7	25.9	.2	.2	2

We are interested in μ, which is the population difference between the gas mileage of the formulas. The differences in gas mileage are listed in the fourth column; their absolute values are listed in column 5. The ranks of the absolute values of the differences, which are given in the last column, can be used to test $H_0 : \mu = 0$ against $H_a : \mu > 0$. The rank sum (S) equals the sum of the signed ranks over all ranks having $X_i > 0$. That is, we sum the signed ranks over all automobiles except automobile 3 and automobile 5, which have negative differences, to obtain S=(7+4+5+6+8+2)=32. From Table 3 we see that the right-tail p-value is .0273. Since we are performing a one-tailed test with $\alpha = .05$ we will reject the null hypothesis. If we were using a two-tailed test the p-value would be 2(.0273)=.0546 and we would not reject the null hypothesis. ■

Figure 5.6 gives the SAS computer output from PROC UNIVARIATE for the signed rank test. This output gives many descriptive statistics for the differences in gas mileage (New - Old). The p-value for the two-sided test is given on the lower right as $p = .0547$, which agrees with the two-sided p-value from our tables. The signed rank test statistic is not given in the output but the "centered" signed rank statistic is given in the last line as $S - \frac{n(n+1)}{4} = 14$, which agrees with our computation of $S = 32$.

Figure 5.6. SAS output for signed rank test.

```
                    Univariate Procedure

                          Moments

    N                   8   Sum Wgts              8
    Mean           1.0125   Sum                 8.1
    Std Dev      1.144474   Variance        1.309821
    Skewness     0.221747   Kurtosis        -1.26529
    USS             17.37   CSS              9.16875
    CV           113.0345   Std Mean        0.404633
    T:Mean=0     2.502269   Pr>|T|            0.0409
    Num ^= 0            8   Num > 0                6
    M(Sign)             2   Pr>=|M|           0.2891
    Sgn Rank           14   Pr>=|S|           0.0547
```

For larger samples having $n > 10$, it is impractical to tabulate the distribution of S, but a good large sample approximation is available. It can be shown that, if the null hypothesis is true, the test statistic

$$z_{SR} = \frac{S - \frac{n(n+1)}{4}}{\sqrt{\frac{1}{4}\sum_{i=1}^{n} R_i^{+2}}}.$$

is approximately distributed as a standard normal random variable for $n \geq 10$. If there are no ties in the data the denominator will equal $\sqrt{[n(n+1)(2n+1)]/24}$, but since ties are often present the formula for z_{SR} is recommended for general use. The normal approximation to the distribution of the signed rank test statistic will be illustrated with the next example.

Example 5.5

Suppose an experiment is designed to determine which of two diets will produce the greatest weight gain in rats. We obtain 25 pairs of rats with each member of a pair coming from the same liter, and we compute the difference between the weight gains for each pair. The null hypothesis is that the two diets are equally effective, which implies that the population mean difference is zero. If we are interested in detecting a difference in either direction for weight gain, we test $H_0 : \mu = 0$ against the alternative $H_a : \mu \neq 0$. If the signed rank statistic was $S = 298$, and if there were no ties in the data, then the signed rank test statistic would be

$$z_{SR} = \frac{S - \dfrac{n(n+1)}{4}}{\sqrt{\dfrac{1}{4}\sum_{i=1}^{n} R_i^{+2}}} = \frac{S - \dfrac{n(n+1)}{4}}{\sqrt{\dfrac{n(n+1)(2n+1)}{24}}}$$

$$= \frac{298 - \dfrac{25(26)}{4}}{\sqrt{\dfrac{25(26)(51)}{24}}} = \frac{135.5}{37.165} = 3.646$$

Since this is a two-tailed test, the p-value is the area under the normal curve to the right of 3.646 and to the left of -3.646. From Table 1 we find that the curve to the right of 3.646 is approximately .0001 so the two-tailed p-value is approximately .0002, which would cause us to reject the null hypothesis for any reasonable level of α. ∎

Controversies in Statistics

Why Would Anyone Use a Rank Test?

At first glance there appears to be no good reason to use a rank test instead of a t test. Common sense would seem to indicate that ranking would be undesirable, because information could be lost by ranking. Besides, the t-test is popular and is thought to be reasonably robust. So why bother with a signed rank test when the t-test works?

There are several good reasons to use the signed rank test. The main reason is that, because the ranks are not as sensitive to outliers, the signed rank test may be much more powerful than the t-test if the distribution is long-tailed. Also, little information is lost by ranking if the number of observations is reasonably large.

Consequently, with large samples the little information that is lost through the ranking process is more than made up by the robustness that is gained by ranking. For a detailed comparisons of the powers of these two test see O'Gorman, T. W. "Applied Adaptive Statistical Methods: Tests of Significance and Confidence Intervals," SIAM press, 2004, Ch.5

Should the signed rank test always be used instead of the t-test? Probably not. The t-test is easy to compute and is slightly more powerful than the signed rank test if the observations are normally distributed. The properties of these tests and some recommendations for the selection of the best test is given in Section 5 of this Chapter.

If there are ties in the absolute values the usual procedure is to compute the average ranks of the tied observations, which are called the **midranks**. For example, if the absolute value of the differences are

$$\{3, 11, 11, 12, 15, 15, 15\}$$

then the midranks would be

$$\{1, 2.5, 2.5, 4, 6, 6, 6\}.$$

The presence of ties is not a serious problem for large samples if the test statistic z_{SR} is used with the normal approximation. However, for small samples that have many tied values there is no easy way to compute an accurate p-value. Typically, researchers use the normal approximation, even if $n < 10$, if there are a few ties.

Sometimes there are several zero differences in a paired comparison data set. In this case the best approach appears to be to remove the zero differences from the data and to adjust the sample size so that n equals the number of non-zero differences. That is, if we have 3 differences that equal zero in a data set having 20 differences, then we should compute the signed rank test statistic on the remaining $n=17$ non-zero differences.

Exercise Set D

1. An experimental supplementary reading program was evaluated using $n = 7$ pairs of below grade level readers who were matched on their reading ability prior to their admission to the program. One subject within each pair was randomly assigned to a behavioral program that was designed to deal with certain problems of these readers, while the other subject in the pair did not receive any special tasks or reinforcement. (Based on Heiman, J. R., Fischer, M. J., and Ross, A. O. "A supplementary behavioral program to improve deficient reading performance," *Journal of Abnormal Child Psychology* 1 (1973) pp. 390-399.) The reading score differences for these subjects is given in the following table:

Pair	Experimental	Control	Difference
1	1.8	0.8	1.0
2	1.4	0.6	0.8
3	1.1	0.7	0.4
4	1.2	-0.5	1.7
5	1.8	0.2	1.6
6	0.9	0.3	0.6
7	1.9	2.6	-0.7

a) State the null and alternative hypotheses for a one-tailed test.

b) Calculate the signed rank test statistic and find the p-value.

c) State your conclusion.

2. Twenty subjects were used in an experiment to determine the effectiveness of sodium valproate in reducing the frequency of seizures with chronic uncontrolled epilepsy. The researchers compared the number of fits observed over an 8 week period when they used the placebo to the number of fits observed when they used valproate. (Based on Richens, A. and Ahmad, S. "Controlled trial of sodium valproate in severe epilepsy," *British Medical Journal* 4 (1975) pp. 255-256) The results, along with the differences (Placebo - valproate) are given in the following table:

Subject	Placebo	Valproate	Difference
1	37	5	32
2	52	22	30
3	63	41	22
4	2	4	-2
5	25	32	-7
6	29	20	9
7	15	10	5
8	52	25	27
9	19	17	2
10	12	14	-2
11	7	8	-1
12	9	8	1
13	65	30	35
14	52	22	30
15	6	11	-5
16	17	1	16
17	54	31	33
18	27	15	12
19	36	13	23
20	5	5	0

a) State the null and alternative hypotheses that would be appropriate for performing a two-sided test.

b) Perform the signed rank test and determine the p-value.

c) Assuming that you used a level of significance of $\alpha = .05$ for your test, state your conclusion.

3. An intensive one-week reading course was developed for 7th grade students that promised to increase the performance of the students on certain reading tests. The improvement was measured by the difference between the reading test score given after the course and the reading test score given before the course. The researchers were interested in testing $H_0: \mu=0$ versus $H_a: \mu>0$ using the improvement scores. For the $n=8$ students in this course the improvement scores were:

$$5, -3, 7, 6, -4, 15, -1, 2$$

a) If the null hypothesis were true, how many of these differences would you expect to be positive?

b) Compute the signed rank test statistic.

c) Should the normal approximation be used to estimate the p-value? Why or why not?

d) Give the p-value.

e) State your conclusion.

4. Consider the following data sets for testing $H_0: \mu=0$ versus $H_a: \mu>0$. Which data set would produce the largest signed rank test statistic? Explain.

(a)   (b)   (c)

```
o oo o o o  o o          oo o ooo o              ꝏ o oo  o
─┬─────────┬──          ─┬───────┬──            ─┬───────┬──
 0        100            0       10              0        2
```

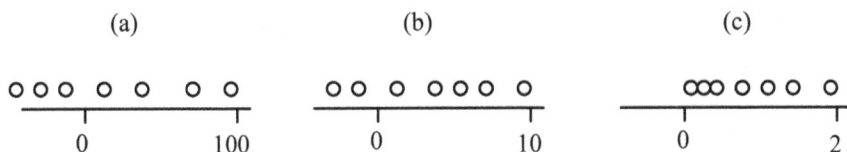

5. Consider the following data sets for testing $H_0 : \mu = 0$ versus $H_a : \mu > 0$. Would the two data sets produce equal signed rank statistics? Why or why not?

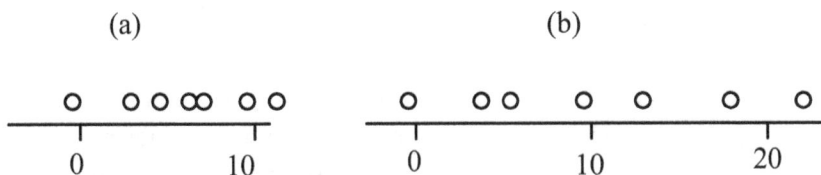

(a)   (b)

```
 o  o o ꝏ o o              o    o o   o  o     o   o
─┬─────────┬──           ─┬────────┬────────┬──
 0        10              0       10        20
```

6. A nutritionist wanted to determine if rats would live longer if they were kept on a low calorie diet. Ten pairs of rats were obtained, with the members of the pairs being males from the same liter. One member of the pair was randomly assigned to a normal diet while the other member was assigned to a low calorie diet. All rats were followed until they died a natural death. The observations that are to be analyzed are the lifetimes of the low calorie rats minus the lifetimes of the normal diet rats. The differences were recorded in days as follows:

33, 18, 22, 9, 42, -197, 35, 48, 15, 27, 36

a) Make a dot plot for the differences. Based on the plot do you believe that the differences are normally distributed? Why or why not?

b) Compute the signed rank statistic and find the p-value.

c) If you use the signed rank statistic would you reject the null hypothesis if you were testing $H_0 : \mu = 0$ versus $H_a : \mu > 0$?

d) Compute the t-test statistic for these observations. Using Table 2, estimate the p-value from the t-test. Would you reject the null hypothesis if you were testing $H_0 : \mu = 0$ versus $H_a : \mu > 0$?

e) Why do you believe there a difference between the results of these tests?

7. When we used a t test to analyze the corn yield data in the previous section we obtained a p-value between $p = .025$ and $p = .01$. Compute the signed rank statistic for this data set and, using the normal approximation, compute the approximate p-value for a one-tailed test. [Note the tied values in this data set.] Compare your result to that obtained in the previous section. If you were using a significance level of $\alpha = .05$, would you reject the null hypothesis using the signed rank test?

8. An investigator wanted to determine if a new diet would improve the weight gain in growing rats. In order to increase the efficiency of the experiment he used a paired comparison design with 25 pairs of rats. Each pair consisted of two rats from the same liter. After completing the experiment and computing the 25 differences he noted that there were two differences that were zero and there were no ties in the non-zero differences. Among the non-zero differences he obtained a signed rank of $S = 155$. Compute the approximate p-value for testing $H_0 : \mu = 0$ against $H_a : \mu > 0$. State your conclusion.

5.5   Guide to One-Sample Tests for Paired Comparisons

In the last two sections we have described the t-test and the signed rank test. Although either test can be used to analyze paired data, one test may be preferable to the other in some circumstances. In this section we will compare the performance of these two tests and will suggest a strategy for test selection.

Two tests can be compared based on their ability to maintain their level of significance near the stated value of α and on their power. Fortunately, both the t-test and the signed rank test do have a chance of rejecting the null hypothesis that is near α, even if the observations are from a non-normal population. The real difference between the two tests can be seen when we compare the power of the tests. That is, we want to compare the tests based on their chance of rejecting the null hypothesis

$H_0 : \mu = \mu_0$, when the population mean does not equal μ_0. Since the usual objective in using these tests is to reject the null hypothesis, we want to use the test that has the best chance of rejecting the null hypothesis if the null hypothesis really is false.

It can be shown that if the observations are normally distributed, the t-test is slightly more powerful than the signed rank test. For example, if a value of μ is chosen so that the power of the t-test is 50%, the power of the signed rank test will be around 47%, which is a difference of only 3%. However, if the observations are from an outlier distribution, which has generally the same shape as the normal except it has longer tails, then the signed rank test is much more powerful than the t-test. For example, if $n > 10$ the signed rank test may be 20% to 30% more powerful than the t-test for outlier distributions. Thus, if we suspect that the observations could have come from an outlier distribution, there is much to gain and little to lose by using the signed rank test. On the other hand, if we are reasonably confident that the population is normally distributed, the t-test will always be the more powerful than the signed rank test.

Another important practical consideration is the sample size. If $n < 10$ we should use the tables to obtain an accurate p-value for the signed rank test, but even these are of little value if there are ties. Furthermore, with small sample having $n < 10$ the signed rank test does not have as great a power advantage over the t-test with long-tailed distributions as it does with larger samples. Therefore, for small samples having $n < 10$ the t-test would almost always be recommended unless you were confident that the observations came from a long-tailed distribution.

It is important to keep in mind that both the t-test and the signed rank tests are fair tests, in the sense that they both have actual levels of significance that are close to the stated value of α. It is also important to remember that both tests have good power to detect large effects. However, the magnitude of the effect is usually unknown and the choice of a test does make a difference when the effect that you are trying to observe is not large. For these situations the following recommendations may be helpful:

- For large samples having $n \geq 10$ consider using the signed rank test unless you are confident that the observations are approximately normally distributed.

- For small samples having $n < 10$ consider using the t-test unless you believe that the data is from a long-tailed distribution.

Exercise Set E

1. Suppose an experimenter used 33 pairs of rats in a paired comparison experiment. The experimenter felt that the observations could have come from a long-tailed distribution. Which one-sample test would be recommended. Why?

2. Suppose you were involved with a research project that used data from a number of experiments on animals. If you wanted to use either the t-test or the rank sum test for each experiment and you usually used 6 to 10 animals for each experiment, which test would you use? You may assume that there are no outliers in the data.

3. Suppose a researcher uses $n=100$ observations from a paired comparison experiment. If the researcher suspects that the distribution of the differences has long tails, which test would be recommended? Justify your answer.

5.6 Sample Size Calculations for the One-Sample Test

In sections 5.3 and 5.4 we discussed two ways of performing a test of significance, based on the data that was collected in an experiment. In this section we consider how many subjects we need in our experiment in order to have a good chance of detecting an effect. Our discussion will be confined to paired data, and we will assume that we have decided to do a t-test.

We note that in our testing procedures we have kept the level of significance α fixed, while we have allowed the probability of a Type II error, which is the probability of not rejecting the null hypothesis when the null hypothesis is false, to vary. We now want to determine how many pairs we need to obtain a probability of a Type II error that we specify. Since the probability of a Type II error is the probability of not rejecting the null hypothesis when the null hypothesis is false, the power of the test is computed as $Power = 1 - P(Type\ II\ error)$. In this section we will compute the number of subjects (n) that we will need to achieve a power that we specify. These sample size calculations are based on the level of significance that we specify before we perform our experiment and on an estimate of the standard deviation of the differences (σ).

For example, suppose a dietician wanted to determine if a "food pyramid" diet, without caloric restriction, would cause a decrease in weight for dormitory residents. What sample size should the dietician use to achieve 80% power for a one-tailed test with $\alpha = .05$? Although it is difficult to calculate exactly how many sample would be needed, the formula that provides an approximate answer is easy to use. The sample size formula for a one-sided test is

$$n = \frac{\sigma^2 (z_\alpha + z_\beta)^2}{\Delta^2},$$

where σ is the standard deviation of the differences, Δ is the true difference that you want to observe, and $\beta = 1 - Power$. In practice we do not know σ^2 so we usually would use an estimate of σ^2 obtained from a prior study. Also, we don't know the exact value of Δ; we usually are interested in calculating the required sample size for a value of Δ that we believe to be important to detect. If we calculate the sample size for the smallest values of Δ that is important to detect then we will have greater power to detect larger values of Δ.

In our example, if we want 80% power then $\beta = 1 - Power = 1 - .8 = .2$, which gives, from Table 1, $z_\beta = .84$. If we use a level of significance of $\alpha = .05$, then $z_\alpha = 1.65$. Now suppose, from a previous study, that we expect the standard deviation of the weight changes, over the period of the study, to be approximately $\sigma = 5$ pounds. If we believe that a difference of $\Delta = 2$ pounds would be important to detect then we would need

$$n = \frac{\sigma^2 \left(z_\alpha + z_\beta\right)^2}{\Delta^2} = \frac{25(.84 + 1.65)^2}{2^2} = 39$$

subjects for the experiment, in order to have an 80% chance of determining a difference of $\Delta = 2$ pounds. However, if we needed to detect a difference of $\Delta = .5$ pounds we would need

$$n = \frac{\sigma^2 \left(z_\alpha + z_\beta\right)^2}{\Delta^2} = \frac{25(.84 + 1.65)^2}{.5^2} = 620$$

subjects to achieve a power of 80%. As you can observe from the formula, you will need a large sample to detect small differences.

Exercise Set F

1. A pharmaceutical company wanted to determine if a certain drug, which might reduce the craving for cigarettes, would be effective in reducing the number of cigarettes smoked per day by smokers. They planned a paired-difference experiment using a one-tailed test with $\alpha = .05$. Based on a previous study, they believed that the standard deviation of the differences (After - Before) in the number of cigarettes used per day is approximately $\sigma = 4$. If they want a 90% chance of detecting a difference of $\Delta = 2$ cigarettes, how many subjects do they need in the experiment?

2. Refer to the drug experiment in the previous exercise. If the company was willing to proceed with the experiment if it had an 80% chance of detecting a difference of $\Delta = 3$ cigarettes, how many subjects would they need in the experiment?

3. Researchers wanted to determine if exercise could lower the level of Triglycerides, which were thought to contribute to heart disease. Their plan was to measure the level of Triglycerides (in mmol/l) before and after an intense 4 week exercise program. They planned to use a one-tailed t-test to analyze the data with a level of significance of $\alpha = .05$. The researchers wanted a 80% chance of detecting a difference of $\Delta = .2$ mmol/l.

a) If they used an estimated standard deviation of $\sigma = .3$ mmol/l, how many samples would be required to meet their objectives?

b) If they used an estimated standard deviation of $\sigma = .5$ mmol/l, how many samples would be required to meet their objectives?

c) Does the sample size appear to be very sensitive to the estimate of the standard deviation? Explain.

4. An exercise scientist wanted to determine if caffeine increased the performance of sprinters. She designed a paired-difference experiment to see if caffeine would decrease the amount of time to run 100 meters. Based on some prior data she estimated the standard deviation of the differences to be near $\sigma = .4$ seconds. She wanted a 90% chance of detecting a difference using a one-tailed test with $\alpha = .05$.

a) If she wanted to detect a difference of $\Delta = .05$ seconds, how many runners would be required for this experiment?

b) If she wanted to detect a difference of $\Delta = .2$ seconds, how many runners would be required?

c) Why is there a large difference between the sample sizes that were calculated in (a) and (b)?

Chapter Review Exercises

1. Do people gain weight after they quit smoking? To answer this question researchers followed 373 asbestos-exposed smokers for one year to determine if they had quit smoking during that year and to determine how much weight they had gained or lost over that time period. In the study $n = 25$ smokers, who had continued smoking for 3 months but had subsequently quit smoking, had a 3.76 pound weight gain, on average, with a standard deviation of 10.03 pounds. (Based on Coates, T. J., and Li, V. C. "Does smoking cessation lead to weight gain? The experience of asbestos-exposed shipyard workers," *American Journal of Public Health*, 73 (1983) pp. 1303-1304.)

a) Perform a one-tailed test to detect weight gain using $\alpha = .05$. State your conclusion.

b) If you thought they might gain or lose weight, would you use a two-tailed test with $\alpha = .05$? Perform a two-tailed test using $\alpha = .05$ and state your conclusion.

c) Did you come to the same conclusion in (a) and (b)? Explain.

2. In some municipalities that obtain drinking water from wells, the presence of radium is a health concern. Suppose people in a town believe that their drinking water may exceed the limit of 5 picocuries of radium per liter. They obtain $n = 7$ water samples at random from their water supply and find the concentration of radium, in picocuries per liter, to be:

$$8.2, \quad 7.7, \quad 9.2, \quad 5.6, \quad 6.4, \quad 5.9, \quad 8.1$$

Because they are interested in detecting a potential hazard, they use a one-sided test of $H_0 : \mu = 5$ against $H_a : \mu > 5$. You may assume that the distribution of radium concentration is normal.

a) Compute $\bar{x}$ and s.

b) Compute the test statistic for a t-test.

c) If you used $\alpha = .05$ for your test, would you reject the null hypothesis?

d) Does it appear that the radium content exceeds the limit?

3. As part of a larger study, $n = 9$ coffee drinkers abstained from drinking coffee for ten weeks to determine if their serum cholesterol would decrease. After 10 weeks they experienced an average decrease in serum cholesterol of 45 mg per 100 ml with a standard deviation of 28 mg per 100 ml. (Based on Førde, O. H., Knutsen, S. F., Enesen, E., and Thelle, D. S. "The Tromsø heart study: coffee consumption and serum lipid concentrations in men with hypercholesterolaemia: a randomized intervention study," *British Medical Journal*, 290 (1985) pp. 893-895.)

a) Suppose the researchers would be interested in detecting either a decrease or an increase in serum cholesterol. Give the null and alternative hypotheses for an appropriate test.

b) Perform the test and roughly estimate the p-value.

c) Would you reject the null hypothesis using $\alpha = .05$?

d) State your conclusion.

4. A quality control inspector at a food processing plant wanted to determine if the "16 ounce" cheese packages really had 16 ounces of cheese in them. He selected $n=25$ cheese packages at random from a warehouse containing a very large number of packages and found that the sample mean was $\bar{x} = 15.1$ ounces and the standard deviation was $s = 1.2$ ounces. Suppose the quality control inspector decided to perform a test of significance because he believed that the mean weight is really less than the 16 ounces that is indicated on the package.

a) State the null and alternative hypotheses.

b) Compute the t-test statistic.

c) Determine the p-value.

d) State the conclusion of the t-test using a level of significance of $\alpha = .05$.

5. Eight adults enrolled in an eight week diet and exercise program at a local university. The director of the program believed that the program should be effective in reducing weight. The weights of the participants before and after were recorded as follows:

	Weight (pounds)	
Subject	Before	After
1	202	204
2	168	159
3	192	188
4	173	170
5	186	175
6	153	154
7	168	164
8	214	209

    a) State the null and alternative hypotheses.

    b) What statistical test would be most appropriate? You may assume that outliers were not expected to be observed in this data set.

    c) Compute the test statistic for the test that you have chosen.

    d) Determine the p-value.

    e) State your conclusion.

6. The director of a rehabilitation program for alcohol abusers wanted to determine if an experimental treatment program was effective in reducing alcohol consumption. Twenty individuals were selected from a large population of alcohol abusers and their alcohol consumption was recorded before and after the treatment program. The differences (After-Before) in alcohol consumption (in ounces of alcohol per week) were computed for each individual. For this data a negative number indicates improvement. The differences are:

    -2.1,   -6.2,   3.1,   -7.1,   -11.3, -5.6,   -1.9,   -4.3,
    3.2,   -4.5,   1.8,   2.7,   -3.4,   -7.9   -9.3,   4.6,
    -6.1,   -8.2   -2.7,   1.8

    a) State the null and alternative hypotheses.

    b) What statistical test would be most appropriate? Explain.

    c) Compute the test statistic for the test that you have chosen.

    d) Compute the p-value.

    e) State your conclusion.

7. A weight trainer wanted to determine if a low intensity workout program would increase the weight lifting performance of male high school basketball players. She selected 11 male basketball players at random who agreed to participate in the program. The effectiveness of the program was determined by the number of repetitions of a bench press exercise. The number of repetitions before and after the program were:

Participant	Before	After	Difference
1	8	14	6
2	12	21	9
3	13	11	-2
4	19	20	1
5	14	25	11
6	6	9	3
7	7	15	8
8	9	14	5
9	12	6	-6
10	7	11	4
11	11	18	7

Use a signed rank test to test the $H_o : \mu = 0$ versus $H_a : \mu > 0$. Compute the p-value and state the conclusion.

8. Refer to the previous exercise. Use the data in the previous exercise to compute a t-test of $H_o : \mu = 0$ versus $H_a : \mu > 0$. Compute the p-value and state your conclusion.

9. Compare your answers to the two previous exercises. Were the conclusions the same for both exercises? Were there large differences in the p-values?

10. Consider the following data sets:

Data Set A

Data Set B

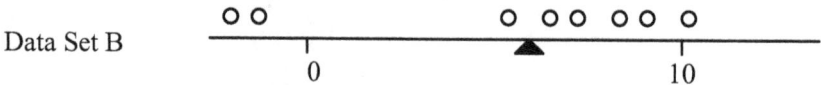

Suppose we compute the t-test statistics for these two data sets to test $H_0 : \mu = 0$ versus $H_a : \mu > 0$ for each of these data sets. Which of the following is true? Justify your answer.

a) The test statistic for data set A is greater than the test statistic for data set B.

b) The test statistic for data set B is greater than the test statistic for data set A.

c) The test statistics should be equal.

d) From the data given it is impossible to determine the relationship between these two test statistics.

11. Refer to the previous exercise and the dot plots that describe the data in the two data sets. Suppose we compute the signed rank test statistic for these two data sets. Which of the following is true? Justify your answer.

a) The test statistic for data set A is greater than the test statistic for data set B.

b) The test statistic for data set B is greater than the test statistic for data set A.

c) The test statistics should be equal.

d) From the data given it is impossible to determine the relationship between these two test statistics.

12. Do right-handed individuals have greater strength in their right arms? To determine if there is a meaningful difference in strength between the right and left arms a scientist designed a paired-difference experiment. In the experiment, the scientist recorded the number of repetitions of a certain arm exercise that could be done by each individual with their right hand and the number that could be done with their left hand. She then computed the difference (Right - Left) for each individual. The scientist believed that the standard deviation of the differences would be near $\sigma = 1.7$ repetitions. If the experimenter is planning to perform a test of $H_0 : \mu = 0$ versus $H_a : \mu > 0$, using a level of significance of $\alpha = .05$, how many subjects would be needed to achieve a power of 90% ?

Chapter 6. Confidence Intervals and Tests for Two Samples

Introduction

The two previous chapters concerned confidence intervals and tests of hypotheses using data obtained from a single sample. This chapter also concerns confidence intervals and tests, but now we will analyze data obtained from two samples. Two-sample confidence intervals can be used to estimate the difference between population means. For example, they can be used to estimate the effectiveness of a new drug compared to an older drug, or to estimate the difference in crop yield between two varieties of wheat. They could also be used to estimate the effectiveness of a new surgical procedure. For example, if surgeons develop a new surgical technique, it is important to compare the new technique to the old technique. We can use a confidence interval to estimate the difference in the proportion of the surgeries that were successful between the new and old techniques. In this case we would use a confidence interval for the difference between proportions. Or we may want to estimate the difference in the mean recovery time between the new and the old techniques. In this situation we would use a confidence interval for the difference between mean recovery times.

We will also develop two-sample tests of significance, which will be used to compare two population means. For example, if a surgeon has developed a new surgical procedure we may want to perform a statistical test to demonstrate that the new procedure is superior to the old procedure, even before we consider calculating a confidence interval. These two-sample tests have proven to be invaluable for researchers in the sciences who need to know if one population mean is really different than another population mean.

Two-sample tests and confidence intervals can often give the same information. Researchers in the basic sciences tend to use tests of significance while those working with business or engineering applications tend to use confidence intervals. Which should you use? In most cases either procedure could be used, your choice should be based primarily on which method will communicate the results most effectively to your audience.

216

All the procedures described in this chapter assume that we have obtained information from two samples and that the samples in one group are independent of the samples in the other group. We will begin by discussing several commonly used confidence intervals.

6.1 Confidence Intervals

In this chapter we will assume that we have obtained a random sample of n_1 observations from the first population, and a random sample of n_2 observations from the second population. We will also assume that the first sample is independent of the second sample. The general procedure will be to calculate the sample means from both samples and use the sample means to estimate the difference between the population means.

For example, suppose a psychiatrist wanted to more accurately quantify the amount of caffeine that people consume. Since most people consume either percolated or drip coffee she decided to take a random sample of $n_1 = 20$ consumers of percolated coffee and a second random sample of $n_2 = 30$ consumers of drip coffee. The individuals who brewed and drank percolated coffee had a sample average caffeine content of 115mg per 6 ounce cup, while the individuals who brewed and drank drip coffee had a sample average caffeine content of 81mg per 6 ounce cup. The analysis will naturally use the difference between the sample averages, which is $115 - 81 = 34\,mg$, as an estimate of the difference between the brewing methods. The populations and samples are illustrated in Figure 6.1.

Figure 6.1 Two independent random samples.

Population 1   Population 2

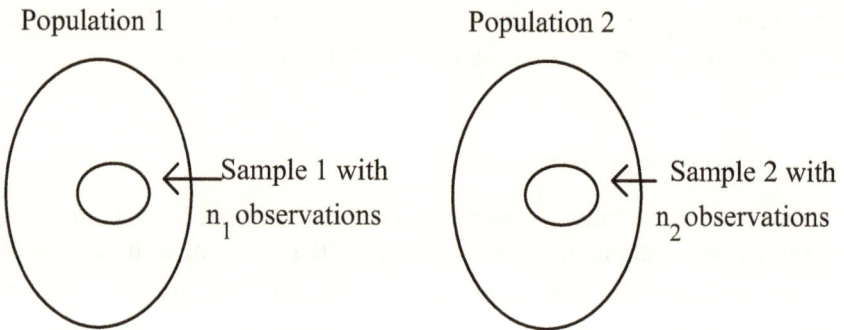

Sample 1 with n_1 observations

Sample 2 with n_2 observations

Our objective in this section is to estimate the difference between the population means. Naturally, we will use the difference between the sample means as our estimate of the difference between the population means. The confidence intervals that we will compute will give us some idea of the accuracy of the point estimate of the population difference.

Large sample confidence intervals for the difference between means

For the first population we will use the sample mean $\bar{x}_1$ to estimate the population mean μ_1. Similarly, for the second population we will use the sample mean $\bar{x}_2$ to estimate the population mean μ_2. Let s_1 and s_2 be the sample standard deviations in the first and second populations. Our objective in this section is to make a confidence interval for $(\mu_1 - \mu_2)$, which is the difference between the population means. We begin by noting that the point estimate of $(\mu_1 - \mu_2)$ is $(\bar{x}_1 - \bar{x}_2)$, which is the difference between the sample means. The next step is to estimate the precision of $(\bar{x}_1 - \bar{x}_2)$. If we have estimates of the variance of $\bar{x}_1$ and $\bar{x}_2$, the variance of $(\bar{x}_1 - \bar{x}_2)$ can be estimated because the variance of a difference is the sum of the variances, provided the two statistics are independent. We know that $\dfrac{s_1}{\sqrt{n_1}}$ is an estimate of the standard error of $\bar{x}_1$ in the first population, and that $\dfrac{s_2}{\sqrt{n_2}}$ is an estimate of the standard error of $\bar{x}_2$ in the second population. Since the estimated variance of $\bar{x}_1$ is $\dfrac{s_1^2}{n_1}$ and the estimated variance of $\bar{x}_2$ is $\dfrac{s_2^2}{n_2}$, we can compute the estimated variance of $(\bar{x}_1 - \bar{x}_2)$ as

$$Estimated\ Variance(\bar{x}_1 - \bar{x}_2) = \frac{s_1^2}{n_1} + \frac{s_2^2}{n_2}.$$

Using this estimate of the variance of $(\bar{x}_1 - \bar{x}_2)$, we can compute confidence intervals for the difference $(\mu_1 - \mu_2)$. The central limit theorem states that the distribution of $\bar{x}_1$ is approximately normal if n_1 is large and that the distribution of $\bar{x}_2$ is approximately normal if n_2 is large. It can be shown that if $\bar{x}_1$ and $\bar{x}_2$ are both approximately normally distributed then $(\bar{x}_1 - \bar{x}_2)$ will be approximately normally distributed. Hence, a large sample 95% confidence interval is

$$(\bar{x}_1 - \bar{x}_2) \pm 1.96 \sqrt{\frac{s_1^2}{n_1} + \frac{s_2^2}{n_2}},$$

and the general formula for the $100(1-\alpha)\%$ confidence interval for $(\mu_1 - \mu_2)$ is

$$(\bar{x}_1 - \bar{x}_2) \pm z_{\alpha/2} \sqrt{\frac{s_1^2}{n_1} + \frac{s_2^2}{n_2}}.$$

This formula is valid for large samples having $n_1 \geq 30$ and $n_2 \geq 30$. Because the central limit theorem does not require the distributions to be normal, the confidence interval formula is valid for large samples even if the observations are not normally distributed.

Example 6.1

A psychiatrist took a sample of $n_1 = 30$ consumers of drip coffee and $n_2 = 20$ consumers of percolated coffee in order to determine the caffeine content of a 6 ounce cup. Suppose from the first sample she obtained a sample average of $\bar{x}_1 = 115\,mg$ of caffeine with a sample standard deviation of $s_1 = 33\,mg$. From the second sample she obtained a sample average of $\bar{x}_2 = 81\,mg$ with a sample standard deviation of $s_2 = 28\,mg$. The point estimate of the difference between the population means is $(\bar{x}_1 - \bar{x}_2) = 115 - 81 = 34\,mg$. A 95% confidence interval for the difference between the population means is

$$(\bar{x}_1 - \bar{x}_2) \pm z_{\alpha/2} \sqrt{\frac{s_1^2}{n_1} + \frac{s_2^2}{n_2}}$$

$$= (115 - 81) \pm 1.96 \sqrt{\frac{33^2}{30} + \frac{28^2}{20}}$$

$$= 34 \pm 17 = (17, 51)$$

This is a 95% confidence interval for the difference between the caffeine content of drip and percolated coffee. That is, if we took many samples, each with n_1=30 and n_2=20, from these populations, approximately 95% of the confidence intervals would cover the difference between the population means $(\mu_1 - \mu_2)$. ∎

Small Sample Confidence Intervals for the Difference Between Population Means

The large sample confidence intervals that we have discussed can be used with skewed or long-tailed data. In contrast, small sample confidence intervals for the difference between population means $(\mu_1 - \mu_2)$ require the observations to be normally distributed. If we have small samples from populations that have skewed distributions with either $n_1 < 30$ or $n_2 < 30$, then the sampling distribution of $(\bar{x}_1 - \bar{x}_2)$ may not be normal and it may be necessary to make additional assumptions in order to construct a confidence interval. However, if the distributions of the observations in both samples are normal, the sampling distribution of $(\bar{x}_1 - \bar{x}_2)$ will be normal, even if small samples were taken. If the standard deviations are equal it is possible to calculate a confidence interval for $(\mu_1 - \mu_2)$ by computing a pooled estimate of the common variance using

$$s_p^2 = \frac{(n_1 - 1)s_1^2 + (n_2 - 1)s_2^2}{n_1 + n_2 - 2}.$$

(Since this pooled estimate of the common variance (s_p^2) is a weighted average of s_1^2 and s_2^2, s_p^2 should approximate s_1^2 and s_2^2). Since s_p^2 is an estimate of the variance in both populations, the standard error of

$(\bar{x}_1 - \bar{x}_2)$ is $\sqrt{\dfrac{s_p^2}{n_1} + \dfrac{s_p^2}{n_2}} = s_p \sqrt{\dfrac{1}{n_1} + \dfrac{1}{n_2}}$. Therefore, the $100(1-\alpha)\%$ confidence interval for $(\mu_1 - \mu_2)$ is

$$(\bar{x}_1 - \bar{x}_2) \pm t_{\alpha/2, n_1+n_2-2}\, s_p \sqrt{\frac{1}{n_1} + \frac{1}{n_2}}.$$

An example will illustrate the calculation of the pooled estimate of the common variance and the confidence interval.

Example 6.2

Morton's Neuroma is an enlarged nerve in the foot that can cause pain near the toes. Two types of surgery are sometimes used to correct the problem. A surgical procedure that uses a dorsal (top) incision is believe to heal quicker than the procedure that uses a plantar (bottom) incision. To see if the dorsal incision promotes quicker healing, patients were randomly assigned to one of two surgical procedures and the number of days before the patient returned to work was determined. The $n_1 = 10$ patients who had the dorsal incision had an average recovery time of $\bar{x}_1 = 34.2$ days with a standard deviation of $s_1 = 8.6$ days. The $n_2 = 12$ patients assigned to the plantar incision had an average of $\bar{x}_2 = 47.5$ days with a standard deviation of $s_2 = 9.1$ days. In order to compute a confidence interval for this small sample we must make the assumption that the scores were normally distributed and that the variances were equal. Using these assumptions we compute the pooled variance as

$$s_p^2 = \frac{(n_1-1)s_1^2 + (n_2-1)s_2^2}{n_1+n_2-2} = \frac{9(8.6)^2 + 11(9.1)^2}{20} = 78.82,$$

which implies that $s_p = 8.88$.   With these assumptions the 95% confidence interval for $(\mu_1 - \mu_2)$ is

$$\left(\overline{x}_1 - \overline{x}_2\right) \pm t_{\alpha/2, n_1+n_2-2}\, s_p \sqrt{\frac{1}{n_1} + \frac{1}{n_2}}$$

$$= (34.2 - 47.5) \pm 2.086(8.88)\sqrt{\frac{1}{10} + \frac{1}{12}}$$

$$= -13.3 \pm 7.93 = (-21.23,\ -5.37).$$

Note that the validity of the confidence interval depends on the validity of the normality assumption and the equal variance assumption. If the distributions are not normal and either $n_1 < 30$ or $n_2 < 30$ then, in repeated samples, 95% of the confidence intervals may not cover $(\mu_1 - \mu_2)$.

Large Sample Confidence Intervals for the Difference Between Population Proportions

We will now turn our attention to confidence intervals for the difference between the proportions in two populations. We will assume that we have binary variables, which can take on only two values, in each population.   For example, suppose a researcher is interested in the difference in the proportion of men and women who drink coffee.   Let p_1 be the population proportion in the first population (men) who drink coffee and p_2 be the population proportion in the second population (women) who drink coffee.   We will use the proportion of coffee drinkers that we observe in the first sample ($\hat{p}_1$) to estimate p_1 and will use the proportion that we observe in the second sample ($\hat{p}_2$) to estimate p_2.   In general, the proportion of some outcome in the first sample will be denoted by $\hat{p}_1$ and the proportion of the same outcome in the second sample will be denoted by $\hat{p}_2$.

To compute a confidence interval for $(p_1 - p_2)$, we use $\dfrac{\hat{p}_1(1-\hat{p}_1)}{n_1}$

as our estimator of the variance of the $\hat{p}_1$ and use $\dfrac{\hat{p}_2(1-\hat{p}_2)}{n_2}$ as our

estimate of the variance of $\hat{p}_2$. Using these estimates the large sample $100(1-\alpha)\%$ confidence interval can be written as

$$(\hat{p}_1 - \hat{p}_2) \pm z_{\alpha/2} \sqrt{\frac{\hat{p}_1(1-\hat{p}_1)}{n_1} + \frac{\hat{p}_2(1-\hat{p}_2)}{n_2}}.$$

For the large sample confidence interval to be valid for the coffee example, the number of people who drink coffee and the number of people who do not drink coffee in both groups must equal or exceed 5. To express the sample size requirements in a general form, we note that, in the first group, the number of people who have a certain outcome is $n_1\hat{p}_1$ and the number who do not have that outcome is $n_1(1-\hat{p}_1)$. For the second group the number of people who have a certain outcome is $n_2\hat{p}_2$ and the number who do not have that outcome is $n_2(1-\hat{p}_2)$. For the large sample formulas to be valid $n_1\hat{p}_1$, $n_1(1-\hat{p}_1)$, $n_2\hat{p}_2$, and $n_2(1-\hat{p}_2)$ must equal or exceed 5.

Example 6.3

Before obtaining permission to market a drug, pharmaceutical companies must estimate the proportion of users who will experience certain adverse events, or side-effects. A company that wanted to market an anti-depressant drug did a randomized double-blind experiment by giving 200 subjects a new drug and another 200 subjects a placebo. After the subjects used the drugs for 3 weeks they were asked if they had experienced drowsiness. Of the 200 subjects who received the drug 16 reported drowsiness, so that $\hat{p}_1 = .08$. Of the 200 subjects who took the placebo only 9 reported drowsiness, so that $\hat{p}_2 = .045$. The difference between the sample proportions $(.080 - .045) = .035$ is the point estimate of $(p_1 - p_2)$. We can use the large sample confidence interval because the number who reported drowsiness and who did not report drowsiness exceeded 5 in both groups. The 95% confidence interval for the difference between the population proportion $(p_1 - p_2)$ is

$$(.080 - .045) \pm 1.96 \sqrt{\frac{.08(1-.08)}{200} + \frac{.045(1-.045)}{200}}$$

$$= .035 \pm .047 = (-.012, .082).$$

Note that the 95% confidence interval is rather wide, which indicates that we have only a rough estimate of the proportion of adverse events due to the antidepressant drug. Since the confidence interval covers zero we are not confident that the new drug causes drowsiness. A much larger sample would be required to obtain a more accurate answer. ■

Exercise Set A

1. As part of a larger study of diet and occupation, researchers compared the caloric intake of a group of civil servants to a group of bus conductors on London double-decker buses. The 83 conductors in the first group consumed an average of 2844 calories with a standard deviation of 437 calories. The 68 civil servants in the second group consumed an average of 2704 calories with a standard deviation of 355 calories. (Based on Marr, J. W. and Heady, J. A. "Within- and Between-person variation in dietary surveys: number of days need to classify individuals", *Human Nutrition: Applied Nutrition*, 40A (1986), pp. 347-364.)

    a) Compute the pooled estimate of the common variance (s_p^2).

    b) Compute a 95% confidence interval for the difference in calories that are consumed between the conductors and the civil servants.

2. Pediatricians investigated the effect of inositol supplementation on mortality in premature infants who had respiratory problems. In a randomized, double-blind trial, 114 infants were assigned to the placebo group and 119 were assigned to the inositol group. In the placebo group 24.3% of the infants died within 28 days compared to 11.4% who died in the inositol group during the same period. (Based on "Inositol supplementation in premature infants with respiratory distress syndrome", *The New England Journal of Medicine*, vol. 326, no. 19(1992) pp. 1233-1239)

a) Compute a 95% confidence interval for the difference between the population proportions.

b) Based on this confidence interval would you conclude that inositol has a beneficial effect? Why or why not?

3. A surgeon wanted to determine if a new surgical procedure is more effective than the standard surgery for patients who had a certain diagnosis. Twenty hospital patients who had a certain disease volunteered for an experiment. They were assigned at random to either a standard surgery group or to an experimental surgery group. The 10 patients who had the standard surgery recovered in an average of 8.2 days with a standard deviation of 4.6 days. The 10 patients who had the experimental surgery recovered in an average of 6.6 days with a standard deviation of 3.7 days. You may assume that the recovery time is normally distributed. Compute a 95% confidence interval for the difference in the population means.

4. Two researchers (A and B) took samples from two populations in order to estimate the difference between the population means $(\mu_1 - \mu_2)$. The observations for both researchers are shown in the dot diagrams. You may assume that the observations are normally distributed, and that the scales are the same for all the plots.

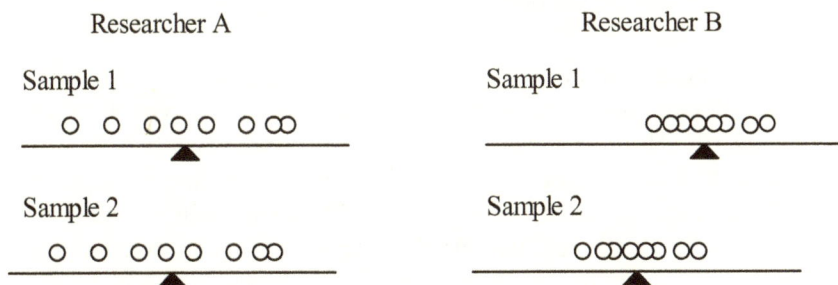

Researcher A   Researcher B

Sample 1   Sample 1

Sample 2   Sample 2

a) Which of the researchers will have a point estimate of $(\mu_1 - \mu_2)$ closest to zero?

b) Which of the researchers will have the narrowest confidence interval? Explain.

5. In the Epidemiologic Catchment Area survey, 3058 people were interviewed in New Haven, Connecticut to determine if they met the criteria for certain psychiatric disorders. Of the 1292 males who were interviewed, 49 met the criteria for a diagnosis of "simple phobia." Of the 1766 females who were interviewed, 150 met the criteria for "simple phobia." (Based on Robins, L. N. *et al.* "Lifetime prevalence of specific psychiatric disorders in three sites," *The Archives of General Psychiatry*, 41 (1984) pp. 949-958.)

   a) Are these samples large enough to justify the use of the large sample formula for the confidence interval for the difference between the population proportions? Explain.

   b) If possible, compute the 95% confidence interval for the difference between the population proportions.

6. A psychiatrist did a survey of 100 males and 100 females in a community to determine who met the criteria for "simple phobia." She found that 3 males and 7 females met the criteria.

   a) Are these samples large enough to justify the use of the large sample formula for the confidence interval for the difference between the population proportions? Explain.

   b) If possible, compute the 95% confidence interval for the difference between the population proportions.

7. An oenologist wanted to compare the alcohol content of a certain type of wine produced at one vineyard to the same type at another vineyard. He took a random sample of 25 white bottles from one vineyard and a sample of 22 red wine bottles from another vineyard. The average alcoholic content was 10.5% for the white wine bottles and 11.7% for the red wine bottles. The oenologist wanted to compute a 95% confidence interval for the difference between the alcohol content of the red and the white wine.

   a) Are the observations continuous or discrete?

   b) Is this a confidence interval for a difference in proportions or for a difference in means?

c) Compute a confidence interval if you have enough information to do so. If you do not have enough information state what additional information would be required.

6.2 The t-test For Comparing Two Means

In the last section we used a confidence interval to describe the difference between two populations. In this section we will take a different approach; we will attempt to determine if there is any difference between the population means or proportions. This approach is often taken in the sciences because of the need to test research hypotheses.

While there are many statistical procedures that can be used to compare two means, the t-test is the most popular. It is often used in observational studies to compare the means of two populations. In experimental work, the t-test is used to compare the means of two treatment groups. In this section we will assume that we have obtained two independent random samples from two populations. We will concentrate on the one-sided test of $H_0 : \mu_1 = \mu_2$ versus the alternative $H_a : \mu_1 > \mu_2$. The two-sample t-test will be performed in a similar manner to that used in the last chapter for the one-sample test. For this one-sided test we will consider rejecting the null hypothesis if $\bar{x}_1$ is much greater than $\bar{x}_2$. We will reject the null hypothesis if the p-value is less than α.

In order to estimate the p-value for a test statistic we need to know the exact distribution of the test statistic. But the exact distribution for the two-sample t-test statistic can only be determined if the distributions of the observations in both populations are normal and if the variability in the first population equals the variability in the second population. Consequently, we will proceed by assuming that $\sigma_1^2 = \sigma_2^2$ in order to compute the pooled estimate of the common variance as

$$s_p^2 = \frac{(n_1 - 1)s_1^2 + (n_2 - 1)s_2^2}{n_1 + n_2 - 2}.$$

This is the same variance estimator that we used in the previous section for small sample confidence intervals.

If the distributions are normal, have equal variances, and if the null hypothesis is true, it can be shown that the statistic

$$t = \frac{\bar{x}_1 - \bar{x}_2}{s_p \sqrt{\dfrac{1}{n_1} + \dfrac{1}{n_2}}},$$

will be distributed as a t distribution with $v = n_1 + n_2 - 2$ degrees of freedom. Knowledge of the distribution of this statistic allows us to find the p-values for any value of the test statistic. The distribution of the t statistic, under the null hypothesis, is shown in Figure 6.2 along with a critical value.

Figure 6.2 Distribution of the t statistic under the null hypothesis.

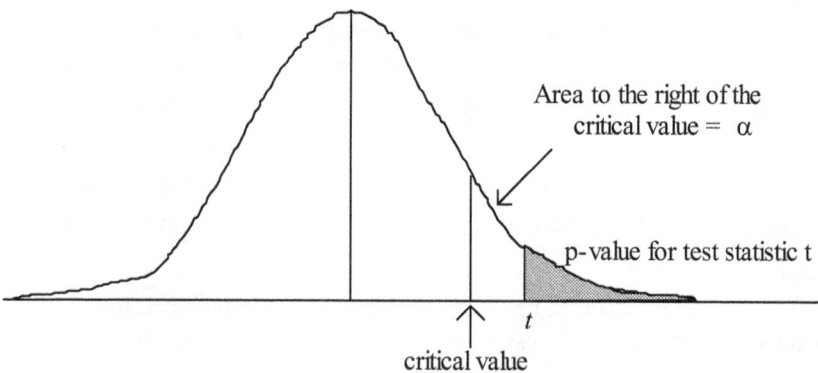

After the test statistic is computed the general procedure for performing the test is similar to that used in the previous chapter for one-sample tests. For a one-sided test of $H_0: \mu_1 = \mu_2$ versus $H_a: \mu_1 > \mu_2$ we calculate s_p and the test statistic t. We then use Table 2 or statistical software to estimate the p-value and reject the null hypothesis if the p-value is less than α. Another approach that can be used is to use Table 2 to obtain a critical value of $t_{\alpha, n_1 + n_2 - 2}$ and then reject the null hypothesis if the test statistic exceeds the critical value.

Example 6.4

A researcher designed an experiment to determine if a certain kind of high protein diet could produce weight gain in rats. The researcher randomly assigned 24 rats to either a high protein diet or to a low protein diet. Because she believed that the high protein diet would increase weight gain she did a one-tailed test. Using the high protein diet as the first population she would like to reject $H_0: \mu_1 = \mu_2$ in favor of $H_a: \mu_1 > \mu_2$. The weights that each rat attained at the 8th week after birth were recorded as follows:

High Protein   494,   526,   687,   707,   718,   877,   925,

                         967,   1087,   1292,   1318,   1466

Low Protein   227,   277,   418,   553,   572,   612,   692,

                         782,   808,   847,   1113,   1367

Based on these observations she obtained a sample average weight of $\bar{x}_1 = 922$ grams with a sample standard deviation of $s_1 = 316.8$ grams for the $n_1 = 12$ rats in that sample. For the $n_2 = 12$ rats assigned to the low protein diet, she obtained a sample average weight of $\bar{x}_2 = 689$ grams with a sample standard deviation of $s_2 = 327.8$ grams.

She began the calculation of the t-test by computing the pooled standard deviation

$$s_p^2 = \frac{(n_1 - 1)s_1^2 + (n_2 - 1)s_2^2}{n_1 + n_2 - 2} = \frac{11(316.8)^2 + 11(327.8)^2}{22} = 103908,$$

which implies that $s_p = \sqrt{103908} = 322.3$. The two-sample test statistic was

$$t = \frac{\bar{x}_1 - \bar{x}_2}{s_p\sqrt{\dfrac{1}{n_1} + \dfrac{1}{n_2}}} = \frac{922 - 689}{322.3\sqrt{\dfrac{1}{12} + \dfrac{1}{12}}} = \frac{233}{131.6} = 1.77.$$

If the weight were normally distributed and if $\sigma_1 = \sigma_2$, she could estimate the p-value from a t distribution.   She referred to the t distribution, in Table 2, having $v = n_1 + n_2 - 2 = 12 + 12 - 2 = 22$ degrees of freedom to estimate a p-value of approximately $p = .04$.   Consequently, if the null hypothesis were true, there was only a 4% chance of obtaining a t statistic as large or larger than 1.77. If she used $\alpha = .05$ she would reject the null hypothesis because $p < \alpha$.

The t-test can also be performed by using the data analysis tool in the EXCEL spreadsheet software.   By entering the data in two rows and selecting "T-test: Two-samples assuming equal variances" she could obtain the output shown in Figure 6.3.   Note that this output gives the one-tailed and two-tailed p-values.   The one-tailed p-value of $p = .0452$ on the computer output is more accurate than the p-value that we estimated from Table 2.   The EXCEL output also includes means, variances, and the pooled variance (s_p^2). ∎

Figure 6.3. Output from the t-Test data analysis tool in EXCEL.

t-Test: Two-Sample Assuming Equal Variances		
	Variable 1	*Variable 2*
Mean	922	689
Variance	100363.8182	107453.2727
Observations	12	12
Pooled Variance	103908.5455	
Hypothesized Mean Difference	0	
df	22	
t Stat	1.770540696	
P(T<=t) one-tail	0.045247153	
t Critical one-tail	1.717144187	
P(T<=t) two-tail	0.090494305	
t Critical two-tail	2.073875294	

It should be noted that if the observations are not normally distributed or if the variances are not equal then the sampling distribution of t may not follow a t distribution with $v = n_1 + n_2 - 2$ degrees of freedom. Fortunately, although the p-values may not be exact if the assumptions are not satisfied, statisticians have shown that the significance level of the test is not greatly affected by moderate departures from normality or by moderate inequality of variance. That is, if the null hypothesis is true and we set α=.05 we will find that, in repeated samples of the same size, the null hypothesis will be rejected in about 5% of the samples, even if the distributions are skewed or there are moderate differences in variability between the groups. Thus, the t-test is a valid procedure that can honestly be used in a wide variety of circumstances, even if the assumptions are violated. However, this does not imply that the t-test is as powerful as other tests for non-normal data. Since we want to use the most powerful test that maintains its significance level, the t-test may not be the best test for many circumstances when the assumptions are not met. Another test that often has higher power will be discussed in the next section.

It should also be remembered that the two sample t-test should only be used with independent samples. In our example the 24 rats were randomly assigned to the two groups, which guarantees their independence. The independence of the observations implies that if a rat in the first sample had a higher than average weight gain then that rat would not influence the weight gain for any of the other rats in the study. If we had used a **repeated measures** experimental method we might have assigned 12 rats to a high protein diet treatment for two weeks and recorded their weight gain over a two week period, and then assigned those same rats to the low protein diet for two weeks and recorded their weight gain over the second two week period. The two-sample t-test should not be used to analyze repeated measures data or paired data because the observations are not independent. A rat that is growing more rapidly than average on one diet may also grow more rapidly than average on the second diet. The repeated measures experiment is an example of a paired-comparison experiment and the methods described in the previous chapter must be used to analyze it. Before you attempt to analyze data you should decide if it comes from two independent samples or from a paired comparison. The two-sample t-test that we described in this section should only be used to analyze data from two independent samples.

Exercise Set B

1. In an effort to reduce iron deficiency in infants, pediatricians sometimes recommended iron supplementation of infant formula. As part of a larger study of iron status during the 1st year of life, 25 infants where randomly assigned to receive an iron fortified formula for 4 months while 20 infants were randomly assigned to receive the same formula without added iron for 4 months. (Based on Picciano, M. F., and Deering, R. H. "The influence of feeding regiments on iron status during infancy," *The American Journal of Clinical Nutrition,* 33 (1980), pp. 746-753) Among the many measurements made

on these infants was body weight at three months, which is summarized in the following table:

	Iron fortified	No iron
Number of infants	25	20
Sample Mean Weight	6.4 Kilograms	6.0 Kilograms
Sample Standard Deviation	0.8 Kilograms	0.7 Kilograms

a) Compute the pooled estimate of the common standard deviation.

b) Perform a two-sample t-test using $\alpha=.05$. Use a two-sided test.

c) Is there enough evidence to conclude that iron supplementation
increases body weight at three months? What conclusion would
you draw?

2. A social science instructor wanted to boost the social science test
scores of eighth grade students. She randomly assigned 30 students
in a social studies class to one of two groups. The 15 students who
were assigned to the treatment group received a home subscription to
a weekly news magazine and were encourage to read it at home. The
15 students who were assigned to the control group received no
subscription. (Consider the subscription group to be sample 1 and
the control group to be sample 2.)   At the end of the school year the
results of the final exams were summarized as follows:

Subscription Group	Control Group
$n_1 = 15$	$n_2 = 15$
$\bar{x}_1 = 72.3$	$\bar{x}_2 = 75.7$
$s_1 = 11.6$	$s_2 = 12.4$

a) Is this a paired comparison or a two-sample experiment?

b) If possible, perform a one-sided t-test and estimate the p-value.
Will the instructor reject the null hypothesis using $\alpha=.10$?

c) State the conclusion.

3. In order to investigate the relationship between coronary heart disease and histamine concentration, researchers measured the histamine concentration in the arteries that were obtained from a post-mortem analysis. The coronary arteries of 7 patients who died from coronary heart disease were compared to the coronary arteries of 11 patients who died of accidents. The data in the table below summarizes the results. (Based on Kalsner, S. and Richards, R. , "Coronary arteries of cardiac patients are hyperreactive and contain stores of amines: A mechanism for coronary spasm," *Science,* 223 (1984) pp. 1435-1437.)

	Histamine Concentration (ng/g)	
	Cardiac Patients	Noncardiac Patients
number of patients	7	11
sample average	8619	4544
sample standard deviation	1274	754

a) Compute the pooled estimate of the common standard deviation.

b) Compute the t statistic and roughly estimate the p-value for a two-tailed test of the null hypothesis.

c) State your conclusion.

4. At one university Researcher A has assigned 16 dogs at random to two treatment groups in order to investigate the relationship between the treatment variable and the response variable. At a second university Researcher B has also assigned 16 dogs at random to two treatment groups to investigate the same relationship. Their results can be displayed in the following dot diagrams. (You may assume that the same scale was used for all of the diagrams.)

Researcher A   Researcher B

Sample 1   Sample 1

O O OOO O ∞   O O OOO O ∞

Sample 2   Sample 2

O O OOO O ∞   O O OOO O ∞

a) Which researcher would have the largest two-sample t statistic? Why? No calculations should be necessary.

b) Which researcher would obtain the smaller p-value? Why? No calculations should be necessary.

5. A researcher wanted to determine the effect of caffeine on the mathematical skills of high school students. She assigned the 400 volunteers at random to either a high caffeine group or a no-caffeine group. Students in the high caffeine group ingested 12 ounces of a high caffeine soft drink 10 hours before the mathematics test. The students in the no caffeine group ingested the same amount of a no-caffeine soft drink. Neither the students nor the test administrators knew which soft drinks had the caffeine. The results of the mathematics test were tabulated as follows:

Treatment	Students	Average	Standard Deviation
High caffeine	196	89.6	20.1
No caffeine	204	85.3	22.7

a) Compute a pooled estimate of the common standard deviation.

b) Use the t-test to test $H_0: \mu_1 = \mu_2$ against $H_a: \mu_1 > \mu_2$. Use $\alpha = .05$.

c) State your conclusion.

6. Suppose a new drug, which is being developed to treat allergies, may
    have some adverse events, or side effects. One of the adverse events
    is that the drug may cause a patient to sleep more than usual. To
    investigate this possible adverse event, 20 volunteers were randomly
    assigned to either a drug group or a control group. The volunteers
    did not know if they were taking the new drug or a placebo. The
    results are shown in the following dot diagram.

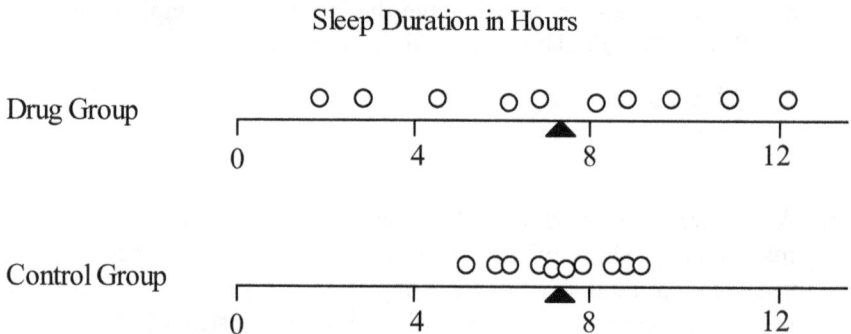

Sleep Duration in Hours

a) Does it appear that the drug has an effect on sleep? If it has an
    effect, describe it.

b) If you had computed a t-test statistic what would it be? No
    computations should be necessary.

c) Does it appear that the assumptions of the two-sample t-test have
    been met? Explain.

6.3 The Rank-Sum Test

It can be shown that the t-test is the most powerful test if the distributions are normal and the variances are equal. However, in practical
situations we really don't know that the distributions are normal and the
variances are equal. If the distributions are skewed, the t-test may not be
the most powerful test.

Fortunately, there is an alternative to the t-test, called the **rank-sum
test**, which is often more powerful than the t-test. It is based on the
ranks of the observations, which are not very sensitive to outliers. The

null hypothesis is that the distributions in the two populations are the same. The alternative hypothesis is that the distributions are shifted relative on one another. If the null hypothesis is true the p-values computed from **rank-sum test statistic** do not depend on the distribution of the observations, hence the rank-sum test is sometimes called a **distribution-free test statistic**. The rank-sum test goes by several names, including the **Wilcoxon rank-sum test** and the **Wilcoxon-Mann-Whitney** test. A closely related test is the **Mann-Whitney U test,** which produces the same p-value as the rank sum statistic.

The rank-sum test proceeds by first combining the n_1 observations in the first sample with the n_2 observations in the second sample, and ranking over all the observations. The ranks of the n_1 observations in the first sample are $Q_1, Q_2, \ldots, Q_{n_1}$ and the ranks of the n_2 observations in the second sample are denoted by $R_1, R_2, \ldots, R_{n_2}$. The rank-sum test is based on the sum of the ranks of the observations in the second sample, which can be expressed as $W = \sum_{i=1}^{n_2} R_i$. Since the sum of the ranks over both samples must equal

$$1 + 2 + \ldots + (n_1 + n_2) = \frac{(n_1 + n_2)(n_1 + n_2 + 1)}{2} \ ,$$

the average rank over the $(n_1 + n_2)$ observations must equal $\overline{R} = \frac{(n_1 + n_2 + 1)}{2}$. If the null hypothesis is true the average rank in the second sample should be near $\overline{R}$, so that the sum of the ranks in the second sample will be near $n_2 \overline{R}$. However, if the observations in the second sample tend to be larger than those in the first sample, then the ranks in the second sample will be larger than those in the first sample, and the rank sum W will exceed $n_2 \overline{R}$, which will cause us to reject the null hypothesis.

The one-sided alternative is that the distribution of the random variables in the second population is shifted to the right of the distribution in the first population. Our general procedure will be that if we test $H_0 : \mu_1 = \mu_2$ versus $H_a : \mu_1 < \mu_2$ we will reject the null hypothesis if W is so large that we obtain a p-value less than α. For small samples a table must be used to estimate these p-values. Table 4 gives the p-values for $n_1<10$ and $n_2<10$.

Example 6.5

A geneticist wants to compare two varieties of corn. The null hypothesis is that the two varieties of corn have the same yield. The alternative hypothesis is that Variety B has a greater mean yield than Variety A. Seven plots are selected at random for variety A and six plots are selected for variety B. The yields, in bushels per acre, are as follows:

Yield (bushels per acre)
Variety A: 126,   130,   110,   137   123,   128,   131,
Variety B: 143,   144,   129,   139,   132,   141.

We combine the observations before ranking to obtain the ranks:

Ranks
Variety A: 3,   6,   1,   9,   2,   4,   7,
Variety B: 12,   13,   5,   10,   8,   11.

The rank sum is the sum over the ranks in the second sample, which is $W=12+13+...+11=59$. From Table 4, with $n_1=7$ and $n_2=6$ we find an exact p-value of $p=.0070$ for $W=59$. Therefore, we would reject the null hypothesis using a level of significance of $\alpha=.05$. The conclusion would be that variety B has a greater mean yield. One advantage of this test is that it is not necessary to assume that the distributions are normal. Also, the test is not sensitive to the presence of an outlier because it is based on ranks.

The p-value gives the chance of obtaining a rank sum as large or larger than that observed if the null hypothesis is true. In our example the null hypothesis is that the varieties have the same yield, and the p-value of .0070 indicates that the chance of obtaining a rank sum of 59 or larger is .0070. Since the chance is so small we reject the null hypothesis. ∎

For large values of n_1 and n_2 it is not practical to publish a table containing p-values for all possible values of W. Fortunately, a normal approximation can be used that gives reasonably accurate p-values for

$n_1 \geq 10$ and $n_2 \geq 10$. Recall that the expected value of W, under the null hypothesis, is $n_2\overline{R}$. It can be shown that the variance of W is

$$Var(W) = \frac{n_1 n_2}{n_1 + n_2} s_{ranks}^2,$$

where s_{ranks}^2 is the sample variance of the ranks over both samples. This variance formula incorporates a correction for ties so that it is can be used even if there are a large number of ties in the data. If there are no ties in the data the expression for the variance of W can be simplified to

$$Var(W) = \frac{n_1 n_2 (n_1 + n_2 + 1)}{12}.$$

By using the expected rank sum and the variance of the rank sum we can compute the rank-sum test statistic

$$z_{RS} = \frac{W - n_2\overline{R}}{\sqrt{Var(W)}},$$

where $\overline{R} = \frac{(n_1 + n_2 + 1)}{2}$ is the average rank. The normal distribution is used to approximate the p-value as the area to the right of z_{RS} for a one-sided test. For a two-sided test the p-value must be multiplied by two to obtain the correct two-sided p-value.

We can illustrate this method of approximating p-values by using the corn yield data, although we would ordinarily use the exact p-values found in Table 4 for $n_1 = 7$ and $n_2 = 6$. Since there are no ties we can compute the variance of W as

$$Var(W) = \frac{n_1 n_2 (n_1 + n_2 + 1)}{12} = \frac{6(7)(14)}{12} = 49.$$

The average rank is $\overline{R} = \frac{(n_1 + n_2 + 1)}{2} = 7$, so that the large sample rank-sum test statistic is

$$z_{RS} = \frac{W - n_2 \overline{R}}{\sqrt{Var(W)}} = \frac{59 - 6(7)}{\sqrt{49}} = 2.429.$$

From the normal distribution in Table 1 we find an approximate p-value of p=.0076, which is close to the exact value of .0070 that was obtained from Table 4. Although the exact p-values can ordinarily be used for $n_1 < 10$ and $n_2 < 10$, the large sample approximation yields reasonably accurate p-values for $n_1 \geq 6$ and $n_2 \geq 6$.

Another way to obtain the p-value is to use the NORMSDIST function in an EXCEL worksheet. The only parameter in this function is z_{RS}, and this function returns the lower tail of the standard normal distribution. Thus, to obtain the p-value, which is the upper tail, we would enter "=1-NORMSDIST(2.429)" in a cell to find $p = .00757$, which is a more accurate estimate than that obtained from the table.

The rank-sum test procedures need to be modified if there are ties in the data. The general procedure is to assign midranks to the tied values and then to compute the rank sum in the usual manner. However, Table 4 cannot be used to compute the p-value because the table was computed assuming that there were no ties. Fortunately, the formula for the large sample approximation can also be used when there are ties, as illustrated with the following example.

Example 6.6

Suppose that, in the corn yield example, the researcher had obtained the following data:

                      Yield (bushels per acre)
Variety A: 124, 130, 110, 137, 124, 128, 130
Variety B: 143, 144, 130, 139, 132, 141

In this example there are two observations tied at 124 bushels per acre and three tied at 130 bushels per acre. The researcher would assign the average rank to tied observations to obtain the midranks:

                      Midranks
Variety A: 2.5,   6,   1,   9,   2.5,   4,   6
Variety B: 12,   13,   6,   10,   8,   11

The rank sum is $W = 12 + \ldots + 11 = 60$. Table 4 cannot be used to compute the exact probability because it was constructed for data that does not contain ties. However, the normal approximation can be used to approximate the p-value. The sample variance of the ranks is $s^2_{ranks} = 14.958$, so the large sample rank-sum statistic is

$$z_{RS} = \frac{W - n_2\overline{R}}{\sqrt{Var(W)}} = \frac{W - n_2\overline{R}}{\sqrt{\dfrac{n_1 n_2}{n_1 + n_2} s^2_{ranks}}} = \frac{60 - 42}{\sqrt{\dfrac{7(6)}{13} 14.958}} = \frac{18}{6.952} = 2.589.$$

By referring to a normal table the researcher would find an approximate p-value of $p = .0048$. Note that the midranks that were used to calculate $Var(W)$ reduced the variance of W from 7.0 to 6.952. ∎

The procedures that we have discussed to compute one-sided p-values can easily be modified to perform a two-sided test of H_0: $\mu_1 = \mu_2$ versus H_a: $\mu_1 \neq \mu_2$. Under the null hypothesis the distribution of W is symmetric about the mean $n_2\overline{R}$. Thus, to calculate a two-sided p-value we must double the one-sided p-value. In our last example, if we were testing $H_0 : \mu_1 = \mu_2$ against $H_a : \mu_1 \neq \mu_2$ the   p-value   would   be $p = 2(.0048) = .0096$.

Figure 6.4 gives SAS output for the analysis of the corn yield data. The rank-sum test, which is a nonparametric procedure, can be performed by using the NPAR1WAY procedure in SAS. The output shows the sum of the ranks, which are the "scores" in the rank-sum test, and the expected ranks under the null hypothesis. The last line in the table gives the two-tailed p-value for the rank-sum test, which agrees with our two-sided p-value of $p = .0096$. The other p-value should be ignored; it is based on an approximation that is not used in this text.

Figure 6.4. SAS output for rank-sum test.

```
N P A R 1 W A Y   P R O C E D U R E

Wilcoxon Scores (Rank Sums) for Variable YIELD
Classified by Variable VARIETY

                     Sum of     Expected      Std Dev          Mean
      VARIETY    N   Scores     Under H0      Under H0         Score

       A         7    31.0        49.0      6.95175683     4.4285714
       B         6    60.0        42.0      6.95175683    10.0000000

Average Scores Were Used for Ties

Wilcoxon 2-Sample Test (Normal Approximation)
(with Continuity Correction of .5)

S =  60.0000           Z =  2.51735           Prob > |Z| = 0.0118

T-Test Approx. Significance = 0.0270

Kruskal-Wallis Test (Chi-Square Approximation)
CHISQ =   6.7043             DF =  1           Prob > CHISQ = 0.0096
```

Up to this point we have limited our discussion to one-sided tests of $H_o : \mu_1 = \mu_2$ versus $H_a : \mu_1 < \mu_2$. Now suppose we want to perform a one-sided test with H_a: $\mu_1 > \mu_2$. The easiest way to perform the test with the other alternative is to switch the labels on the populations so that, after the switch, the alternative is H_a: $\mu_1 < \mu_2$. This will permit the use of Table 4 and the normal approximation without further modifications.

The rank-sum test can be viewed as a simple and effective alternative to the t-test. At first glance, it may appear that a rank test could not be as powerful as the t-test because it is based on the ranks of the observations rather than the observations themselves. However, as we shall see in the next section, the rank test can be more powerful than the t-test with many distributions.

Exercise Set C

1. A nutritionist enlisted 15 volunteers for an experiment designed to evaluate the effectiveness of two weight loss programs. Seven subjects were assigned at random to the "diet" treatment and eight subjects were assigned at random to the "exercise" treatment. Subjects in the "diet" treatment group were instructed to modify their diets to reduce their caloric intake while the subjects in the "exercise" treatment group were instructed to exercise at least 20 minutes per day. One year later the nutritionist weighed the subjects and calculated the weight loss over the year for each subject. A positive value indicated weight loss and a negative value indicated weight gain. The weight losses for the subjects were recorded as follows:

Diet Treatment	$n_1=7$	-3, -2, 2, 5, 6, 9, 15
Exercise Treatment	$n_2=8$	3, 7, 8, 11, 13, 17, 22, 25

a) Compute the rank sum.

b) A one-tailed test would be used if the nutritionist believed that the exercise treatment would prove to be more effective than the diet treatment. Use the rank-sum test to compute the p-value for $H_0: \mu_1=\mu_2$ versus $H_a: \mu_1<\mu_2$.

c) A two-tailed test would be used if the nutritionist did not know which treatment would be most effective. Use the rank-sum test to compute the p-value for the two-sided test of $H_0: \mu_1=\mu_2$.

d) For the two-sided test would you reject the null hypothesis using $\alpha=.05$? State your conclusion.

2. A researcher performed an experiment with 4 animals in each sample and an experiment with 6 animals in each sample. If the researcher tests $H_o : \mu_1 = \mu_2$ against $H_a : \mu_1 < \mu_2$ for both experiments. Compute the p-values for (a) and (b).

a) n_1=4, n_2=4

Sample 1

Sample 2

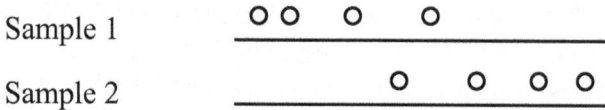

b) n_1=6, n_2=6

Sample 1

Sample 2

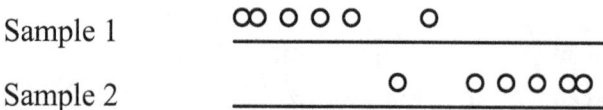

c) Which of the two experiments would have more convincing results? Why?

3. Students who were registered to take an introductory statistics course were randomly assigned to either a traditional statistical methods course or to an enriched course that utilized a special statistical calculator. Except for the use of the special calculator, the enriched course was identical to the traditional course and was taught by the same instructor. The traditional course had n_1=32 students and the enriched course had n_2=36 students. The rank sum was computed to be W=1302 for the students in the enriched course.

a) Under the usual null hypothesis, what is the expected sum of the ranks for the students in the enriched course?

b) Using the rank-sum test, compute an approximate p-value for a one-sided test of H_o: μ_1=μ_2 versus H_a: μ_1<μ_2 . You may assume that there were no ties in the data.

c) State your conclusion.

4. In an effort to measure the exposure of infants to tobacco smoke, researchers measured the concentration of cotine, which is the major metabolite of nicotine, in urine samples of $n_1 = 18$ infants who were not exposed to tobacco smoke and $n_2 = 28$ infants who were exposed. (The data was estimated from Figure 1 in Greenberg, R. A., *et al* "Measuring the exposure of infants to tobacco smoke," *The New England Journal of Medicine* 310 (1984) pp. 1075-1078.) The urinary cotine concentration in ng/mg were estimated to be :

Non-exposed   0, 0, 0, 0, 0, 0, 0, 0, 0, 7, 11, 14, 14, 14,

                     16, 19, 35, 104

Exposed   21, 38, 42, 43, 82, 95, 102, 118, 122,

                     123, 141, 145, 162, 186,

                     192, 203, 212, 215, 247, 291, 332,

                     341, 344, 618, 982, 1117, 1462, 1832

a) Under the usual null hypothesis, what is the expected sum of the ranks for the rats in the exposed group.

b) Compute the sum of the ranks in the exposed group. To facilitate computations use the fact that $26 + 27 + ... + 46 = 756$.

c) Use a rank-sum test to compute a p-value for a one-tailed test.

d) State your conclusion.

5. Ten animals were randomly assigned to two groups such that each group had five animals. The experimenter wanted to test $H_0: \mu_1 = \mu_2$ versus $H_a: \mu_1 < \mu_2$ using a level of significance of $\alpha = .005$.

a) In order to reject the null hypothesis, what value of W would the experimenter need to observe?

b) Based on your answer in part (a), draw a dot diagram for any set of observations in the second sample that would produce that value of W.

Group 1

Group 2

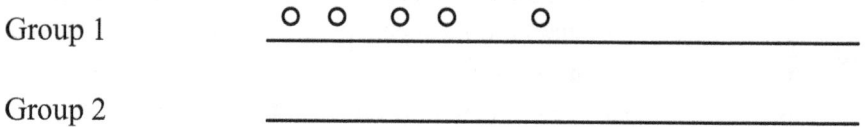

6. In an effort to quantify the anxiety that people may experience prior to hypnosis, researchers obtained data on 16 subjects who were randomly assigned to either a hypnosis treatment or to a control treatment. (Based on Agosti, E. and Camerota, G. "Some Effects of Hypnotic Suggestion on Respiratory Function" *Intern. J. Clin. Expt. Hypnosis* (1965):149-156) A measure of their ventilation before the hypnosis was recorded as follows:

Control group:	3.99	4.19	4.21	4.54	4.64	4.69
	4.84	5.48				
Hypnotized group	4.36	4.67	4.78	5.08	5.16	5.20
	5.52	5.74				

a) Under the usual null hypothesis, what is the expected sum of the ranks for the subjects in the hypnotized group?

b) Compute the sum of the ranks in the hypnotized group.

c) Use a rank-sum test to compute a p-value.

d) State your conclusions.

7. An investigator used rabbits in an experiment designed to investigate the effect of a certain drug on blood pressure. The experimenter was reluctant to use any more rabbits than were absolutely necessary because the total cost per rabbit, including laboratory time and materials, was $1000.00 per rabbit. He planned to randomly assign four rabbits to each of two treatment groups, which he thought should be sufficient because he believed that the drug would be so effective that there would be no overlap between the groups. He intended to reject the null hypothesis if the p-value was less than α=.01. What is wrong with this experiment? How could the experiment be modified to achieve the experimenter's objective.

6.4 A Guide to the Selection of Two-Sample Tests

Comparison of the t-test to the rank-sum test

We noted in the last chapter that the assumptions that were used in the development of a test do not always provide an adequate guide for test selection. The two-sample t-test requires the distributions to be normal, but we seldom, if ever, are certain that the distributions are normal. But if we use a rank-based test that does not assume normality, we may lose some power by ranking the data. In order to decide whether to use a rank-sum test or a t-test we need to compare these tests for a variety of sample sizes and distributions.

A simulation study was used to compare the t-test to the rank-sum test. (The full simulation study included sample sizes configurations from $n_1 = n_2 = 6$ to $n_1 = n_2 = 400$ and fifteen distributions. For details see O'Gorman, T. W. "A Comparison of an Adaptive Two-Sample Test to the t-test and the Rank-Sum Test" *Communications in Statistics-Simulation and Computation* (1997):1393-1411) Many sample sizes and distributions were used in the full simulation but, to save space, we will only look at a results for the four distributions shown in Figure 6.5 .

Figure 6.5 Distributions Used in the Simulations

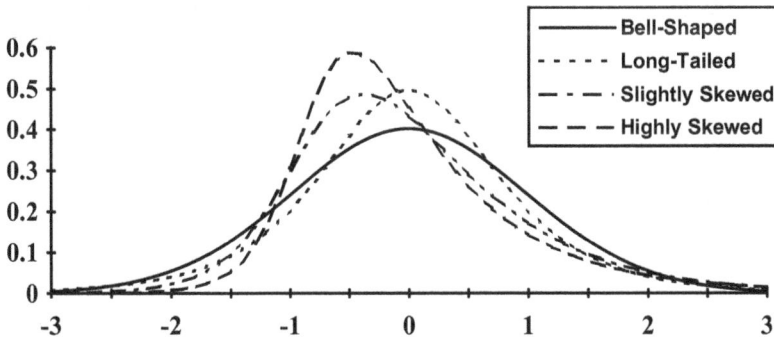

The simulation study demonstrated that both the t-test test and the rank-sum tests are valid tests because they maintained their size close to the desired level. That is, the proportion of times that the null hypothesis was rejected was close to α. Consequently, we can say that both tests are honest in the sense that if we set $\alpha = .05$, we have a 5% chance of rejecting the null hypothesis if the null hypothesis is true. This is an important property, but power is also important, and in this regard these tests do differ.

The simulation studies were also designed to compare the power of these tests, which is their ability to detect a difference between the populations. Since we want a test to be able to detect a difference between the two distributions, we want a test to have high power. In these simulations the second distribution was shifted to the right a certain amount, 1000 data sets were generated using this shift, and we counted the number of data sets that led to rejection of the null hypothesis. The power is the percent of data sets that were rejected. Figure 6.6 gives the power comparisons for a few simulations that used equal sample sizes of $n_1 = n_2 = 8$ and $n_1 = n_2 = 15$.

Figure 6.6 Power of the t-test and rank-sum test.

a) $n_1=n_2=8$

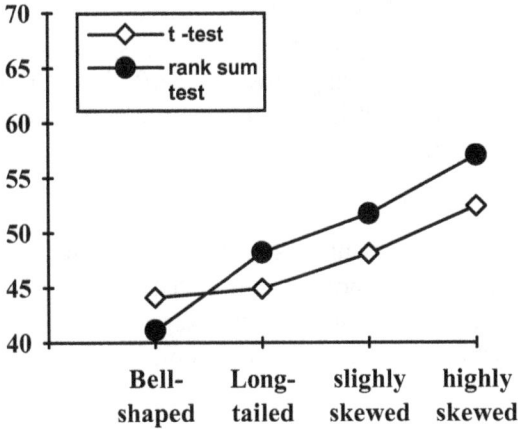

b) $n_1=n_2=15$

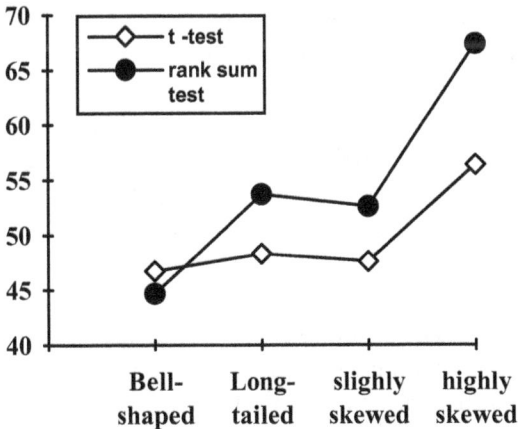

The results for $n_1 = n_2 = 8$ indicate that there is little difference in power between the tests. The results for $n_1 = n_2 = 15$ indicate that the rank-sum test is more powerful for distributions that are not bell-shaped, and that there is little difference in power between the tests for a bell-shaped distribution.   For sample sizes that exceed 15, the power advantage of the rank-sum test is slightly larger than that shown in Figure 6.6 (b) for distributions that are not bell-shaped. Power estimates for unequal sample sizes have not been shown but they are similar to those shown for equal sample sizes. Based on the results from the full simulation study, the t-test is nearly as powerful or is more powerful than the rank-sum test for studies having fewer than 10 observations in each sample, while the rank-sum test has a considerable power advantage for larger samples from non-normal distributions.

Choosing a test

We have seen that the rank-sum test is more powerful than the t-test for larger samples from many non-normal distributions. We also know that the t-test is always the most powerful test if the distributions are normal.   Is there anything else that we should consider before selecting a test?

Unfortunately, neither of these tests are appropriate if there are large differences in variability between the groups, so our first task is to make a judgment about the difference in variability.   If we see large differences in variability, as indicated by histograms or by the sample standard deviations, then we should use one of the techniques that we will describe in the next section.   There are statistical tests for differences in variability, but these tests are seriously flawed because they do not perform well with non-normal data. (See Controversies in Statistics in this chapter.) Hence, it seems best to use graphical techniques and the sample standard deviations to arrive at a judgment about the differences in variability between the groups. In the simulation study cited above, some power comparisons were performed with differences in variability; these showed that differences in variability are not much of a problem if the population standard deviation in one group is less than 30% larger than the population standard deviation in the other group. Consequently, if there is no indication that the variability in one group is more than 30% larger than the variability in the other group then it seems best to choose the t-test or the rank-sum test.

As indicated earlier, the pooled t-test is the most powerful test for location when the observations are normally distributed and the equal variance assumptions are satisfied. However, it is often difficult to know if the observations are normally distributed. Unless the samples are quite large, statistical tests have little power to detect a difference between a normal distribution and some other distributions. (See Controversies in Statistics). If the distribution of the observations is unknown, as is usually the case, it seems prudent to use a rank-sum test if the sample sizes are large because it tends to have greater power than the t-test for most distributions. For small samples, the t-test seems preferable because there is little to gain from using the rank-sum test with non-normal distributions and much to lose if the distributions are normal. These recommendations are summarized in Figure 6.7

Figure 6.7 A Guide to the Selection of Two-sample Tests.

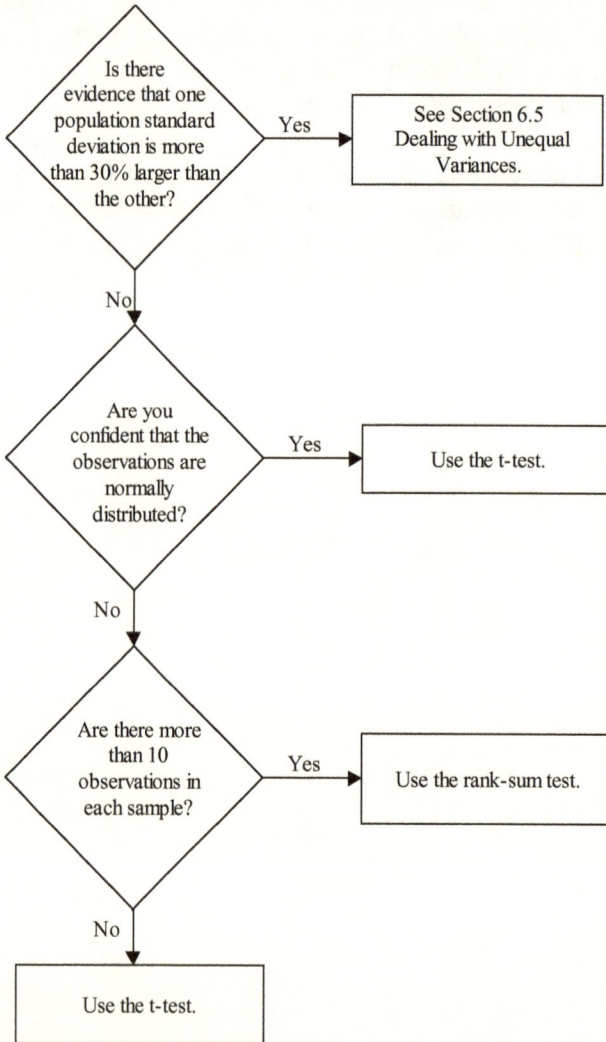

Controversies in Statistics

Choosing a test—What is the best strategy?

Many strategies have been proposed for test selection. We would like to always select the most powerful test but we usually don't know enough about the true situation to know which test would be the most powerful. If we knew that the population distributions were normal and had equal variances then we would use the pooled-t test because it is the most powerful test for that situation.   If we knew the population distributions were normal but the variances were unequal we might use the unequal variance t test. If we knew the population distributions were skewed we might use the Wilcoxon rank-sum test.   However, we typically don't know anything about the shape of the distribution or the population variances; all we usually have is the data to guide us.

One approach is to use statistical procedures to determine if the distributions are normal and the variances are equal. The problem with this approach is that it is not practical for most data sets. For example, consider the following data set on the time to relief of headache pain for two pain relievers:

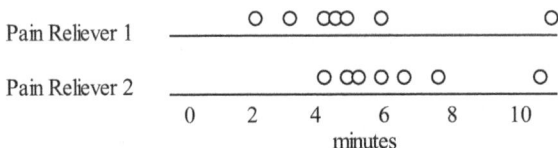

If we attempt to determine if the populations are normal we will find that we have far too little information to determine if they are normal. Both appear to be somewhat skewed to the right, but the populations that these samples came from could have been normal. There are statistical tests for normality but these have very low power when there are few observations.

Even if we could somehow determine that the distributions were normal we would still face the problem of determining if the variances are equal. The dot plot and the sample standard deviations of $s_1 = 2.92$ and $s_2 = 2.02$ suggest that there may some difference in variability. An F test for the equality of variance is often recommended to detect unequal variances. However, if we used this test, which is not covered in this text, for this data we would obtain a two-tailed p-value of $p = .39$, so we would be unable to reject the null hypothesis $H_o : \sigma_1^2 = \sigma_2^2$. Since the p-value is not very small we might be tempted to conclude that the variances are equal, but we must remember that this test has low power to detect moderate differences with small samples. Thus, we still don't know that the variances are equal. A further complication arises because the F test itself is not a valid test if the distributions are not normal. So, while it may appear that we could easily detect differences in variability, in many practical situations graphical techniques should be used to decide if the population standard deviations are not too different.

Some authors recommend using an unequal variance t-test for all two-sample tests. The general idea is to avoid the problems associated with the F-test by always using the unequal variance t-test since it should work reasonably well with equal and unequal variances. The problem with this approach is that the unequal variance t-test fails to maintain its significance level for small samples. (See section 6.5 for more information on the unequal variance t-test.) Because of this problem, the unequal variance t-test cannot be recommended for general use, although it can be a valid and powerful test in some situations.

The approach taken in this text is to compare the power of the pooled t-test to the rank-sum test and to recommend the test that appears to have the greatest power overall for samples of that size. Our approach is based on the size of the sample, which we know, instead of the variances and distributions, which are usually unknown.

Exercise Set D

1. A researcher wanted to compare the weight of males living in the United States in 1995 to the weights of males living in the United States in 1975. Separate samples were taken in 1995 and in 1975. If he wanted to compare the 125 males in the 1975 sample to the 167 males in the 1995 sample, what test would be recommended? Explain.

2. In a cardiovascular experiment a researcher randomly assigned 6 dogs to one treatment group and 6 dogs to another treatment group. A continuous variable was recorded on each animal in both groups. The researcher needed to use some two-sample test to compare the groups. Should she consider using a t-test or a rank-sum test. Explain.

3. As part of a larger study of diet and occupation, researchers compared the alcohol intake of a group of civil servants to a group of bus conductors on London double-decker buses. The 83 conductors in the first group consumed an average of .39 grams per day with a standard deviation of 1.00 grams per day. The 68 civil servants in the second group consumed an average of 10.0 grams per day with a standard deviation of 20.6 grams per day. (Based on Marr, J. W. and Heady, J. A. "Within- and Between-person variation in dietary surveys: number of days needed to classify individuals", *Human Nutrition: Applied Nutrition*, 40A (1986), pp. 347-364.)

    a) Do you believe that alcohol consumption is normally distributed in these two groups? Why or why not?

    b) Do you believe that there is little difference in the variability of alcohol consumption between these groups?

    c) Suppose someone used a t-test to compare the alcohol consumption of the two groups. In your opinion is this the best approach? Why or why not?

4. Social scientists often compare the responses of subgroups to questions posed in a telephone or mail interview. For example, a researcher may compare the responses from the males to the responses from the females. If the researcher always has samples of at least 30 in every subgroup, would a t-test or a rank-sum test be recommended for routine use?

*6.5 Dealing with Unequal Variances

Using Transformations with Two-Sample Tests

Scientists who use two-sample tests sometimes find that the sample variances in the two groups are markedly different. Unequal variances can create a major problem in testing H_0: $\mu_1=\mu_2$, because the t-test and the rank-sum test may not be appropriate. Fortunately, this problem can sometimes be solved by transforming the raw data. For example, suppose a botanist counts the number of insects in ten plots of ground where each plot covers a square meter of ground. The plots were assigned, at ranodm, to be treated with either insecticide A or insecticide B. The following insect counts were obtained:

Insecticide A: 27,   93,   102,   150,   819

$\bar{x}_1 = 238.2, \ s_1 = 327.6$

Insecticide B: 12,   32,   50,   70,   250

$\bar{x}_2 = 82.8, \ s_2 = 95.9$

The large differences in the sample standard deviations suggest that the population standard deviations are not equal. Since the t-test assumes that the standard deviations are equal the t-test may not be appropriate. To avoid this problem the botanist may decide to work with the logarithms of the counts, or with the square roots of the counts, rather than with the raw data. The base 10 logarithms of the data, along with the means and standard deviations of the transformed data, are:

Log_{10} Transformed Data

Insecticide A: 1.43,   1.97,   2.01,   2.18,   2.91

$$\bar{x}_1 = 2.10, \quad s_1 = 0.53$$

Insecticide B: 1.08,   1.51,   1.70,   1.85,   2.40

$$\bar{x}_2 = 1.71, \quad s_2 = 0.48$$

Note that the sample standard deviations are approximately equal on the logarithm scale so that the t-test could be used on the transformed data. Transformation is a recommended method of dealing with unequal variances when the transformation makes sense and when it achieves its purpose of making the variances nearly equal.

The tests are performed using the transformed data in the same way that the tests are usually performed. Figure 6.8 gives SAS output from the TTEST procedure for the analysis of the transformed data. The output gives the p-value for the two-sided t-test as $p = .2577$, which would lead us not to reject the null hypothesis. This p-value is on the row for the equal variance t-test.

However, the transformation approach should be used with some caution because the analysis using the transformed data may be difficult to interpret, especially if the transformation appears to be highly artificial. As a rough rule, if the larger standard deviation is less than 30% larger than the other standard deviation, then it may be preferable to proceed with the analysis using the raw data, rather than using the transformed data. However, a problem with the logarithmic transformation is that it is not possible to take the logarithm of negative and zero values.

Figure 6.8 SAS output for a two-sample t-test.

```
TTEST PROCEDURE

Variable: LOG_DATA

INSECTCD    N              Mean              Std Dev              Std Error
-----------------------------------------------------------------------------
A           5         2.10000000         0.53301032           0.23836946
B           5         1.70800000         0.48287680           0.21594907


Variances        T        DF      Prob>|T|
-----------------------------------------------
Unequal       1.2187     7.9      0.2580
Equal         1.2187     8.0      0.2577

For H0: Variances are equal, F' = 1.22   DF = (4,4) Prob>F' = 0.8528
```

The Unequal Variance t-test

The t-test that was discussed in section 6.2 is sometimes called the **pooled t-test** because the variances in the two samples are pooled to provide an estimate of the common variance.   If we are testing $H_0: \mu_1 = \mu_2$ versus $H_a: \mu_1 > \mu_2$, but we believe that $\sigma_1^2 \neq \sigma_2^2$, another approach is to use the **unequal variance t-test** statistic

$$t_{unequal} = \frac{\bar{x}_1 - \bar{x}_2}{\sqrt{\dfrac{s_1^2}{n_1} + \dfrac{s_2^2}{n_2}}}.$$

To perform the unequal variance t-test we compute $t_{unequal}$ and then obtain the p-values from the t distribution with the degrees of freedom equal to

$$\nu = \frac{\left(\dfrac{s_1^2}{n_1} + \dfrac{s_2^2}{n_2}\right)^2}{\dfrac{\left(s_1^2/n_1\right)^2}{n_1 - 1} + \dfrac{\left(s_2^2/n_2\right)^2}{n_2 - 1}}.$$

It has been shown that the test often fails to maintain its significant levels for unequal sized samples from skewed distributions having n_1 or n_2 less than 30. (see O'Gorman, T. W. "A Comparison of an Adaptive Two-Sample Test to the t-test and the Rank-Sum Test" *Communications in Statistics-Simulation and Computation* (1997):1393-1411) For large samples the test does maintain its significance level but, for these sample sizes, the unequal variance t-test has roughly the same power as the pooled t-test. The unequal variance t-test has more power than the pooled t-test with normally distributed observations when the larger sample has the larger variance, and in these situations it is the recommended test. Apart from these situations there appears to be no good reason to use the unequal variance t-test.

Figure 6.8 gives $t_{unequal}$ and the associated p-value for the transformed insect count data. Note that the test statistics for the pooled t-test and the unequal variance t-test are identical if $n_1 = n_2$. The p-values are not equal because the unequal variance t-test has a different number of degrees of freedom.

Tests for Dispersion

Since the pooled t-test requires equal variances it is tempting to use a test for equality of variance before proceeding with the pooled t-test. Such a test, called the F-test for equality of variance, has been developed for normally distributed random variables to test the null hypothesis $H_o : \sigma_1^2 = \sigma_2^2$. However, there are several serious problems associated with the application of this test. This F-test for equality of variances does not work well if the distributions are not normal because it may reject the null hypothesis too often. That is, if we set $\alpha=.05$ and the null hypothesis is true, the test should have a 5% chance of rejecting the null hypothesis, but the actual chance of rejecting the null hypothesis may be much greater than 5% with non-normal data.

The failure of this F-test to maintain its significance level is a serious problem; the other tests used in this book do maintain their significance level. The poor performance of the F-test with non-normal data limits its application to those rare situations where the distributions are known to be normal. For this reason the F-test is not covered in this text.

Conclusion

In conclusion, small differences in variability should not be much of a concern for the user of the pooled t-test and the rank-sum test since they both perform well even if there are small differences in variability between the groups. If graphical techniques and the sample standard deviations suggest that there are large differences in variability, and if transformations cannot be used to reduce the difference in variability, then the researcher is faced with a difficult problem.

If there are larger differences in variability and a test of location is required, the user of a t-test or a rank-sum test should be aware that these tests are far from optimal. The t-test and rank-sum tests may not be very sensitive to differences in location if there are large differences in variability. More sophisticated tests have been developed to detect differences in location and variability, but these are beyond the scope of this text.

Exercise Set E

1. An experiment was performed to determine if the addition of 10 mcg. of vitamin B_{12} per pound of ration would increase the weight gain in swine. Eleven swine were used in the experiment; five were randomly assigned to the control group that did not receive a sup-plement while the other six were assigned to the B_{12} group. The average daily weight gains were recorded as follows:

	Average daily weight gain (pounds per day)					
Control Group	1.15	1.05	1.20	1.00	.85	
B_{12} Group	1.10	1.35	1.05	1.45	1.65	1.20

The sample standard deviation in the control group was .137 and the standard deviation in the drug group was .228 .

a) Compute the ratio of the standard deviation in the drug group to the standard deviation in the control group. Would a transformation be recommended before using a t-test?

b) Use a logarithmic transformation on the data.

c) Compare the standard deviations of the log-transformed data. Compute the ratio of standard deviations for the transformed data. Is a t-test on the log-transformed data appropriate? Explain.

2. A researcher compared the basal metabolism of 6 male college students who usually got more than 6 hours of sleep to the basal metabolism of 6 male college students who usually got less than 6 hours of sleep. The plan was to use a two-tailed t-test to compare the samples. The basal metabolism, which was measured as calories per square meter per hour, was recorded as follows:

hours of sleep ≥ 6 36.8, 32.3, 37.5, 35.2, 29.8

$\bar{x}_1 = 34.32$, $s_1 = 3.22$

hours of sleep < 6 33.1, 37.6, 35.6, 34.7, 36.1, 30.2

$\bar{x}_2 = 34.55$, $s_2 = 2.60$

a) Based on the guidelines given in this section, do you believe that a logarithmic transformation would be necessary before doing a t-test? Why or why not?

b) Calculate the t-test statistic.

c) Would you reject the null hypothesis using $\alpha = .05$? State your conclusion.

3. A health scientist wanted to compare the weighs of 6th graders who participate in athletics to those who do not participate. He took a random sample of 20 participants and 20 non-participants and obtained their weights. Before he tested $H_0 : \mu_1 = \mu_2$ versus the alternative $H_a : \mu_1 \neq \mu_2$ he performed an F-test for the equality of vari-

ance. Why should the scientist be cautious about using the F-test for this data?

4. A farmer wanted to compare two varieties of corn. Twelve plots were selected and the yield (in bushels per acre) from these two varieties were recorded as follows:

variety A	120	131	118	117	121	128
		$\bar{x}_1 = 122.5$		$s_1 = 5.683$		

variety B	129	133	138	132	141	131
		$\bar{x}_2 = 134.0$		$s_2 = 4.561$		

a) Compute the ratio of the standard deviations.

b) Before performing a t-test would it be advisable to transform the data? Why or why not?

*6.6 Sample Size Calculations for the Two-Sample Test

Occasionally there is a need to design an experiment in order to achieve a certain outcome. A medical researcher may want to design an experiment so that she has a 80% chance of detecting a certain difference between two varieties of corn. If she uses too few plots she will have little chance of rejecting the null hypothesis. If she uses too many plots she will be wasting time and money.

We begin by noting that while α, the significance level of the test, is usually fixed, the power of a test, which is the probability of rejecting the null hypothesis, depends on the actual difference between the two groups. If there really are large differences between the population means the power of a test may be quite large, even for a small experiment. On the other hand, if the difference between the population means is quite small a large experiment may be needed to have any chance of detecting the difference. In this section we will assume that the pooled t-test will be used in the analysis of the data and that the variances are equal. Further, our discussion of sample size will be confined to experiments having equal numbers of experimental units assigned to each group.

As we shall see, the sample sizes that are required to provide a specified power are a function of the variability that is present in the two populations. We typically do not know the variability in the two populations, but we will need to use an educated guess or the results from a previous study in order to estimate it. We will assume that $\sigma_1^2 = \sigma_2^2$ and will denote the common variance by σ^2. When designing a survey we must specify the power that we hope to achieve for a specified difference between the groups. For example, suppose we are designing an experiment to determine if one variety of corn has a greater yield than another. We want an experiment large enough so that we have a 80% chance of rejecting the null hypothesis if there is a difference of $\mu_1 - \mu_2 = 5$ bushels per acre. If we believe the standard deviation in each group is around 10, we can calculate the sample sizes that would be required to give 80% power of detecting a difference of 5 bushels per acre. With a power of 80%, the probability of a Type II error is 20%. That is, with 80% power we have a 20% chance of not detecting a difference when there really is a difference of 5 bushels per acre. Let $P(\text{Type II error}) = \beta = 1 - \text{Power}$, which implies that $\text{Power} = 1 - \beta$. The sample size needed to give a power of $1 - \beta$ for a one-tailed test is

$$n_1 = n_2 = \frac{2\sigma^2 \left(z_\alpha + z_\beta\right)^2}{\left(\mu_1 - \mu_2\right)^2}$$

where z_α and z_β are obtained from the normal table. If we use $\alpha = .05$ for our test then $z_\alpha = 1.645$. For a power of 80% we find $\beta = 1 - .8 = .2$ so that $z_\beta = 0.842$. Consequently, the sample size necessary to achieve 80% power would be

$$n_1 = n_2 = \frac{2\left(10^2\right)\left(1.645 + 0.842\right)^2}{5^2} = \frac{1237}{25} = 49.5$$

That is, we will need to take 50 sample plots with each variety to have an 80% chance of detecting a difference of 5 bushels per acre. If the difference between these varieties is larger, say $\mu_1 - \mu_2 = 10$, then the sample sizes would be

$$n_1 = n_2 = \frac{2\left(10^2\right)\left(1.645 + 0.842\right)^2}{10^2} = \frac{1237}{100} = 12.37$$

Thus, only 13 plots would be need for this experiment. As you can see, the sample size requirements are greatly influenced by the estimate of the difference between the varieties.

Exercise Set F

1. A researcher wanted to determine if nonvegetarians consumed more calories per day than vegetarians. From previous studies it was determined that the standard deviations in both groups were around 620 calories per day.

   a) Using a one-sided t-test with $\alpha = .05$, how many samples would be required in each group to give a 90% chance of detecting a difference of 100 calories per day?

   b) Using a one-sided t-test with $\alpha = .05$, how many samples would be required in each group to give a 80% chance of detecting a difference of 400 calories per day?

2. A sociologist wanted to determine if there was gender bias in the awarding of end-of-year bonuses. Suppose she wanted an 80% chance of detecting a difference of $1000 between men and women using a one-tailed t-test with $\alpha = .05$. If she used $2500 as an estimate of the standard deviations in both groups, how many samples would be required?

3. A researcher needed to determine if the drug Propranolol had any effect on blood pressure in rats. He wanted a 90% chance of detecting a difference of 12 mmHg using a one-tailed t-test with $\alpha = .05$. If it is reasonable to assume that the standard deviation is 25 in each group, how many rats would be required in each group?

Chapter Review Exercises

1. Researchers investigated the use of isolation timeout as a behavioral control intervention by placing students into one of two isolation groups based on the severity of their behavioral disorder. Each group was supervised by one teacher and one paraprofessional. Group 1 had a maximum of 12 students and Group 2 had a maximum of 6 students. The total number of hours in the timeout over one school year were recorded for the $n_1 = 100$ students assigned to Group 1 and for the $n_2 = 55$ students assigned to Group 2. (Based on Costenbader, V. and Reading-Brown, M. "Isolation timeout used with students with emotional disturbance," *Exceptional Children,* 61(1995) pp. 353-363) The average and standard deviation of the hours were reported as follows:

Group 1	n_1=100,	$\bar{x}_1 = 20.55$,	s_1=19.84
Group 2	n_2=55	$\bar{x}_2 = 26.56$,	s_2=21.74

   a) Compute a 95% confidence interval for the mean difference $(\mu_1 - \mu_2)$.

   b) Interpret the confidence interval.

   c) Based on the confidence interval, are you justified in concluding that the population mean of the total hours for the students assigned to the second group is greater than the population mean for the students assigned to the first group?

2. Can a physician determine the sex of a fetus by using the fetal heart rate? Some obstetricians believed that a rapid fetal heart rate indicated that the fetus was more likely to be female. To investigate this theory, researchers collected data on 250 male and 250 female births. They found that the males had an average fetal heart rate of 137.21 beats per minute with a standard deviation of 9.78 beats per minute,

while the females had an average of 137.18 beats per minute with a standard deviation of 8.10 beats per minute. (Based on Petrie, B. and Segalowitz, S. J. "Use of fetal heart rate, other perinatal and maternal factors as predictors of sex," *Perceptual and Motor Skills,* 50 (1980) pp.871-874) The researchers used a t-test to analyze the data.

a) Before doing any calculations, do you believe that a two-tailed t-test would produce a p-value less than $\alpha = .05$? Explain.

b) Compute the test statistic for the t-test and roughly estimate the p-value.

c) State your conclusion.

3. An epidemiologist compared the smoking habits of people who had developed bladder cancer to people who were disease-free. The epidemiologist determined if the bladder cancer cases and the disease-free controls have ever been smokers. In this exercise p_1 is the proportion of smokers in the population of all bladder cancer cases, p_2 is the proportion of smokers in the disease-free population. The number of people in each population and the number who were smokers are given below:

	Population	
	Bladder Cancer Cases	Disease-Free Controls
Smokers	110	241
Non-smokers	38	183
Total	148	424

a) Compute $\hat{p}_1$ and $\hat{p}_2$ for the smokers in the two populations.

b) Compute a 90% confidence interval for the difference in the population proportions.

c) Does there appear to be a difference in the population proportion of smokers between the cases and the controls?

4. A sociologist took a survey to determine the opinions that the Republicans and the Democrats have concerning the death penalty. The results of the survey showed that 72% of the 300 Republicans

that were surveyed approved of the death penalty and that 64% of the 300 Democrats approved of the death penalty.

a) Compute a 95% confidence interval for the difference between the population proportions.

b) Does the 95% confidence interval demonstrate that there is a difference in the population proportions?

5. Can certain diets reduce cholesterol?   To find out researchers randomly assigned 20 subject to either an oat-bran diet or to a bean diet.   The results indicated that both of these diets decreased total serum cholesterol. The table below gives summary statistics for the reduction in total serum cholesterol. (Based on Anderson, J. W., *et al* , "Hypocholesterolemic effects of oat-bran or bean intake for hypercholesterolemic men," *The American Journal of Clinical Nutrition* 40, (1984) pp. 1146-1155.)

	Diet	
	Oat-bran	Bean
Patients	10	10
Sample average (mg/dl)	53.6	55.5
sample standard deviation (mg/dl)	31.1	29.4

a) The researchers wanted to test H_0: $\mu_1=\mu_2$ versus H_a: $\mu_1\neq\mu_2$. Before doing a t-test make an educated guess about whether you will reject the null hypothesis.

b) Compute a p-value for a two-sided t-test of H_0: $\mu_1=\mu_2$ versus H_a: $\mu_1\neq\mu_2$. State your conclusion.

c) Compute a 95% confidence interval for the mean difference. Is the interpretation of the confidence interval consistent with the results of the test?

6. In section 6.1, we gave an example of a confidence interval for a population difference in the caffeine content of percolated and drip coffee. That 95% confidence interval had a width of 2(17)=34 with

$n_1=20$ and $n_2=30$. Now suppose we increased the sample size by a factor of 4 so that $n_1=80$ and $n_2=120$.

a) Would the value of s_p increase by a factor of 4? Why or why not?

b) The width of the 95% confidence interval would be closest to which of the following?

$$136, \quad 8.5, \quad 17, \quad 68$$

7. At one university Researcher A has assigned 16 dogs at random to two treatment groups in order to investigate the relationship between the treatment variable and the response variable. At a second university Researcher B has also assigned 16 dogs at random to two treatment groups to investigate the same relationship. Their results are displayed in the dot diagrams. Without doing any calculations, which researcher would have the larger two-sample t statistic? (You may assume that the scales are the same on all the plots and that the difference in sample means that Researcher A observes equals the difference in sample means that Researcher B observes.)

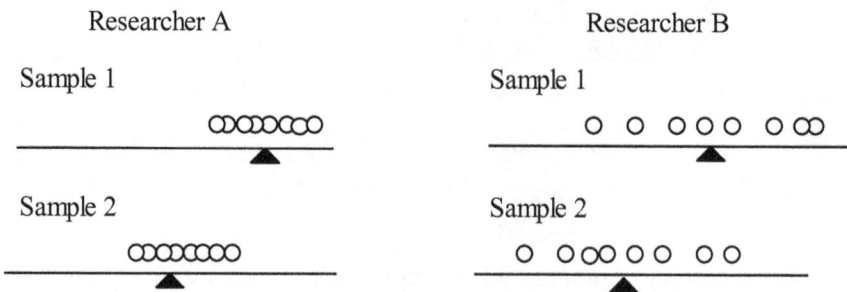

Researcher A

Sample 1

Sample 2

Researcher B

Sample 1

Sample 2

8. Can biofeedback methods help to reduce blood pressure? To investigate the effectiveness of biofeedback methods for the employees of a large industry, researchers randomly assigned employees who had several coronary risk factors into a biofeedback treatment group or a control group. The average reduction in systolic blood pressure for the 95 subjects in the biofeedback group was 15.3 mm Hg with a standard deviation of 15.1 mm Hg. The 84 subjects in the

control group had an average reduction of 6.1 mm Hg and a standard deviation of 14.3 mm Hg. (Data from Patel, C., Marmot, M. G., and Terry, D. J., "Controlled trial of biofeedback-aided behavioural methods in reducing mild hypertension," *British Medical Journal* 282(1981) pp. 2005-2008.)

a) Perform a one-sided t-test with $\alpha = .05$.

b) State your conclusion.

c) Based solely on the data given in this exercise, can you determine if the observations are symmetric or skewed?

d) Do you believe that it would be better to use the rank-sum test for the analysis of this data? Why or why not?

e) Is it possible to compute the rank-sum test with the information given above? Explain.

9. A quality control inspector compared the lifetimes of two types of incandescent bulbs. The lifetimes of the bulbs were recorded in the following dot diagram:

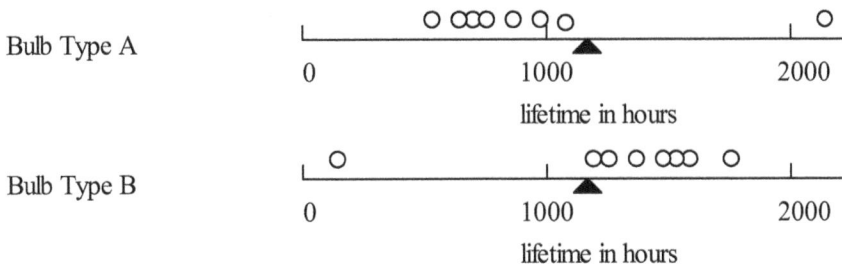

Bulb Type A

Bulb Type B

a) Without doing any calculations, roughly estimate the approximate value of the t-statistic. What would be the p-value, based on that t-statistic, for a one-sided test of H_0: $\mu_A = \mu_B$ versus H_1: $\mu_B > \mu_A$?

b) Compute the value of W for the rank sum statistic. Give the p-value for the one-sided test of H_0: $\mu_A = \mu_B$ versus H_1: $\mu_B > \mu_A$.

c) In your opinion, which test result seems to best summarize the results of this study?

10. In certain parts of the United States, the western fence lizards often contract malaria from a parasite. As part of a larger study to determine if the oxygen transport capacity of the infected lizards had been reduced, researchers obtained data on the distances the lizards could run in 2 minutes for 15 infected lizards and 15 noninfected lizards. The infected lizards had a mean running distance of 26.9 meters with a standard deviation of 6.81 meters. The noninfected lizards had a mean running distance of 32.2 meters with a standard deviation of 8.10 meters. (Based on Schall, J. J., Bennett, A. F., and Putnam, R. W., "Lizards Infected with Malaria: Physiological and Behavioral Consequences," *Science,* 217 (1982) pp. 1057-1059.)

a) Is the population mean running distance in the two groups equal? Compute a t statistic for a two-sample test and roughly estimate the p-value.

b) Would you reject the null hypothesis using $\alpha = .01$? State your conclusion.

c) In their publication the experimenters used a test that is equivalent to the Wilcoxon rank-sum test. In your opinion was it reasonable for them to use a this approach? Why or why not?

11. A researcher wanted to determine the long-term effect of high doses of alcohol on the liver function of Gorillas. Six gorillas were randomly assigned to receive very high doses of alcohol daily and six gorillas were assigned to a control group that did not receive any alcohol. All gorillas were fed the same diet. After two years the liver function was evaluated on a scale of 0 to 100, with 100 being a fully functioning liver. The results were recorded as follows:

| Alcohol Group: | 34, | 22, | 18, | 30, | 99, | 15 |

$$\bar{x}_1 = 36.33, \quad s_1 = 31.53$$

| Control Group: | 81, | 98, | 95, | 93, | 36, | 92 |

$$\bar{x}_2 = 82.5, \quad s_2 = 23.50$$

Perform a test of $H_o : \mu_1 = \mu_2$ versus $H_a : \mu_1 < \mu_2$ by using the rank-sum test. If you use a significance level of $\alpha = .05$, would you reject the null hypothesis? State your conclusion.

12. Refer to the data given in the previous exercise. Perform a test of $H_o : \mu_1 = \mu_2$ versus $H_a : \mu_1 < \mu_2$ by using the pooled t-test test. If you use a significance level of $\alpha = .05$, would you reject the null hypothesis? State your conclusion.

13. Refer to data and results from the two previous exercises.

a) Did you reach the same conclusions with the t-test and the rank-sum test using a significance level of $\alpha=.05$?

b) If you had used a significance level of $\alpha=.01$, would you have reached the same conclusions with the t-test and the rank-sum test?

14. A researcher wanted to determine if a low fat diet would reduce serum cholesterol in rats. In this experiment 5 rats were randomly assigned to a high fat diet and 5 were assigned to a low fat diet. The serum cholesterol was recorded at the end of the study and is shown in the dot diagrams:

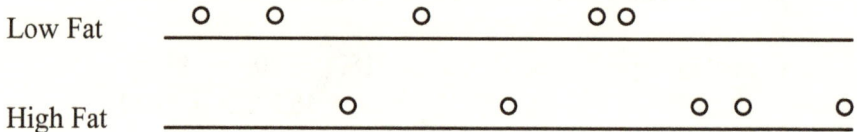

Low Fat

High Fat

a) Compute the rank-sum statistic.

b) Compute the p-value for a one-tailed test H_o: $\mu_1=\mu_2$ versus H_a: $\mu_1<\mu_2$ using the rank-sum test.

c) The researcher feels that if one more "high fat" animal were used the p-value would decrease. If the next "high fat" rat has a cholesterol value that is greater than the largest value in the "high fat" group, would the p-value decrease or increase?

d) If the next "high fat" rat has a cholesterol value that is less than the smallest observation in the "low fat" group, what would the p-value be?

15. A cardiologist, who wanted to determine if an exercise program would reduce systolic blood pressure in the middle-aged males, enlisted 40 volunteers for an experiment. The volunteers were randomly assigned to an exercise program or to a control program. After 6 weeks the systolic blood pressures were recorded for the 20 males who were assigned to each of the programs.

a) Should a test for paired data be used or should a two-sample test be used?

b) Suppose it is known from large surveys that the distribution of systolic blood pressure is slightly skewed. Which test would be most appropriate? Give at least one reason for your choice.

16. A petroleum company needed to determine if their new low-friction oil additive would reduce gasoline consumption in mid-sized automobiles. A local utility company had 37 identical automobiles that were used in an experiment designed to test the effectiveness of the additive. The researchers randomly assigned 18 of the vehicles to use the standard motor oil and assigned 19 of the vehicles to use the low-friction motor oil. After the appropriate motor oil was given to each vehicle, the gasoline consumption, in miles per gallon, was recorded for the next 2000 miles for each vehicle. The vehicles were then ranked on the basis of their average gasoline consumption. The sum of the ranks for the 19 vehicles that used the low-friction oil was $W = 472$. You may assume that there were no ties in the data.

a) What value of W would you expect if the low-friction oil additive did not change the fuel consumption?

b) Compute the large sample test statistics for the rank-sum test and compute the p-value for a one-sided test. Hint: Use the fact that there were no ties.

c) State your conclusion about the effectiveness of the low-friction additive.

Chapter 7. Correlation and Regression

7.1 Working with Bivariate Data

In the first chapter we computed descriptive statistics for a single variable. In the last two chapters we calculated confidence intervals and tests for a single variable. In this chapter we will analyze two continuous variables at a time. The analysis of this data usually begins with a **scatterplot**, which represents the data for the two variables as points in the plane. The scatterplot for this **bivariate** data, which is shown in Figure 7.1, displays the height and weight values for five students. In this scatterplot height is plotted on the horizontal axis and weight is plotted on the vertical axis. The coordinates for the first two points are indicated on the scatterplot. Note that each of the points corresponds to a student and that students who are shorter and lighter are represented by points on the lower left. Students who are taller and heavier are displayed on the upper right. It is traditional to label the horizontal axis the "x-axis" and to label the vertical axis the "y-axis." The values that are plotted on the x-axis are called x-values and those that are plotted on the y-axis are called y-values.

Figure 7.1 A scatterplot of height and weight for five students.

Student	Height (in.)	Weight (lbs.)
1	68	165
2	61	110
3	72	180
4	62	130
5	71	200

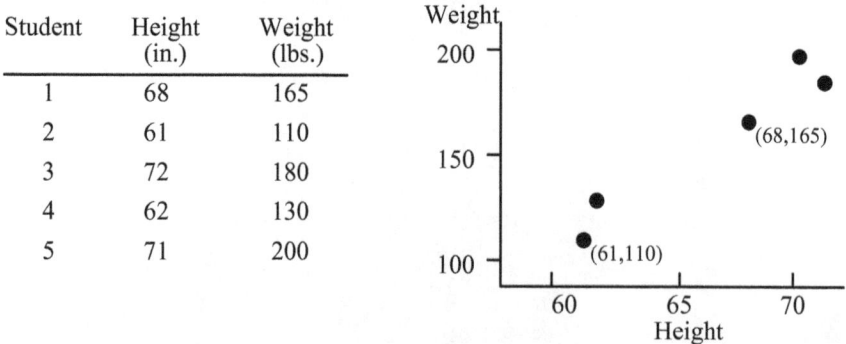

This chapter concerns the closely related topics of correlation and regression. **Correlation** is used when we want to describe the strength of the relationship between two variables. With our small sample, if we were interested in the magnitude of the relationship between height and weight we would compute a correlation coefficient. **Regression** would be used if we wanted to predict the y-values based on the x-values. If we were interested in predicting the average weight for a given height we would base our prediction on a regression line. Correlation will be developed first and regression will be developed later as an extension of regression.

Exercise Set A.

1. Ten eighth grade students were selected at random to take a standardized test on mathematics and language skills. The scores were recorded on a scale of 0 to 100 on both tests. A scatterplot using the mathematics scores as the x-values and language scores as the y-values is:

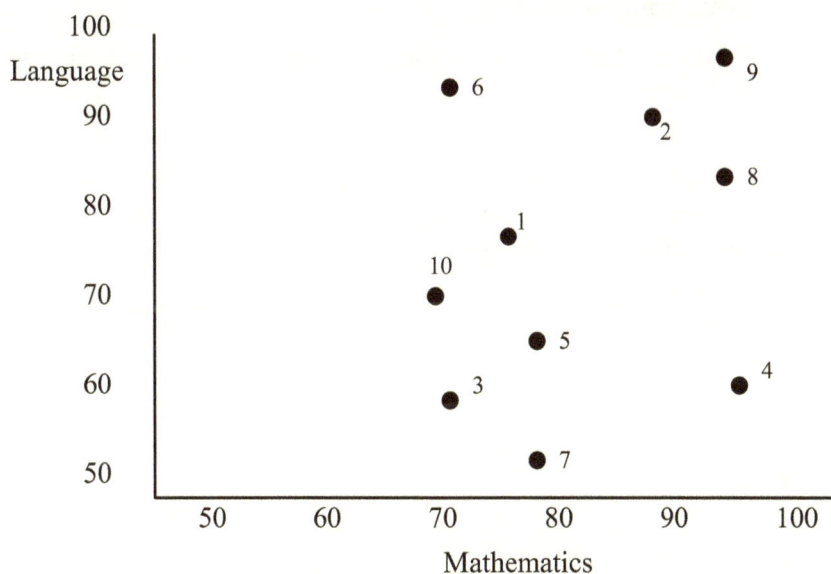

a) The average mathematics score is approximately

62   72   82

b) The average language score is approximately

77   87   97

c) Which of the following statements are true?

    i) The mathematics scores are more variable than the language scores.

    ii) The language scores are more variable than the mathematics scores.

    iii) The language and mathematics scores are equally variable.

d) Describe the scores that were received by student number 4 on this test.

e) Which students scores higher than average on both exams?

f) Which students scores lower than average on both exams.

g) Which students had language scores that were greater than their mathematics scores?

h) Which students had mathematics scores that were greater than their language scores?

2. The following scatterplot gives the age and systolic blood pressure for 20 females. The average age is indicated by a vertical dashed line and the average blood pressure is indicated by a horizontal dashed line.

Blood
Pressure

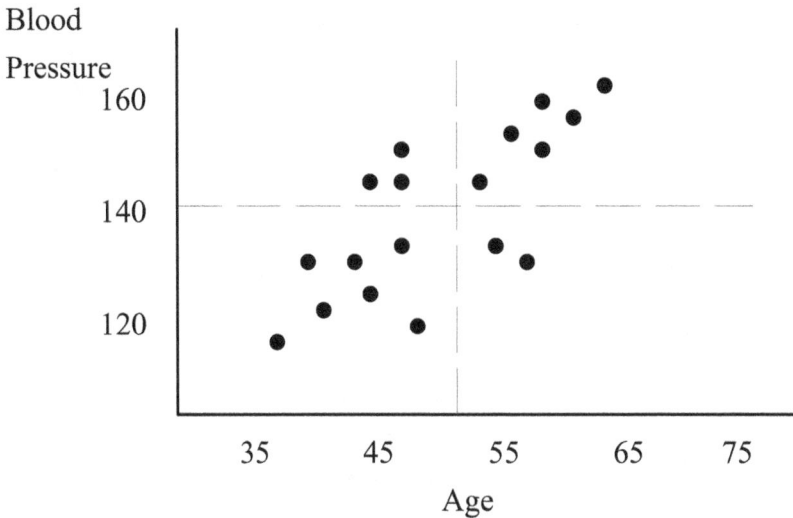

a) How many women who were above average in age had a blood pressure reading that was above average?

b) How many women who were above average in age had a blood pressure reading that was below average?

c) How many women who were below average in age had a blood pressure reading that was above average?

d) How many women who were below average in age had a blood pressure reading that was below average?

e) Does it appear that age is associated with blood pressure? Use the answers from (a) to (d) to justify your claim.

7.2 Computing the Correlation Coefficient

In our small sample of five students shown in Figure 7.1, the taller students tended to be the heavier students and the shorter students tended to be lighter students. We say that height and weight are positively associated in Figure 7.1 because the height values tend to increase as the weight values increase. However, not all scatterplots show such strong positive associations. In Figure 7.2(a) we see that the x-values and y-values are negatively associated because the smaller x-values tend to have the larger y-values. In Figure 7.2(b) we observe little association between the x-values and the y-values because those points having larger x-values have approximately the same average y-values as those points that have smaller x-values.   In Figure 7.2(c) we see some positive association because the larger x-values tend to have larger y-values, but the association is not strong because there are a few large x-values with small y-values and a few small x-values that have large y-values.

Figure 7.2. Three Scatterplots Showing Three Levels Of Association.

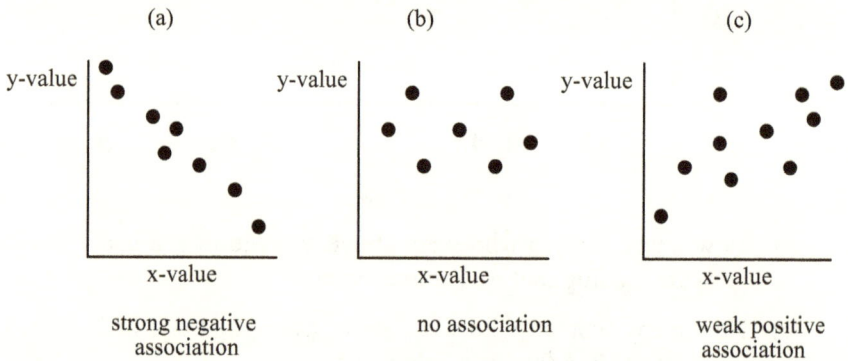

strong negative association   no association   weak positive association

A measure of the association between the x-values and the y-values is the correlation coefficient, which we will learn how to compute.   We will not compute the correlations directly from the data. Instead, we will **standardize** the x-values and the y-values and will then use the standardized values to compute the correlation coefficient.

Standardized values

A standardized value represents the number of standard deviations a value is above or below the average. For example, suppose we have height data for a sample of female undergraduates and we find that $\bar{x} = 64$ inches and $s_x = 3$ inches, where s_x is the standard deviation of the x-values. Note that a student whose height is 67 has a standardized height of +1, because she has a height that is one standard deviation above the mean. To see this, note that $67 = 64 + 3 = \bar{x} + s$. A student whose height is 58 inches has a standardized height of -2 because her height is two standard deviations below the mean. A general formula can be used to compute the standardized value for any height. To compute the standardized height (x_i') for the ith value in the sample we use

$$x_i' = \frac{x_i - \bar{x}}{s_x}.$$

Using this formula, a student whose height is 57 has a standardized height of

$$x_i' = \frac{x_i - \bar{x}}{s_x} = \frac{57 - 64}{3} = \frac{-7}{3} = -2.33.$$

We can standardize the y-values in the same way. Let $\bar{y}$ be the mean of the y-values and let s_y be the standard deviation of the y-values. The standardized y-values use the formula

$$y_i' = \frac{y_i - \bar{y}}{s_y}.$$

Since we subtract the mean value from each observation when we standardize, we obtain standardized values that have a mean of zero. By dividing by the standard deviation we force the standard deviation of the standardized values to equal one. We will use the pairs of standardized values (x_i', y_i'), $i = 1,...,n$ to compute the correlation coefficient.

To understand the calculation of the correlation it is helpful to mark the average of the x-values and the average of y-values on the scatterplot. In Figure 7.3 we have indicated the average x-values by a dashed vertical line and the average y-value by a dashed horizontal line. Note that there are two students who were above average on height and weight, two students who were below average on height and weight, one student who was taller than average that had a below average weight, and one student who was shorter than average that had an above average weight.

Figure 7.3 Scatterplot of height and weight for seven students.

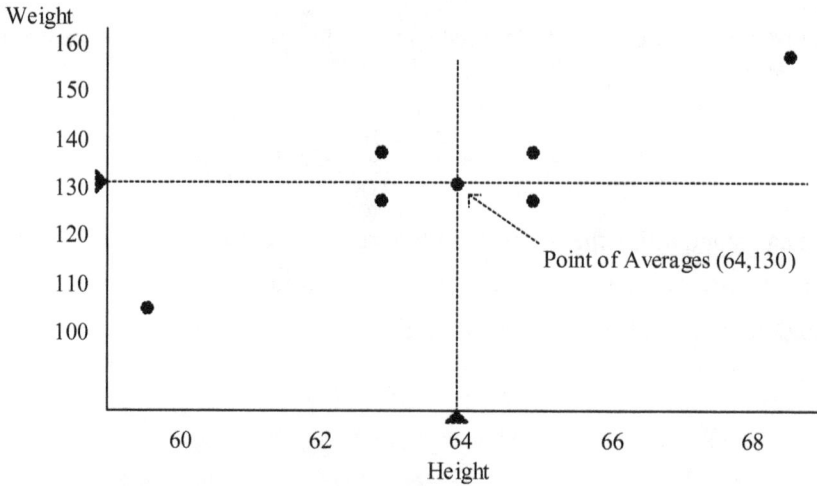

The following table gives the standardized values of height and weight. The standardized values indicate the number of standard deviations above or below the average. A positive standardized value indicates an observation that is above the average and a negative value indicates an observation that is below the average. The standardized value for the height of the first student is -1.67, which indicates that it is 1.67 standard deviations below the average.

	Original Data		Standardized data	
Student	x_i	y_i	$x'_i = \left(\dfrac{x_i - \bar{x}}{s_x}\right)$	$y'_i = \left(\dfrac{y_i - \bar{y}}{s_y}\right)$
1	59	105	$\dfrac{59 - 64}{3} = -1.67$	$\dfrac{105 - 130}{15} = -1.67$
2	63	125	$\dfrac{63 - 64}{3} = -.33$	$\dfrac{125 - 130}{15} = -.33$
3	63	135	$\dfrac{63 - 64}{3} = -.33$	$\dfrac{135 - 130}{15} = .33$
4	64	130	$\dfrac{64 - 64}{3} = 0$	$\dfrac{130 - 130}{15} = 0$
5	65	125	$\dfrac{65 - 64}{3} = .33$	$\dfrac{125 - 130}{15} = -.33$
6	65	135	$\dfrac{65 - 64}{3} = .33$	$\dfrac{135 - 130}{15} = .33$
7	69	155	$\dfrac{69 - 64}{3} = 1.67$	$\dfrac{155 - 130}{15} = 1.67$

Figure 7.4 gives the scatterplots for the raw data and the scatterplots for the standardized values. Note that the scatterplots have the same overall appearance. In fact, the only difference between the scatterplots are the horizontal and vertical scales. Since we are interested in the association between the two variables this association can be seen as clearly with the standardized values as with the raw data. The advantage of the standardized values is that they can be directly used to compute the correlation coefficient because they have a mean of zero and a standard deviation of one.

Figure 7.4 Scatterplots for original and standardized values.

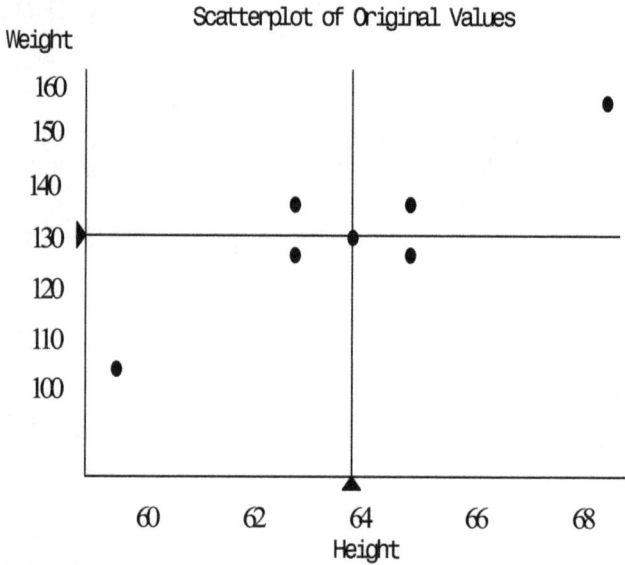

Scatterplot of Original Values

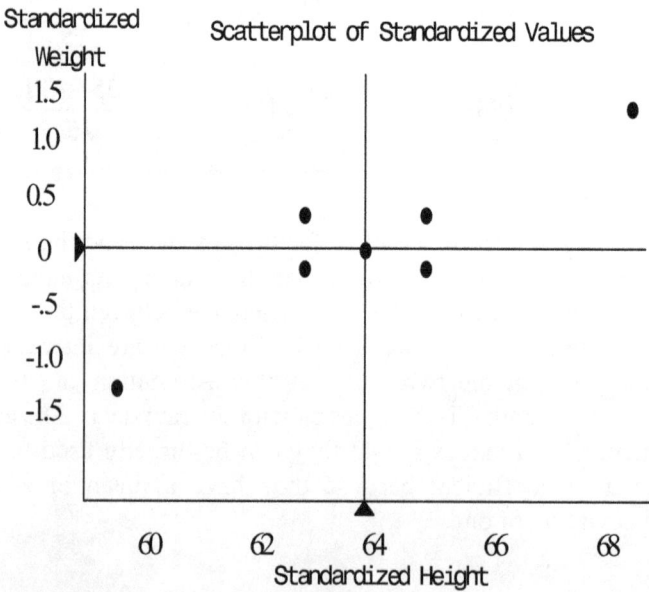

Scatterplot of Standardized Values

The correlation coefficient

After we have computed the standardized values we can easily compute the correlation coefficient. The correlation coefficient (r) can be computed from the standardized x-values (x_i') and the standardized y-values (y_i') with the formula

$$r = \frac{1}{n-1}\sum_{i=1}^{n} x_i' y_i'.$$

Note that, for large samples, the correlation coefficient is approximately equal to the average of the product of the standardized values.

In order to compute the correlation coefficient we need only add a column that has the product of the standardized data $(x_i' y_i')$ to the previous table. Note, in our example, that most of the products are small, with only the first and last observations producing large products. The last two students had heights and weights that were above average and so these values, when standardized, were positive and hence their product was positive. The first two students were below average on height and weight and so their standardized values were negative, but since both were negative their product was positive.

Original Data			Standardized data		Product
Student	x_i	y_i	$x'_i = \left(\dfrac{x_i - \bar{x}}{s_x}\right)$	$y'_i = \left(\dfrac{y_i - \bar{y}}{s_y}\right)$	$x'_i\, y'_i$
1	59	105	$\dfrac{59-64}{3} = -1.67$	$\dfrac{105-130}{15} = -1.67$	2.79
2	63	125	$\dfrac{63-64}{3} = -.33$	$\dfrac{125-130}{15} = -.33$	.11
3	63	135	$\dfrac{63-64}{3} = -.33$	$\dfrac{135-130}{15} = .33$	-.11
4	64	130	$\dfrac{64-64}{3} = 0$	$\dfrac{130-130}{15} = 0$	0
5	65	125	$\dfrac{65-64}{3} = .33$	$\dfrac{125-130}{15} = -.33$	-.11
6	65	135	$\dfrac{65-64}{3} = .33$	$\dfrac{135-130}{15} = .33$	.11
7	69	155	$\dfrac{69-64}{3} = 1.67$	$\dfrac{155-130}{15} = 1.67$	2.79

$$\sum_{i=1}^{n} x'_i\, y'_i = 5.58$$

To compute the correlation we need only sum the products in the last column and divide by $n-1$. Therefore, the correlation is

$$r = \frac{1}{n-1}\sum_{i=1}^{n} x_i' y_i' = \frac{1}{7-1}(5.58) = .93.$$

It can be shown that correlations are always in the interval [-1, +1]. A value of $r = +.9$ indicates positive association. A value of $r = -.9$ indicates a strong negative association. If there is no association between the x-values and the y-values the correlation coefficient should be near zero. The range of correlations is displayed in Figure 7.5.

Figure 7.5 The range of correlations.

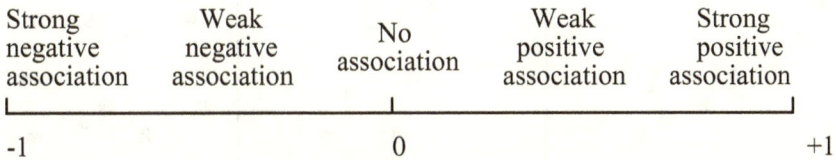

Strong negative association	Weak negative association	No association	Weak positive association	Strong positive association

-1   0   +1

Figure 7.6 gives some scatterplots and their correlations. Note that the magnitude of the correlation coefficient is a measure of the amount of clustering around a line. If the x-values and the y-values are almost on a line and are positively associated they have a correlation near +1. Note also that the correlation is negative if the x-values that are larger than average tend to have y-values that are smaller than average. This can be seen in Figure 7.6(a).

Figure 7.6 Scatterplots for several values of the correlation.

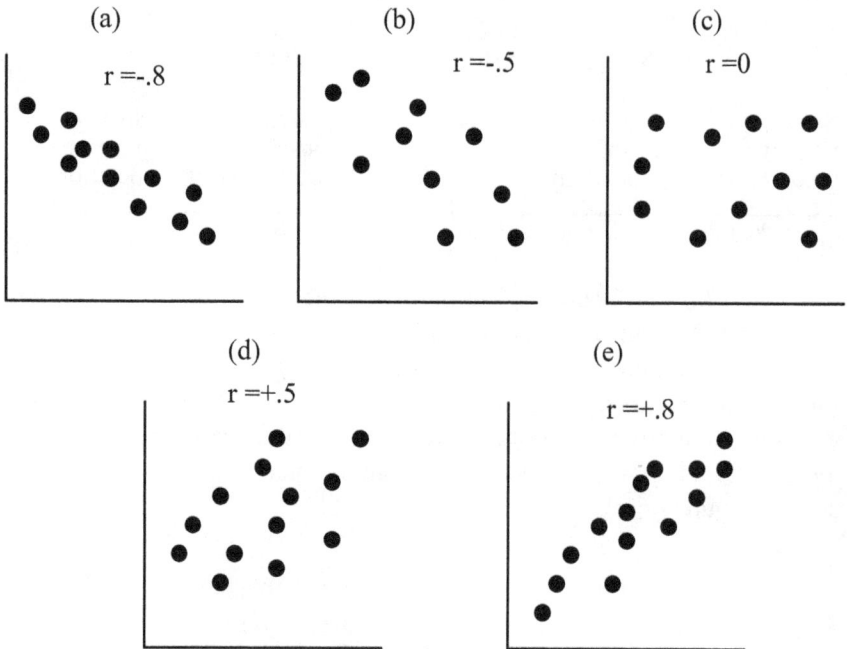

The correlation coefficient is a good measure of the association about a straight line but it does not always give an accurate assessment of the association if the points cluster about a curved line. For the scatterplots shown in Figure 7.7 there is a clear association between X and Y. In Figure 7.7(a) the correlation is zero even though there is a strong association around the curved line. The correlation coefficient is not a good measure for the curved relationship between these two variables. In Figure 7.7(b) the correlation coefficient may be near $r = .3$, which is the correct sign of the association, but it fails to accurately reflect the strong association between the two variables because the association is not linear. Consequently, before computing a correlation it is prudent to make a scatterplot of the data to verify that the association is linear.

Figure 7.7 Scatterplots for non-linear associations.

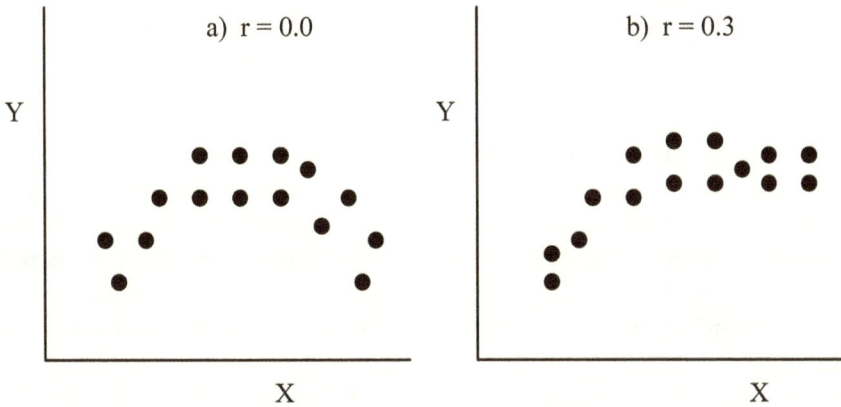

An important property of the correlation coefficient is that it will not change if the scale of measurement is changed. This property of the correlation can be demonstrated by changing the height data, which was originally measured in inches, to centimeters and observing that the standardized value $x_1' = \frac{(x_1 - \bar{x})}{s_x}$ will not change. To see this, note that, since 1 inch = 2.54 centimeters, the first student is $59 \times 2.54 = 149.86$ centimeters in height and the average height is 64 inches, which is $64 \times 2.54 = 162.56$ centimeters. Thus, $(x_1 - \bar{x})$ will be 2.54 times larger when measured in centimeters. To calculate the effect of the change of scale on the standard deviation we recall the formula for the standard deviation is

$$s_x = \sqrt{\frac{\sum_{i=1}^{n}(x_i - \bar{x})^2}{n-1}}.$$

The deviations $(x_i - \bar{x})$ will be 2.54 times larger when centimeters are used than when inches are used. Since these deviations are squared before the square root is taken, it follows that the standard deviation will be 2.54 times larger using centimeters than using inches. Because the standard deviation on the inch scale is 3 inches the standard deviation on the centimeter scale is $3 \times 2.54 = 7.62$ centimeters. Therefore, using the centimeter scale, the standardized value for the height of the first student would be

$$x_i' = \frac{x_i - \bar{x}}{s_x} = \frac{149.86 - 162.56}{7.62} = \frac{2.54(59) - 2.54(64)}{2.54(3)} = -1.67.$$

Note that the factor 2.54 appears in the numerator and denominator and it follows that the standardized value will not change because 2.54 will cancel. This makes sense because the standardized value equals the number of standard deviations above or below the mean, and this should not depend on what scale is used. The same argument may be applied to the standardization of the y-values. We can summarize our result as follows:

> Since the correlation is a function of the standardized values, the correlation will not change if there is a change of scale to either the x-values or y-values (or both).

Take another look at the last column in the table containing the products of the standardized values. The products $(x_i' y_i')$ are large for the first and the last students because the x-values and the y-values are far from the point of averages. In general, correlation coefficients are determined largely by those points that are far from the point of averages. This point can be illustrated with the following figure. Figure 7.8(a) contains scatterplots of the observations in a data set that contains one outlier. Figure 7.8(b) shows the same data with the single outlier removed.

Figure 7.8 The sensitivity of correlations to outliers

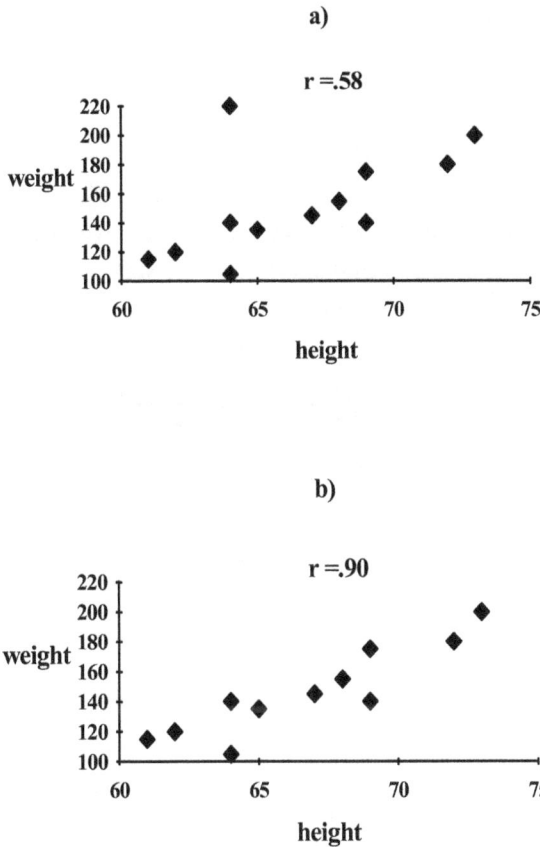

a)

b)

Note that the correlation in Figure 7.8(a) is not very large even though there appears to be a clear relationship in this population between height and weight. When a single observation is removed from the data set the correlation jumps from $r = .58$ to $r = .90$. The low correlation of $r = .58$ is due to the presence of an observation which has a smaller than average x-value and a larger than average y-value which produces a large negative product of the standardized values. Consequently, this outlier had a large effect on the correlation.

In general, the presence of a single outlier may raise or lower the correlation observed with the other points. A scatterplot should always be used to determine if the correlation actually represent the association between the x-values and the y-values or if an outlier is present. Usually, outliers should not be removed from the data unless an error was made in the collection of the data. If outliers are present then a rank correlation may be more suitable than the product-moment correlation that we have described. The rank correlation will be discussed later in this chapter.

Some people prefer to work with the raw data directly rather than compute the standardized values. By substituting the expressions for the standardized values into the formula for the correlation coefficient, we can write the correlation coefficient as

$$r = \frac{1}{n-1}\sum_{i=1}^{n} x_i' y_i' = \frac{1}{n-1}\sum_{i=1}^{n}\left(\frac{x_i - \bar{x}}{s_x}\right)\left(\frac{y_i - \bar{y}}{s_y}\right)$$

$$= \frac{1}{(n-1)s_x s_y}\sum_{i=1}^{n}(x_i - \bar{x})(y_i - \bar{y})$$

This equation may be preferred for manual calculation.

Exercise Set B

1. These scatterplots have correlations of $r = -.3$, $r = 0$, and $r = .3$. Match the correlation with the scatterplot.

(a)   (b)   (c)

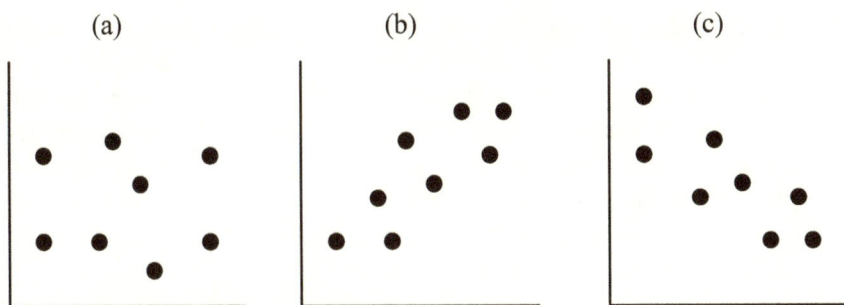

2. Consider the college admission scores (x-values) and the number of years attended in college (y-values).

Student	admission score (x-value)	years attended (y-value)
1	20	0
2	22	4
3	20	0
4	22	2
5	21	4

a) The average x-value is _____.

b) The average y-value is _____.

c) Make a rough scatterplot and identify the point of averages.

d) The standard deviation of the x-values is _____.

e) The standard deviation of the y-values is _____.

f) Compute the standardized values of the admission scores and the years attended.

g) Compute the correlation coefficient.

h) If you considered the admission scores to be the y-values and the years attended to be the x-values, would the correlation be the same or would it be different? Explain.

3. A researcher determined the height (x-values) and the weight (y-values) for 50 high school wrestlers. After she completed her study and computed her correlation she realized that the scale that was used to measure the weights was biased. It consistently recorded weights 3 pounds lower than they actually were so that people who actually weighted 135 pounds were recorded as weighing 132 pounds. Does the correlation need to be computed again based on the actual weights? Why or why not?

4. Suppose a researcher obtained data from 15 subjects on coffee consumption (x-value) and weight gain over the last year (y-value). Based on an observed correlation of $r = .3$ the researcher came to the conclusion that there was an association between coffee consumption and weight gain. A scatterplot of the data is shown below. Do you believe that there really is an association between coffee consumption and weight gain? That is, if you were to do another study of 15 other subjects do you believe that you would obtain a correlation around $r = .3$? Why or why not?

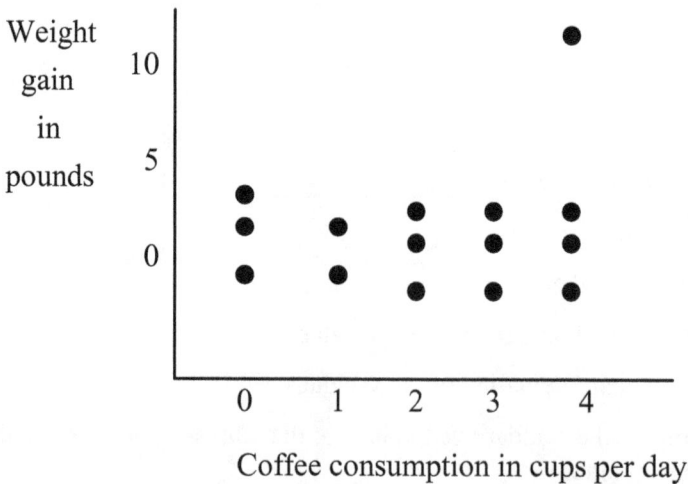

Coffee consumption in cups per day

5. An environmental group wanted to determine the relationship between the weight of automobiles (x-value) and their gas mileage (y-value). The scatterplot of the 15 vehicles that were selected is shown with the mean values indicated by dashed lines.

Mileage in
miles per
gallon

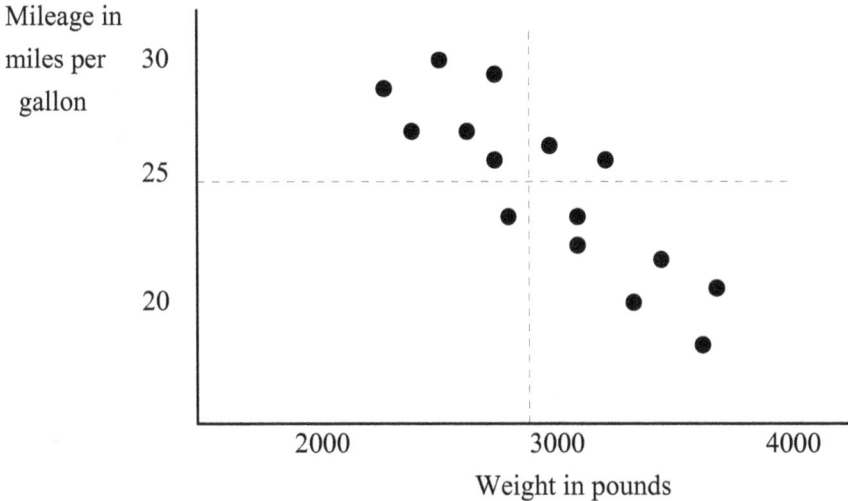

Weight in pounds

a) How many observations had positive products of their standard-ized variables? (That is, how many had $x'_i y'_i > 0$?

b) How many of the products were negative?

c) Without doing any calculations estimate the correlation.

$$-.5 \qquad\qquad 0 \qquad\qquad +.5$$

d) If the weight was recorded in kilograms and the mileage was recorded in kilometers per liter would the correlation change? Why or why not.

6. A soil scientist measured the amount of rainfall during the year (x-value) and the soil loss (y-value) in six plots in various parts of the country. The data, which was measured in inches per year, is given in the following table:

Plot	Rainfall	Soil Loss
1	34	0.1
2	75	0.2
3	25	0.08
4	88	0.02
5	47	0.15
6	60	0.18

a) Make a scatterplot.

b) Compute the correlation.

c) Does the correlation describe the association between most of the points in the data set?

7. In an experiment designed to improve the performance of students on a trigonometry examination, students were assigned at random to 1, 2, 3, 4, or 5 hours of computer assisted instruction. Their scores on the exam (y-values) and the computer assisted instructional times (x-values) were recorded and displayed in a scatterplot. A correlation of $r=.35$ was computed.

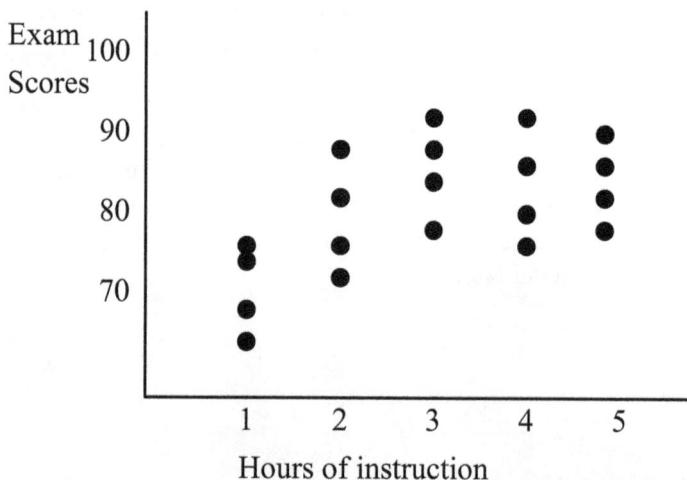

When asked to summarize the results, the researcher stated that "The computer assisted instruction was effective in improving the per-

formance of the students on the exam. Student performance increased with each additional hour of instruction." In your opinion,
was this a fair summary of the results?

7.3 Using Correlations To Make Conclusions

Possible explanations for large correlations

After a little practice, you will find that correlations are not too difficult to compute for small data sets. Spreadsheet programs or specialized
statistical software are usually used to compute correlations for large data
sets. Consequently, correlations are generally easy to calculate.
However, after the correlation is computed, there is often considerable
controversy surrounding the interpretation of the correlation. Problems of
interpretation most often arise when a large correlation is erroneously
interpreted to mean that the x-variable caused the y-variable. In this
section we discuss methods that can be used to aid in the interpretation of
correlation coefficients.

In scientific research the word "cause" can be used in two ways. The
classical definition of cause implies that a cause always produces an
effect. However, in scientific work we often refer to a cause as
something that often, but not always produces an effect. When we say
that "X causes Y" we mean that a change in X will very often produce a
change in Y. For example, we consider cigarette smoking a cause of
lung cancer because, by smoking cigarettes, a person increases the
chance of getting lung cancer. We do not imply that everyone who
smokes will get lung cancer. In this text we will use cause in this
probabilistic sense and say that X causes Y if a change in X alone will
increase the chance of producing a change in Y. In the medical literature
the expression "risk factor" is often used to express this idea.

If we do observe a positive correlation between X and Y, then we
must determine if the correlation could reasonably have been due to
chance. This point will be addressed later in this chapter where we will
give methods for computing the significance level of a correlation. That
is, we will test the null hypothesis that the population correlation is zero.
If there is a large chance of observing a sample correlation of this
magnitude under the null hypothesis, then we have not established that a
true association exists between X and Y. We can only say that we

observed an association in the sample, that may or may not exist in the population.

If, on the other hand, we compute a small significance level, we will reject the null hypothesis of no association and we will conclude that there really is an association in the population. However, even though we are quite certain that an association exists between X and Y, we cannot, on the basis of the test of significance alone, determine if there is a causal relationship between X and Y.

If we want to make a statement about the relationship between X and Y when the correlation is large, we generally need to know something about the possible relationships that could exist between the variables. Suppose we want to investigate the relationship between rainfall and crop yield for a certain variety of corn. If we took a sample of many plots of corn and measured the crop yield (Y) and the rainfall in the growing season (X) we might find a correlation of $\rho = .8$. This large positive correlation demonstrates a strong association between the two variables. However, should the interpretation be that increased rainfall caused increased crop yields or should the interpretation be that the increased crop yield caused increased rainfall? Most people would use their knowledge of science to conclude that the increased rainfall caused the increased crop yield. It is important to understand that, in this example, the correlation could not be interpreted intelligently without a knowledge of agronomy and meteorology. Without knowledge of the scientific principles involved, the direction of the causation cannot be determined.

In many situations it is very difficult to interpret a high correlation between X and Y because the possible relationships between X and Y are unknown. It may be that X caused Y or that Y caused X or that X and Y influenced each other. For example, if X is a measure of alcohol abuse and Y is a measure of antisocial behavior, it may be reasonable to believe that the alcohol abuse caused the antisocial behavior or that the antisocial behavior caused the alcohol abuse. In this situation, unless some other information was available to determine which behavior came first, it may be impossible to make a statement about the causal relationship, even if the correlation coefficient is large.

Even if the association between X and Y is strong, as determined by the large correlation coefficient, there may be no direct relationship between X and Y. It is possible that a third variable, say Z, caused both

X and Y. For example, suppose we measure the height and the reading ability of the students in a grade school, display the results in Figure 7.9, and find that the correlation is $r = .8$. It is clear that height does not cause reading ability and that reading ability does not cause height. In this example the correlation can easily be interpreted because the older students, who tend to be taller, also tend to have greater reading ability. That is, there is a third variable, age(Z), that is the cause of height (X) and is associated with reading ability(Y). Thus, there may not be any direct relationship between height and reading ability for a given age group so that if we took a sample of 10 year old students, as shown in Figure 7.9(b), or a sample of 13 year old students, as shown in Figure 7.9(c), the correlation between their height and reading ability may be close to zero.

Figure 7.9 Correlation between height and reading ability

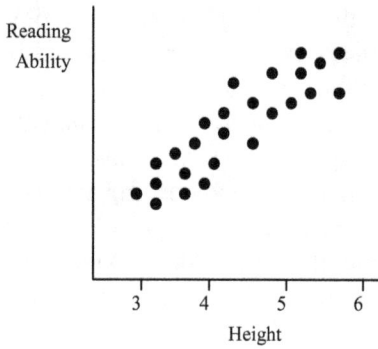

a) r = .8 with all ages included

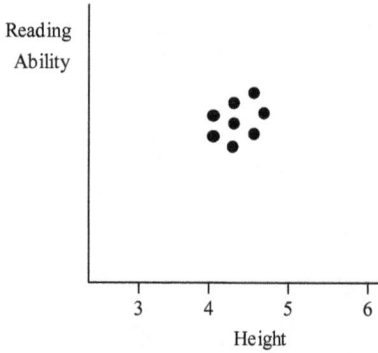

b) r = .04 for students with age=10

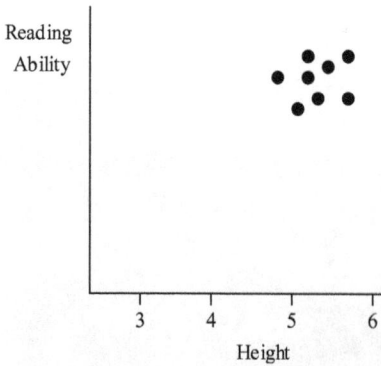

c) r = .1 for students with age=13

Another possible explanation for an association between X and Y is that there is a **confounder** that is associated with X and Y. A confounder is a third variable(Z), that has the effect of distorting the relationship between X and Y, even though it may not directly cause X or Y. For example, suppose we observe a small association between coffee consumption and lung function. The association between X and Y may be due to the presence of a confounder, like smoking, that is correlated with lung function and with coffee consumption. In this situation smoking may not cause coffee consumption, but may be sufficiently correlated with coffee consumption that it distorts the direct relationship between coffee and lung function. Consequently, there may be no direct relationship between coffee consumption and lung function for smokers and for non-smokers. Figure 7.10 shows the possible relationships that may exist when we observe a large correlation between X and Y.

Figure 7.10 Possible explanations for an observed strong association between X and Y.

Explanation	Diagram
X causes Y	X $\longrightarrow$ Y
Y causes X	Y $\longrightarrow$ X
X and Y influence each other	X $\longleftrightarrow$ Y
There is a common cause (Z)	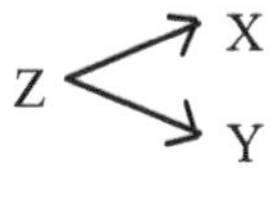 Z with arrows to X and Y
There is a confounder (Z)	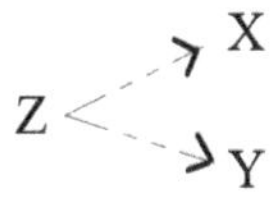 Z with dashed arrows to X and Y

Many correlations cannot be easily interpreted because the presence of a third factor may be suspected, but can neither be proved or disproved. For example, suppose a sociologist determines that the correlation between education and income is $\rho = .4$. Clearly, the education came before the income, but it is not entirely clear that it was the cause. Perhaps motivation is a common cause. Highly motivated individuals may do well in school and may eventually obtain higher paying jobs, but their job success may have little to do with their schooling. Without further analysis it is not entirely clear that education caused the increased salary or if there is a common cause. In this example a positive correlation did not demonstrate that one variable caused the other. Therefore it is prudent to exercise some caution when interpreting correlations.

> Correlation is not causation. The presence of a large positive correlation does not mean that one variable caused the other variable.

Arguments for Causation

We have seen that there are other possible explanations for a large correlation and that a large positive correlation does not imply that a causal relationship exists. However, if we have spent much time and money collecting data on X and Y, and if we are interested in the relationship between X and Y, we want to say as much as we honestly can about the relationship between the variables. The approach presented in this section, which is based on the work of Bradford Hill (Hill and Hill, 1991), is to take several other aspects of the association into account before making a judgment about a causal relationship. If a large correlation exists between X and Y, the biggest obstacle to correct inference is determining if a third factor could be a common cause of X and Y.

The **strength of the association** may give us a clue about the possibility of a common cause. If the association between X and Y is very large, it may be argued that any common cause of X and Y would need to be highly correlated with X and with Y. But a high association between the common cause and either X or Y, if it exists, is very likely to be observed. If no common cause is obvious and if the association is very large, the argument for causation is strengthened.

A **consistent association** may also provide some information about causation. If X really does cause Y then a strong association should be observed by different researchers working in different places. If the association is not consistent, it is likely that some other factors are operating. A consistent association is also useful to rule out the possibility that an error was made in the collection of the data or that the study was biased in some way.

In addition, an argument concerning causation should be **plausible**. If a causal relationship between X and Y appears to be implausible, a prudent person would hesitate to declare that X caused Y. However, this approach is not foolproof because it is possible that a causal relationship really does exist between X and Y even though we, with our limited understanding of nature, now believe it to be impossible.

It is also important to check the **temporal relationship** of X and Y. If X is a recent environmental exposure and Y is a disease that requires 20 years to develop, it does not make sense to state that the recent introduction of an environmental hazard caused the initiation of the disease process. Of course, the environmental hazard may promote the disease process or may be related to the disease in some other way, but it could not be the cause because it did not exist before the initiation of the disease process.

Even though a strong association is consistently observed, is plausible, and has the correct temporal relationship, there may be some other explanation for the observed association. The ideas presented in this section may provide some guidance but cannot be used to prove that a causal relationship exists. In observations studies there is generally some uncertainty about the presence of an alternative explanation for the correlation and this uncertainty is the major weakness of observational studies. Experimental studies are usually more conclusive because there are fewer alternative explanations for the observed association.

Exercise Set C

1. Suppose a sociologist computes a sample correlation of $r=.2$ between the expenditure per pupil on computer equipment in the high school (x-value) and the college entrance exam score (y-value). This correlation was based on a sample of 500 students from many different school districts. Is it clear that additional expenditures on computer equipment in the high school will increase the college entrance exam scores? Name at least one possible common cause.

2. A dietitian took a sample of 300 individuals and determined, from a food frequency checklist, the number of servings of eggs and bacon that each individual consumed in a month. The correlation between the egg consumption and the bacon consumption was $r = .4$. In your opinion, how are these variables related?

3. Suppose a sample of 1000 students was selected for study and their reading ability was measured along with the amount of money per year that was spent on their education. The correlation between reading ability and annual educational expenditure was computed to be $r = .3$. Does this imply that increased school funding in a school district will necessarily cause an increase in the reading score? Name at least one possible common cause or confounder.

4. An educational researcher recorded the reading ability, as measured by a standardized test, and the number of newspapers and news magazines delivered to the home for a sample of n=300 third grade students. The researcher computed a correlation of r=.25 between the number of standardized tests and the number of news magazines and concluded that the amount of reading material in the home increases the reading scores on the test. The parents of a third grader, after reading these results, order a subscription to the *Wall Street Journal* in order to improve their daughter's reading score. Do you believe that this will be effective? Why or why not?

7.4 Regression Analysis

Introduction to regression analysis

We have learned that the correlation coefficient measures the amount of association between the x-values and the y-values. Sometimes the relationship between the x-value and the y-values can be adequately described by the correlation coefficient and, in those situations, no further analysis is necessary. However, there is often a need to predict the average value of y for a given x-value. For example, suppose a researcher, using the heights and weights of 50 female soccer players, wants to estimate the average weight for a given height. This is illustrated in Figure 7.11 where the **regression line** is shown as a solid line going through the middle of the cloud of points. The regression line used to predict the height based on the weight is called the **regression of weight on height**.

Figure 7.11 Regression of weight on height.

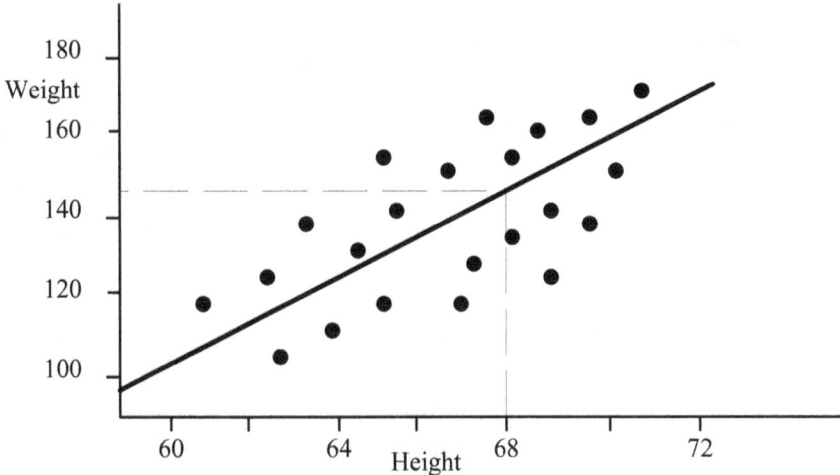

We will learn later how to compute the regression line. For now, we will estimate the average weight for any given height. Suppose we want to estimate the average weight for soccer players who are 68 inches in height. The average height is estimated by the height of the regression line at x=68. That value appears to be approximately 145 pounds. For x=60 inches we estimate the average weight to be around 105 pounds and for x=72 we estimate the average height to be around 165 pounds.

As we shall see, the regression line can easily be computed from the correlation coefficient. However, it is important to carefully distinguish between correlation and regression. The correlation coefficient is a measure of association between two variables and it makes no difference which variable is considered the x-value and which is considered the y-value. In regression analysis, the objective is to estimate an average value of the y-value for a specified x-value, so it does make a difference which variable is considered the x-value.

Computing the regression line

In order to compute the regression line we first need to compute the standard deviation of the x-values (s_x), the standard deviation of the y-values (s_y), and the correlation (r). After these are computed the regression line is defined as the line that :

1) goes through the point of averages $(\bar{x}, \bar{y})$ and

2) has a slope of $b = r\dfrac{s_y}{s_x}$.

Since only one line with slope b goes through $(\bar{x}, \bar{y})$ we have defined the regression line, which allows us to predict the average value of y for any x-value.

We can illustrate the computations of the regression line using the height and weight data shown in Figure 7.11. Suppose we found that the average height was $\bar{x} = 67$ inches and that the standard deviation was $s_x = 2$ inches. For this data we found an average weight of $\bar{y} = 140$ pounds with a standard deviation of $s_y = 20$ pounds. The point of averages is marked on Figure 7.12. We computed the correlation in the usual way as $r=.5$ and determined that the regression line must

1) go through the point of averages (67,140) and,

2) have a slope of $b = r\dfrac{s_y}{s_x} = .5\dfrac{20}{2} = 5$.

This regression line is shown in Figure 7.12. Note that, because the regression line goes through $(\bar{x}, \bar{y})$, the average weight of all players who are 67 inches in height is 140 pounds. To estimate the weight for other values of height we need to remember that the slope equals the rise divided by the run. A slope of $b=5$ indicates that the average weight increases by 5 pounds for each additional inch of height. Thus, the average weight for players who are 68 inches in height is $(140+5)=145$ pounds.

Figure 7.12 Regression of weight on height.

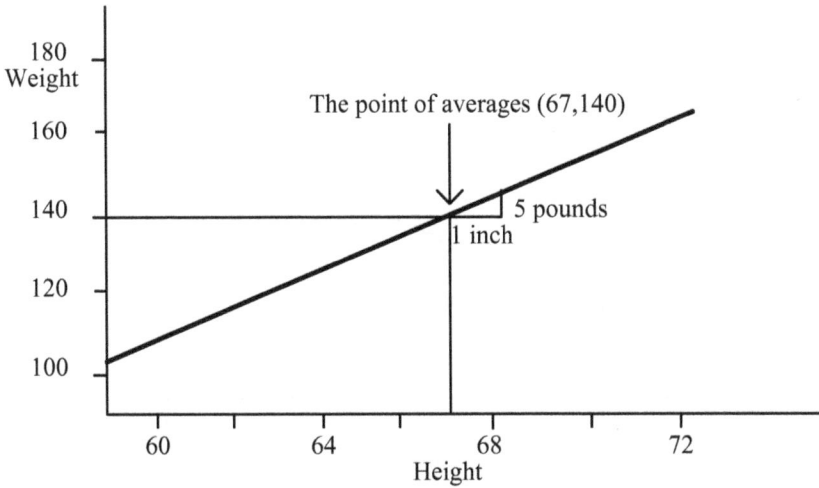

Suppose we wanted to compute the average weight for players who are 70 inches in height, which is 3 inches taller than average. Since the average weight increases at a rate of 5 pounds for each inch, her expected weight would be $3 \times 5 = 15$ pounds more than the average, or 155 pounds. This estimate is shown in Figure 7.13. Now consider a person who is 65 inches in height. Since the height is 2 inches shorter than the average height, the weight will be $2 \times 5 = 10$ pounds less than the average, or 130 pounds.

Figure 7.13 Regression of weight on height.

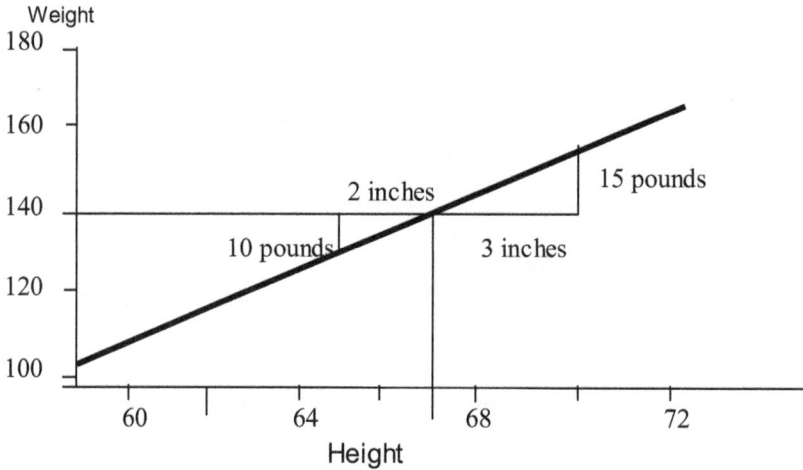

In our example of a regression of weight on height, the slope was positive because the correlation was positive. Regression analysis can also be used with negative correlations. Figure 7.14 shows the association between the number of hours of television viewing per day and the reading scores on a standardized test. Suppose we compute a mean viewing time of $\bar{x} = 3.0$ hours with $s_x = 1.2$ hours and a mean reading score of $\bar{y} = 80$ with $s_y = 9.6$. There is a small negative association between these two variables as measured by a correlation of $r = -.2$. Since the correlation is negative, the slope of the regression line is also negative.   The slope $b = r\dfrac{s_y}{s_x} = (-.2)\left(\dfrac{9.6}{1.2}\right) = -1.6$ indicates that the reading scores drop 1.6 points for every hour of television viewing time. Therefore, based on this data, the average reading score of children who watch 5 hours of television would be $(5 - 3) \times 1.6 = 2 \times (1.6) = 3.2$ below the average of 80, or 76.8.

Figure 7.14 Regression of reading scores on television viewing time.

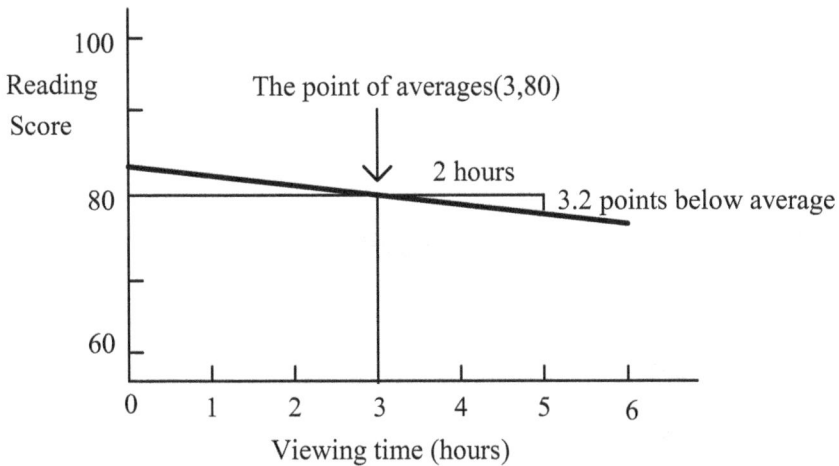

The slope-intercept form of the regression line

Although the procedure that we have described can be used to compute the average y-value for any x-value, it is tedious if a large number of averages are to be computed. A simpler method of describing the regression line is to compute the y-intercept of the line and express the line using the slope and y-intercept. The best way to understand the calculation of the y-intercept is to refer to Figure 7.15. Recall that the general form of a line is

$$y = a + bx$$

where a is the y-intercept and b is the slope. Since the slope can be computed as $b = r\dfrac{s_y}{s_x}$ and since the line must go through $(\bar{x}, \bar{y})$, we can determine the y-intercept by finding the value of the slope (a) such that $\bar{y} = a + b\bar{x}$. Using algebra to rearrange the equation we find that

$$a = \bar{y} - b\bar{x} \ .$$

Figure 7.15 Calculation of the y-intercept

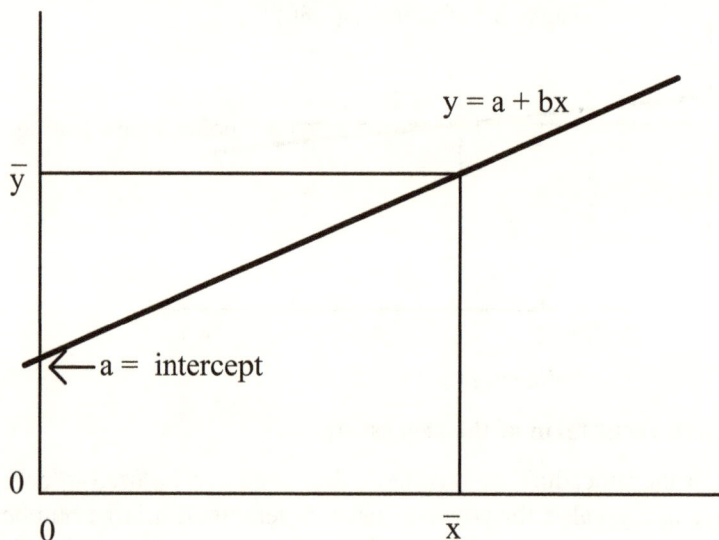

In our example concerning the height and weight of soccer players, we computed a slope of $b = r\dfrac{s_y}{s_x} = .5\left(\dfrac{20}{2}\right) = 5$. Since the regression line goes through $(\bar{x}, \bar{y}) = (67,140)$, the y-intercept is

$$a = \bar{y} - b\bar{x} = 140 - 5(67) = 14 - 335 = -195 \ .$$

Therefore, for this example the regression line is

$$\hat{y} = a + bx = -195 + 5x,$$

where $\hat{y}$ is the predicted value of the y-value for a given x-value. This formula allows us to quickly compute the average weight(y) for any height(x) by substituting the x-values into the equation for the regression line. Thus, for a height of x=70 inches we obtain an average weight of

$$\hat{y} = -195 + 5x = -195 + 5(70) = -195 + 350 = 155 ,$$

which equals the value we found earlier for this value of x. The advantage of using the equation is that it is easy to compute the average weight for any given height because $\bar{x}$ and $\bar{y}$ are not used directly in the calculation.

We can summarize the calculations needed for estimating the slope and intercept in a regression model $y = a + bx$ as follows:

1) Calculate the standardized x-values and the standardized y-values. For the ith observation these will be denoted by x'_i and y'_i.

2) Compute the correlation $r = \dfrac{1}{n-1}\sum_{i=1}^{n} x'_i \, y'_i$.

3) Compute the slope of the regression line using $b = r\dfrac{s_y}{s_x}$.

4) Compute the y-intercept using $a = \bar{y} - b\bar{x}$.

The regression line was developed using a data set consisting of soccer players that were between 60 and 72 inches in height. The regression line should fit the data between 60 and 72 inches but we have no information about what the average weight might be outside this range. This is illustrated in Figure 7.16 which shows the regression line over a range of 0-100 inches. The regression line is extended outside of the range of the data, which is indicated by the dashed lines. Predictions outside of the range of the data are called **extrapolations** and are often incorrect because the regression line may not fit the data outside of the range of the data.

Figure 7.16 The regression of weight on height.

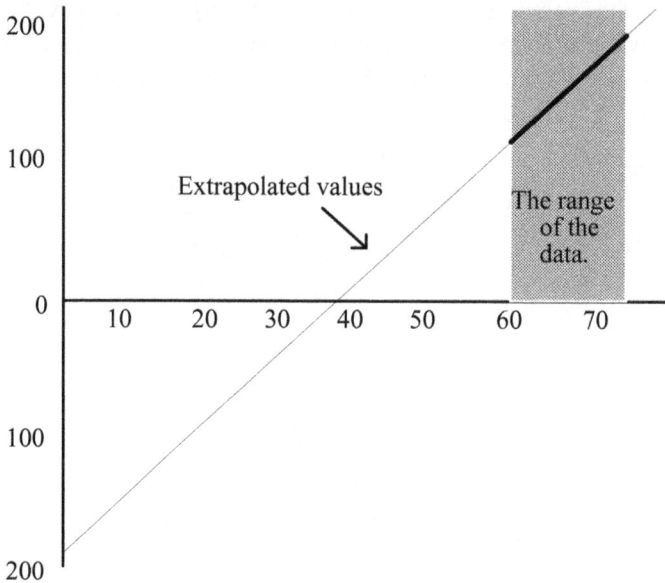

Direct calculation of the regression line

Instead of computing the correlation first, some people prefer to calculate the slope directly from the data. The formula for the slope is easy to derive because

$$b = r\frac{s_y}{s_x} = \left[\frac{1}{n-1}\sum_{i=1}^{n}x'_i\, y'_i\right]\frac{s_y}{s_x} = \left[\frac{\sum(x_i-\bar{x})(y_i-\bar{y})}{(n-1)s_x s_y}\right]\frac{s_y}{s_x}$$

$$= \frac{\sum(x_i-\bar{x})(y_i-\bar{y})}{(n-1)s_x^2}$$

Now, since the variance of the x-values is $s_x^2 = \dfrac{\sum(x_i-\bar{x})^2}{n-1}$ the denominator can be written as $(n-1)s_x^2 = \sum(x_i-\bar{x})^2$, and we can write the formula for the slope as

$$b = \frac{\sum(x_i - \bar{x})(y_i - \bar{y})}{\sum(x_i - \bar{x})^2}.$$

The intercept is then found using the usual formula as $a = \bar{y} - b\bar{x}$.

Exercise Set D

1. A health department took a random sample of n=30 men and recorded their age (X) and their systolic blood pressure(Y). The average age was $\bar{x} = 42$ with a standard deviation of $s_x = 6$. The average systolic blood pressure was $\bar{y} = 130$ with a standard deviation of $s_y = 18$. The correlation between age and blood pressure was $r = .25$.

   a) Make a rough figure indicating the point of averages.

   b) Draw an ellipse around the point of averages indicating where most of the observations should fall. Hint: Use the empirical rule which says that about 95% of the observations should fall with 2 standard deviations of the mean value.

   c) Compute the slope.

   d) Estimate the average systolic blood pressure for a 42 year old male.

   e) Estimate the average systolic blood pressure for a 48 year old male.

2. Using the information provided in the previous exercise, answer the following questions.

   a) Compute the slope and the intercept of the regression line.

   b) Using the slope-intercept form of the regression line, estimate the average systolic blood pressure for a 42 year old male.

   c) Using the slope-intercept form of the regression line, estimate the average systolic blood pressure for a 48 year old male.

3. A soil conservation technician measured the wind velocity and the soil loss in one field once a week. There was a great deal of variability in the average wind velocity and the soil loss from week-to-week. The data was recorded for 44 weeks in one field over an 8 year period. The average wind velocity was 6 miles per hour and the standard deviation was 3 miles per hour. The average soil loss was 85 pounds per acre with a standard deviation of 35 pounds per acre. The correlation was $r = .68$.

a) Make a rough figure that shows the point of averages.

b) Draw an ellipse around the point of averages indicating where most of the observations might fall. Hint: Use the empirical rule which says that about 95% of the observations should fall with 2 standard deviations of the mean value.

c) Compute the slope.

d) Estimate the average soil loss if the wind velocity was 6 miles per hour.

e) Estimate the average soil loss if the wind velocity was 3 miles per hour.

f) Estimate the average soil loss if the wind velocity was 18 miles per hour.

g) Compute the intercept for the regression line and give the regression line in the slope-intercept form.

4. A paper products manufacturer wanted to determine the growth rate of white pine trees in a certain forest. They obtained measurements on 250 white pine trees of varying ages in a forest and found that $\bar{x} = 5.6$ years with $s_x = 2.3$ years. For the heights they found $\bar{y} = 7.2$ feet with $s_y = 3.6$ feet. The correlation between the age and the height was $r = .85$.

a) Make a rough figure that shows the point of averages.

b) Draw an ellipse around the point of averages indicating where most of the observations should fall. Note the high correlation between age and height.

c) Compute the slope.

d) On average, how much does a white pine tree grow in a single year.

e) Estimate the average height for trees that are 5 years of age.

f) Estimate the average height for trees that are 8 years of age.

7.5 Fitting a Line to Points on a Scatterplot

The least squares line

Sometimes we need to fit a straight line to a set of points so that we can describe the relationship between the x-values and the y-values. For example, suppose we have measured the heights of a soybean plant once a week for five weeks. We plot these observations in Figure 7.17 and give a few lines that might be used to fit the data. Clearly, line A does not fit the data well because it greatly underestimates the height for week 6 and week 7. Line B provides a better fit to the points than Line A because the line is closer to the points. However, Line B consistently overestimates the height of the plants. A better fit can be obtained with Line C, which closely approximates the observed values for all weeks.

Figure 7.17 Fitting a line to the heights of soybean plants.

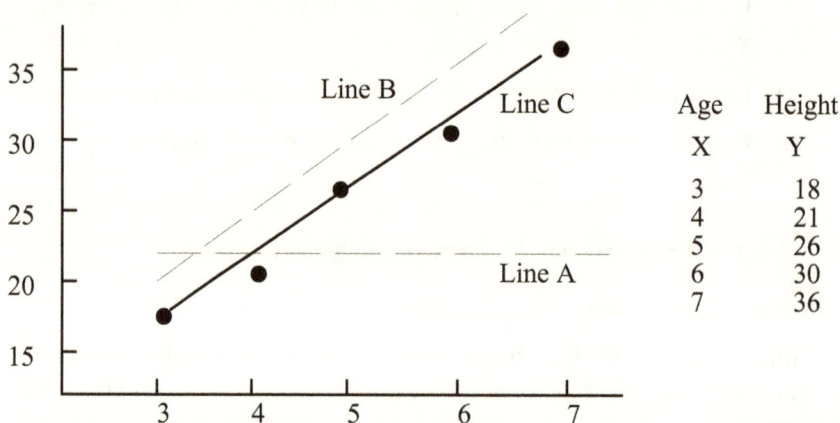

In fitting a line to the ith point (x_i, y_i), we are primarily concerned with the magnitude of the **residual**, or the **prediction error**, which is the observed value (y_i) minus the predicted value $(\hat{y}_i)$. The residual for the ith observation is

$$e_i = y_i - \hat{y}_i.$$

In general, we would like to use a line that make e_i close to zero for all observations. Note that e_i can be positive or negative, depending on whether the observed value is above or below the line. In fitting a line to a set of points we do not care if e_i is positive or negative, so we usually try to minimize e_i^2. Since we cannot minimize e_i^2 for all observations simultaneously, we try to minimize the sum of these squared residuals over all observations. That is, our **least squares line** is the line that minimizes the sum of the squared residuals

$$\sum_{i=1}^{n} e_i^2.$$

Fortunately, it can be shown that the regression line is the least squares line, which we already know how to compute. Thus, when we use a regression line we can be assured that no other line will give a smaller value for the sum of squared residuals.

The following table shows the predicted heights of soybean plants and the residuals for the data shown in Figure 7.17. For this data, the means are $\bar{x} = 5$ and $\bar{y} = 26.2$, and the standard deviations are $s_x = 1.581$ and $s_y = 7.155$. The correlation is $r = .994369$ which implies that $b = r\dfrac{s_y}{s_x} = .994369\left(\dfrac{7.155}{1.581}\right) = 4.5$. The intercept is

$$a = \bar{y} - b\bar{x} = 3.7 - 4.5(5) = 3.7$$

which gives a prediction equation of

$$\hat{y}_i = a + bx_i = 3.7 + 4.5x_i .$$

For the first observation the predicted value is $\hat{y}_i = 3.7 + 4.5(3) = 17.2$ and the residual for the first observation is $e_1 = y_1 - \hat{y}_1 = 18 - 17.2 = .8$. The other predicted values and residuals are entered in the table along with the values of e_i^2 .

Week	Height	Predicted Value	Residual	Squared Residual
x	y	$\hat{y}$	$e_i = y_i - \hat{y}_i$	e_i^2
3	18	17.2	.8	.64
4	21	21.7	-.7	.49
5	26	26.2	-.2	.04
6	30	30.7	-.7	.49
7	36	35.2	.8	.64

The sum of the squared residuals is $\sum_{i=1}^{n} e_i^2 = 2.3$. Since the regression line is the least squares line, no other line can produce a smaller value of $\sum_{i=1}^{n} e_i^2$. In this sense it produces the best fit to the points.

The Root Mean Squared Error

A line that closely fits the points will have a small value of $\sum_{i=1}^{n} e_i^2$ and a line that fits poorly will have a large value of $\sum_{i=1}^{n} e_i^2$. A measure of how closely the line fits the points is given by the **root mean square error** (s) that is calculated using

$$s = \sqrt{\frac{\sum_{i=1}^{n} e_i^2}{n-2}} = \sqrt{\frac{\sum_{i=1}^{n} (y_i - \hat{y}_i)^2}{n-2}} .$$

Note that s is used here to describe the variability of the points about the line, whereas s_x is used the describe the variability of the x-values and s_y is used to describe the variability of the y-values. Because the regression line is the least squares line it minimizes $\sum_{i=1}^{n} e_i^2$ and hence minimizes the root mean square error (s).

The coefficient of determination

Another measure of the fit of a line to a set of points is the **coefficient of determination** (R^2), which is simply the square of the ordinary correlation coefficient. That is, $R^2 = r^2$. Large values of R^2 near 1.0 indicate a good fit while values of R^2 near zero indicate a poor fit. Since it can be shown that

$$R^2 = 1 - \frac{\sum_{i=1}^{n} e_i^2}{\sum_{i=1}^{n} (y_i - \bar{y})^2},$$

and since $\sum_{i=1}^{n} (y_i - \bar{y})^2$ is a constant for any given set of points, the regression line, which minimizes $\sum_{i=1}^{n} e_i^2$, must maximize R^2. The coefficient of determination can be computed using this formula or it can be computed directly from the correlation.

In the table that gives the predicted values for the heights of the soybean plants, the sum of the squared residuals is 2.3 which gives a root mean squared error of

$$s = \sqrt{\frac{\sum e_i^2}{n-2}} = \sqrt{\frac{2.3}{5-2}} = .8756.$$

Since the regression line is the least squares line, the value of s is the smallest such value that can be obtained. The coefficient of determination is $R^2 = r^2 = (.994369)^2 = .988769$, which is the largest such value that can be obtained.

Exercise Set E

1. Consider the three regression lines. Note that the x-values and the y-values are on the same scale.

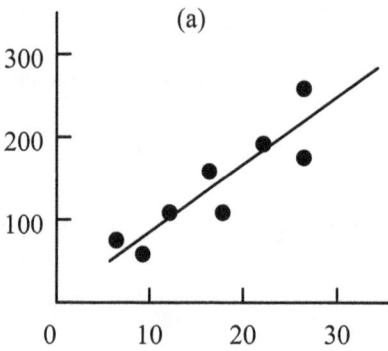

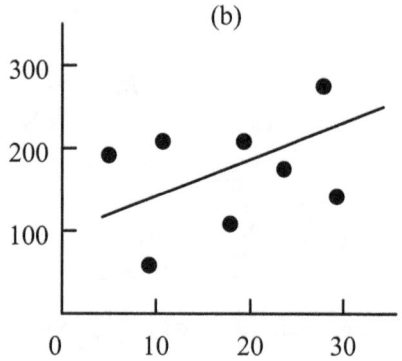

a)   Which has the smaller sum of squared residuals?

b)   Which has the smaller value of s?

c)   Which has the smaller value of R^2

2.   Consider the following regression line. Note the scale that is used in the figure and roughly estimate the residuals from the scale.

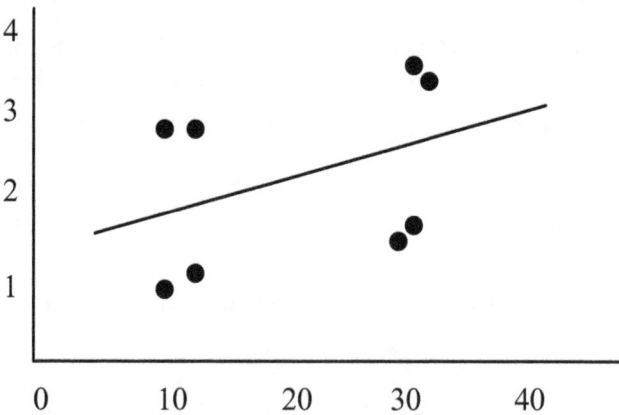

a)   Most of the residuals are

near +10 or -10   near +1 or -1

b)   The root mean squared error is approximately

.1   1   10   20

3. A consumer organization wanted to determine the relationship between the automobile engine size and gasoline consumption. They believe a linear relationship would hold for passenger cars between engine size and number of gallons per mile. Their data for 7 passenger cars were recorded as follows:

Engine Size (liters)	Gasoline Consumption (gallons per mile)
1.6	.026
3.1	.056
4.0	.071
2.2	.032
3.0	.046
3.5	.052
1.8	.029

The least squares estimates are $a = -.00421$ and $b = .017785$.

a) Compute the predicted values $\hat{y}_i$ for the seven observations and record them in a table.

b) Calculate the residuals for the seven observations.

c) Compute the root mean squared error.

d) Using the fact that $\sum_{i=1}^{7}(y_i - \bar{y})^2 = .001632$, compute R^2.

e) Suppose we used $a = -.01$ and $b = .02$. Would $\sum_{i=1}^{7}(y_i - \bar{y})^2$ change or would it equal .001632 ? Explain.

f) Suppose we used $a = -.01$ and $b = .02$. Could the value of R^2 be smaller than that obtained in (d) ? Explain. No calculations should be necessary.

4. In section 7.2, exercise 2 contained a data set that had college admission scores (x) and the number of years attended (y) for five students.

a) Compute the slope of the regression line.

b) Compute the predicted values and residuals.

c) Compute the root mean squared error.

7.6 The Regression Effect and the Regression Fallacy

The regression effect most often occurs in experiments when a measurement is taken for each subject before the treatment begins and another measurement is taken after the treatment. These designs are sometimes called test-retest designs or repeated measures designs. In these designs the first observation on the individual is the x-value and the second observation on that individual is the y-value.

It often happens that the treatment is not effective so that the average of the x-values equals the average of the y-values. In these situations we often find that, for a given x-value, the predicted y-value is closer to the mean than the x-value was. That is, a person who had an above average score on the first test would usually have a lower score on the second test. Also, an individual who had a below average score on the first test would usually achieve a higher score on the second test. Thus, the scores appear to be going back to, or regressing, to the mean value. This effect is called the **regression effect**.

The regression effect is illustrated in Figure 7.18, which is a scatterplot of the scores on equivalent forms of a college admissions test. Suppose that both tests had a mean of 50 and both had standard deviations of 10. Further, suppose that the correlation between the first and second scores was $r = .6$. The scatterplot shows the regression line that goes thru the point of averages and has a slope of

$$b = r\frac{s_y}{s_x} = .6\frac{10}{10} = .6 \ .$$

This value of the slope implies that the average y-value increases by .6 for each unit increase in the x-value. Therefore, individuals who score 70 on the first test should score around

$$50 + (70 - 50) \times .6 = 50 + 12 = 62$$

on the second test. Also, individuals who scored 30 on the first test scored around

$$50 + (30 - 50) \times .6 = 50 - 12 = 38$$

on the second test. This is an example of the regression effect because the second scores tend to be closer to the mean value on average. Note that if there were no measurement errors on these tests then the correlation would be $r = 1.0$ and there would be no regression effect.

Figure 7.18 Scatterplot and regression line for scores from equivalent forms of an admission test.

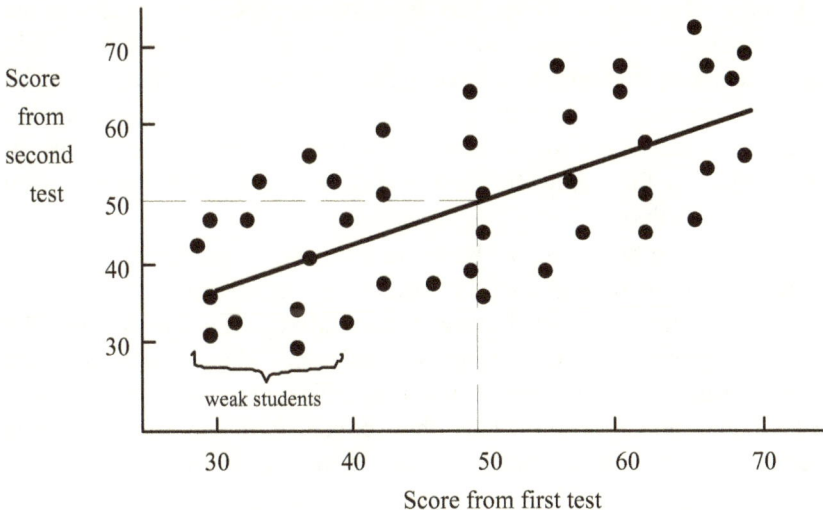

It is incorrect to believe, because the scores on the second test are closer to the average, that there would be less variability with the second test scores than there was with the first test scores. This is clearly not true in this situation because $s_x = s_y = 10$. To see this point more clearly, refer to Figure 7.18 again and note the great variability of the second scores for individuals who scored near the average of 50. These individuals contributed little to the variability of the first scores but contributed a great deal to the variability of the second scores. Therefore, while students who score far from the average tended to regress to the average on the second test, there is enough variability in the second test scores that the overall variability of the second test may equal the variability of the first test scores.

Why does the regression effect occur? To explain this difficult point, consider the students who scored around 30 on the first test. Some of those scores were from individuals who would usually score near 35, but they scored near 30 because they had a bad day. Some of these scores were from individuals who would usually score near 25, but they scored near 30 because they got lucky. These two factors will not cancel out because there are more students who normally score near 35 than there are students who normally score near 25. Therefore, for these two groups, more of the second scores will be nearer to 35 than to 25, so the average will be greater than 30.

In test-retest designs the regression effect can lead to incorrect interpretation of the results. If we had designed an experiment to determine the effectiveness of a preparatory course for the second test we might select the weak students, that have a first test score less than 40, to be given the preparatory course. See Figure 7.18 again. Because of the regression effect, these students will, on average, have better scores on the second test than they did on the first test, even if the preparatory course was worthless !

The mistaken belief that a change in test scores was due to the effectiveness of a treatment, when it could have been due solely to the regression effect, is called the **regression fallacy**. Do not confuse the regression effect with the regression fallacy. The regression effect occurs naturally in test-retest situations. The regression fallacy is a mistake that people make when they conclude that a treatment was effective when the improvement was due to the regression effect.

Exercise Set F

1. Suppose a lengthy achievement test was given to a group of n=100 eighth grade students on two occasions. On the first occasion the students scored an average of $\bar{x} = 500$ with a standard deviation of $s_x = 100$. One month later they took an equivalent form of the test and scored an average of $\bar{y} = 500$ with a standard deviation of

$s_y = 100$. Because these lengthy tests were quite reliable, they had a correlation of $r = .9$.

a) Make a rough figure showing the ellipse where most of the observations would fall.

b) Compute the average of the second test scores for students who scored 500 on the first test.

c) Compute the average of the second test scores for students who scored 700 on the first test.

d) Compute the average of the second test scores for students who scored 300 on the first test.

e) Was there a regression effect for (c) and (d) and, if so, would you describe it as a large or a small effect.

2. Suppose a short achievement test was given to a group of n=100 eighth grade students, as in the previous problem, and they obtained the same means and standard deviations as they did with the lengthy test described in the previous problem. However, because these short tests were quite unreliable, they had a correlation of $r = .3$.

a) Make a rough figure showing the ellipse where most of the observations would fall.

b) Compute the average of the second test scores for students who scored 500 on the first test.

c) Compute the average of the second test scores for students who scored 700 on the first test.

d) Compute the average of the second test scores for students who scored 300 on the first test.

e) Was there a regression effect for (c) and (d) and, if so, would you describe it as a large or a small effect.

3. An educator designed an experiment to determine if six weeks of violin practice could improve the mathematical ability of 4th grade students. From a large school district she selected 600 students who scored in the lowest 20% of those who recently took a lengthy stan-

dardized mathematics test. These 600 students were randomly assigned to either a violin practice group that practiced one hour per day or they were assigned to a control group that did not practice any musical instrument. An equivalent form of the standardized mathematics test was then given immediately after the conclusion of the experiment.

a) After 6 weeks the students in the violin group increased their performance by 5 points. Based on this data is it reasonable to assume that the improvement was solely due to the violin lessons? Explain.

b) After 6 weeks the students in the control group increased their performance by 1 point. Based on the results from both groups what conclusion would you draw?

*7.7 Inference For Correlation and Regression

In this chapter we have computed the sample correlation coefficient and have used it to estimate a regression line. Often, that is all that is necessary. Sometimes however, we need to make an inference about the correlation that exists between the x-values and the y-values in the population. To clarify this idea consider the scatterplot for a large population that is displayed in Figure 7.19(a).

Figure 7.19 Scatterplot for a large population and a sampling distribution for the correlation coefficient.

(a) population correlation= .5

(b) Sampling distribution for repeated samples of size n=50

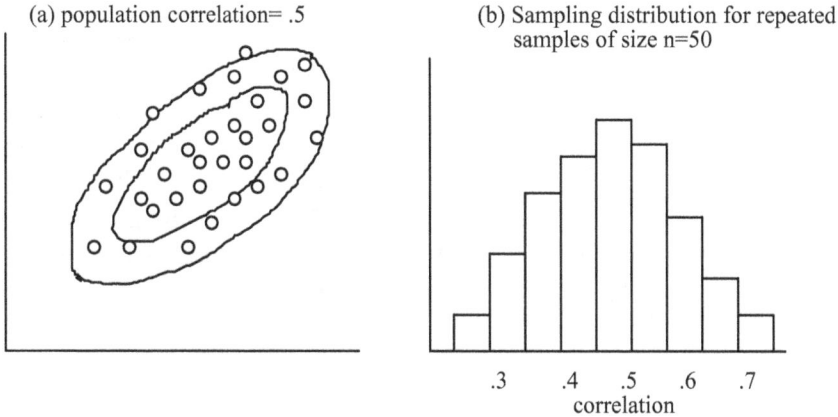

.3   .4   .5   .6   .7
correlation

Now suppose the correlation in the large population is .5. We will use the Greek letter ρ (rho) to denote the population correlation so that $\rho = .5$. If we took one sample of $n=50$ points from this population we might obtain a sample correlation of $r = .45$. If we took another sample of size $n = 50$ we might obtain $r = .53$. Consequently, if we took many samples from this population, each of size $n = 50$, we might obtain a sampling distribution like that shown in Figure 7.19 (b). It is this sampling distribution that enables us to make some inference about ρ. In the usual real-world situation we take one sample from a population and we obtain a sample correlation (r) and then we use this sample correlation to estimate the population correlation ρ or to perform a test of significance about ρ.

A test of significance for the population correlation.

In many studies where correlation is used, it is not clear that the x-values are associated with the y-values. That is, it is not clear that $\rho \neq 0$. Therefore, one of the most common statistical procedures is to test the null hypothesis that $\rho = 0$ against the alternative that $\rho \neq 0$. But to make a test of significance we need to make an assumption. To test H_0: $\rho = 0$ versus H_a: $\rho \neq 0$ we must assume that the population is **bivariate normal**. Bivariate normal population have scatterplots that are elliptical in overall appearance. The population in Figure 7.19 appears to be bivariate normal because most of the points fall in an elliptical region, with the heaviest concentration in the center of the ellipse. Bivariate normal populations have the property that both the x-values and the y-values are normally distributed and, if you pick any x-value, the distribution of the y-values for that x-value will be normally distributed.

Now consider the bivariate normal population in Figure 7.20(a) that has $\rho = 0$. Suppose we take repeated samples of size $n = 50$ from this population and compute the statistic $t = r / \sqrt{\frac{1-r^2}{n-2}}$ for each sample. It can be shown that the sampling distribution of t follows a t distribution with $n - 2$ degrees of freedom, as illustrated in Figure 7.20(b). Thus, to test H_0: $\rho = 0$ versus H_a: $\rho \neq 0$ we compute the sample correlation r and the test statistic

$$t = \frac{r}{\sqrt{\dfrac{1 - r^2}{n - 2}}} \, .$$

We then use Table 2 to estimate the p-value and reject the null hypothesis if the p-value is less than α.

Figure 7.20 Scatterplot for a large population and a sampling distribution for the correlation coefficient.

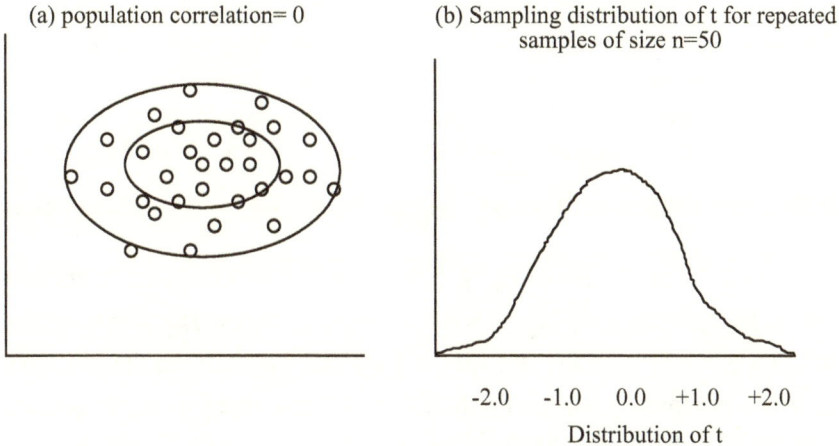

(a) population correlation= 0

(b) Sampling distribution of t for repeated samples of size n=50

-2.0   -1.0   0.0   +1.0   +2.0

Distribution of t

For example, suppose we have a sample of size $n = 27$ of men aged 35-55 and we calculated the correlation between their age and the their oxygen uptake to be $r = -.24$. To determine if the oxygen uptake is related to age we test H_o: $\rho = 0$ versus H_a: $\rho \neq 0$ by computing

$$t = \frac{r}{\sqrt{\frac{1-r^2}{n-2}}} = \frac{-.24}{\sqrt{\frac{1-(-.24)^2}{27-2}}} = \frac{-.24}{.194} = -1.236.$$

We find the two-tailed p-value from the t-distribution in Table 2 with $(27 - 2) = 25$ degrees of freedom to be $p \cong .20$. Thus, we are unable to reject the null hypothesis. We conclude that we have insufficient evidence, with this small sample, to conclude that there is an association between age and oxygen uptake.

A test of significance for the slope of the regression line

Earlier in this chapter we computed the sample slope using the formula $b = r\frac{s_y}{s_x}$. If we take a sample of n observations from a large population the sample slope b will be an estimate of the population slope β. Suppose that the population in Figure 7.19 had a standard deviation of

$\sigma_x = 4$ and $\sigma_y = 16$, and that the population correlation was $\rho = .5$. The population slope can be computed as $\beta = \rho \dfrac{\sigma_x}{\sigma_y} = .5\left(\dfrac{16}{4}\right) = 2$.

However, in a typical study we would not know the population slope β; we would only know the sample slope b for one sample.

We often need to decide, on the basis of one sample slope, if there is any association between the x-values and the y-values. Note that if there really is no association then $\rho = 0$, but since $\beta = \rho \dfrac{\sigma_x}{\sigma_y}$ it also implies that $\beta = 0$. Thus, a test of H_0: $\beta = 0$ versus H_a: $\beta \neq 0$ is equivalent to a test of H_0: $\rho = 0$ versus H_a: $\rho \neq 0$. Therefore, to test H_0: $\beta = 0$ versus H_a: $\beta \neq 0$ we can use the test for the population correlation that we have already described or we can do the following test based on a test statistic that is a function of the slope(b) and the standard error of the slope.

The test of significance for the slope uses the standard error of the slope, which we will denote by s_b. This is calculated by

$$s_b = s \Big/ \sqrt{(x_i - \bar{x})^2} \, ,$$

where s is the root-mean-square error.   To test H_0: $\beta = 0$ versus H_a: $\beta \neq 0$ we first compute the test statistic $t = b/s_b$. We then use the t distribution with $n-2$ degrees of freedom to estimate the p-value, and we reject the null hypothesis if the p-value is less than α.

The assumptions that are needed for inference with regression are somewhat different that those used in correlation. We will assume that the true population model is $y = \alpha + \beta x + \varepsilon$, where ε is the error term in the model. For the t statistic to follow the t distribution with $n-2$ degrees of freedom the following three **inference assumptions** must be met:

1) The error terms in the model have mean zero and have a constant variance for all values of the independent variable x.

2) The error terms are normally distributed for all values of x.

3) The error terms are independent. That is, they are not correlated with other error terms.

A confidence interval for the slope of the regression line

In addition to performing a test of significance it is also possible to compute a confidence interval for the population slope β. The $100(1-\alpha)\%$ confidence interval for β is

$$b \pm t_{\alpha/2}^{n-2} s_b,$$

where s_b is the standard error of the slope. Recall, in our previous example, that the correlation between age and oxygen uptake was $r = -.24$ over the $n = 27$ men who were studied. If we found that the slope of the regression of oxygen uptake on age is $b = -.072$ and that the root mean squared error $s = 1.485$, then the 95% confidence interval for β is

$$b \pm t_{\alpha/2}^{n-2} s_b = -.072 \pm 2.06(1.485) = -.072 \pm 3.059 = (-3.131, 2.987)$$

.

This confidence interval is wide because the correlation between age and oxygen uptake is not large.

Exercise Set G

1. Suppose a researcher takes a sample from a population and computes a least squares line to predict the average y-value for any given x-value.

   a) Would the researcher compute b or β ?

   b) What is the difference between b and β ?

c) Would the researcher compute s or σ ?

d) What is the difference between s and σ ?

2. A survey was conducted of 32 homes in the Cincinnati, Ohio area between 1985 and 1989. These homes were similar in construction and had similar furnaces. The objective of the survey was to determine the actual relationship between house size(in square feet) and annual heating costs (in dollars), and to develop an equation to predict the average annual heating costs for any size home. Some of the summary statistics are:

House size (square feet)	Heating costs (Dollars)
$\bar{x} = 1550$	$\bar{y} = 520$
$s_x = 310$	$s_y = 130$

$$r = .72$$

a) Compute the sample slope.

b) In a sentence, give an interpretation of the sample slope.

c) Compute the sample intercept.

d) Give the least squares line.

e) Compute a 95% confidence interval for the population slope.

3. A national car rental company designed a study to determine the effectiveness of a television advertising campaign. The researchers used 15 regions that were similar in size and varied the amount of local television advertising between the regions. In this study the x-values were the advertising expenditures and the y-values were the increased amount of sales over the previous year. The ordinary correlation was $r = .31$.

a) Does the magnitude of the correlation suggest that the advertising might increase sales?

b) Does this correlation prove that advertising will increase sales? Justify your answer with a test of significance using $\alpha = .05$.

4. A researcher wanted to determine the relationship between body size and blood pressure. Body size was measured by the body mass index (BMI), which is the weight (in kilograms) divided by the square of the height (in meters). Since age might be a confounding factor, the researcher controlled the effect of age by selecting $n=32$ men between the ages of 35 and 37 for the study. The BMI and the systolic blood pressure were measured on each subject. The correlation was calculated to be $r = .45$.

   Even with a correlation of this magnitude, some people might question the association between BMI and blood pressure. Is there sufficient evidence to reject the null hypothesis that the population correlation is zero? Perform a two-sided test of significance and roughly estimate the p-value. What conclusion would you draw if you used $\alpha = .05$?

5. The alcohol consumption (in drinks per week) and the weight (in pounds) was recorded for $n=11$ subjects. The ordinary Pearson correlation between the alcohol consumption and weight was determined to be $r = .863$.

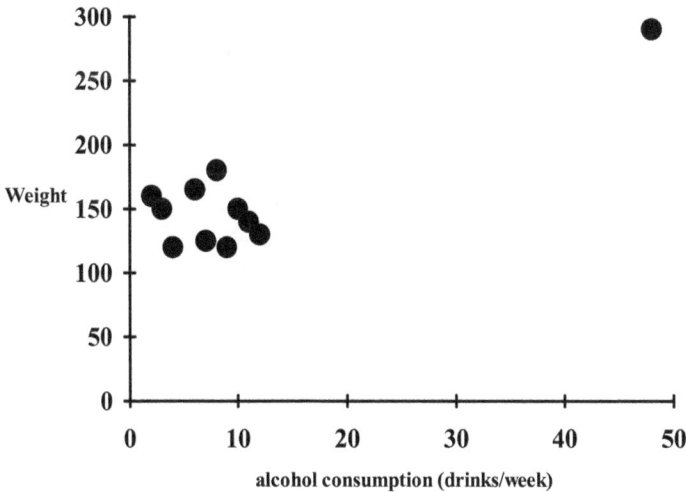

a) In your opinion is a test of $H_0: \rho=0$ appropriate for this data? Explain.

b) In a sentence or two explain the relationship that you observe between the alcohol consumption and the weight.

*7.8 Spearman's Rank Correlation

A major problem with the ordinary Pearson correlation that we computed in section 7.2 is that it is sensitive to the presence of a few outliers. We can avoid this problem by using a correlation that is based on the ranks of the observations, which was proposed by Spearman and which is called Spearman's rank correlation. Spearman's rank correlation can be calculated using the following three steps:

1) Rank the x-values. Use midranks for ties.

2) Rank the y-values. Use midranks for ties.

3) Compute the ordinary Pearson correlation between the ranks for the x-values and the ranks for the y-values to obtain the Spearman rank correlation (r_s).

An example will illustrate this procedure to compute the Spearman rank correlation (r_s) between age and systolic blood pressure for six males. The rank of x_i will be denoted by x^*_i and the rank of y_i will be denoted by y^*_i. Note that $\bar{x}^* = \bar{y}^* = 3.5$ and that $s_{x^*} = s_{y^*} = 1.871$.

Obs.	Age	Blood Pressure	Rank Age	Rank B.P.	Stand. rank x	Stand. rank y	Product
i	x_i	y_i	x^*_i	y^*_i	x'_i	y'_i	$x'_i y'_i$
1	20	125	1	1	-1.336	-1.336	1.785
2	35	140	4	4	.267	.267	.071
3	26	130	3	2	-.267	-.802	.214
4	47	175	5	6	.802	1.336	1.071
5	53	145	6	5	1.336	.802	1.071
6	22	135	2	3	-.802	-.267	.214

It is important to note that if a very large x-value or y-value is present in the data the rank correlation will only be influenced through the rank of that value. Thus, the Spearman rank correlation is not as sensitive as the Pearson correlation to the presence of a few outliers.

The Spearman rank correlation is also used for data that consists of ranks. For example, suppose we have two raters who judge the clarity of 10 computer monitors. The data is shown in the table below. Note that in this situation we do not have the information to compute the Pearson correlation. However, we can use the rank correlation to measure the agreement between the two raters. The ranks given by the first rater are the values of x^*_i and the ranks given by the second rater are the values of y^*_i.

Computer Monitor

	1	2	3	4	5	6	7	8	9	10
Rater 1	2	10	7	8	9	3	6	4	1	5
Rater 2	4	9	8	7	10	2	5	6	1	3

The Spearman rank correlation is not as popular as the ordinary Pearson correlation. The ordinary Pearson correlation is easy to interpret because the sample correlation (r) is an estimate of the population Pearson correlation (ρ), which is approximately the average of the products of the standardized variables. There is no simple interpretation of the Spearman rank correlation. The ordinary Pearson correlation is also popular because it is the basis for the most powerful test of H_0: $\rho = 0$ if the population is bivariate normal.

Exercise Set H

1. Consider the following data on alcohol consumption (in drinks per week) and weight (in pounds). The ordinary Pearson correlation is $r = .921$.

Alcohol Consumption	Weight
2	160
4	145
6	165
8	135
9	125
11	140
48	290

a) Make a scatterplot for this data.

b) Compute the Spearman rank correlation r_s.

c) Compare r to r_s and comment on the difference between the values.

d) In your opinion, which correlation best reflects the relationship between alcohol consumption and weight? Explain.

2. A consumer organization completed a study comparing brands of flour packages for bread machines. They wanted to report a measure of association between the price of the flour packages and the taste of the bread. The panel of experts that evaluated the taste of the

breads was able to rank them from the best tasting to the worst tasting, but was unable to quantify the taste by giving a taste score. What measure of association would be recommended for this data. Describe how this measure of association would be computed.

3. A high school football coach wanted to compare the final standing for the football teams in a six school league to their final standings from the previous year. The standings for the six teams for this year and last year are as follows:

School	Ranking for this year	Ranking for last year
West	1	2
Springfield	2	1
Highland	3	3
Lincoln	4	6
Central	5	4
Jefferson	6	5

To measure the association between this year's ranking and last year's ranking, should the Pearson ordinary correlation or the Spearman rank correlation be used? Should it make a difference? No calculations should be necessary.

*7.9 A Guide to Correlation and Regression

Recommendations for correlations

The ordinary Pearson correlation coefficient (r) is the most popular correlation coefficient because it is easy to compute and is easier to interpret than the rank correlations. The major problem with the ordinary correlation is that it is very sensitive to outliers. The Spearman rank correlation (r_s) is less influenced by outliers because it is based on ranks and should be considered as an alternative. The problem with r_s is that it is not easy to interpret.

Another correlation that is not too sensitive to outliers is Kendall's τ_b. This correlation, which will be discussed in the next chapter, is often used with categorical data. Kendall's t_b correlation is not as difficult to interpret as Spearman's rank correlation and should be considered as an alternative to the ordinary Pearson correlation coefficient. These three correlations (Pearson's, Spearman's, and Kendall's) are available in many large statistical packages.

Recommendations for Regression

The presence of a few outliers can also cause problems with regressions analysis. Consider Figure 7.21 which shows the least squares fit to a set of point with and without an outlier. We say that point A is an **influential observation** because the slope of the line changes dramatically when we remove that observation. On the other hand, Point B is not influential because removal of that point would not change the slope or intercept by a large amount. Generally, outliers that are far from the center of the cloud of points and that have large residuals are influential. However, not all observations with large residuals are influential. Point C has a large residual, but removal of that point would not change the slope much because it is near $\bar{x}$.

Figure 7.21 The influence of one outlier on the slope of a regression line.

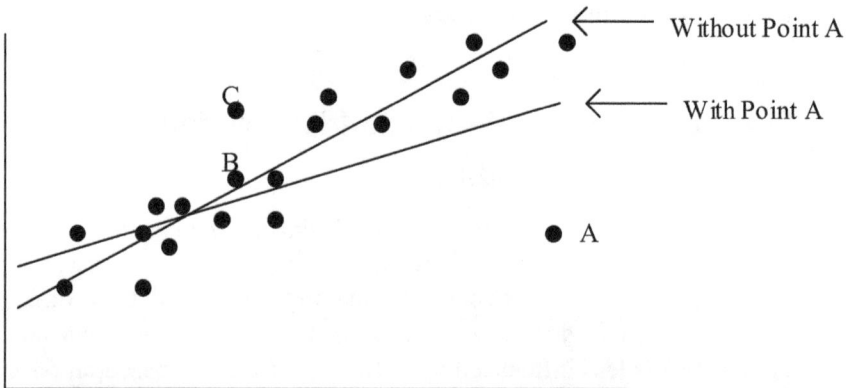

The best approach to regression is to make a scatterplot of the data before you attempt to fit a straight line to the data. You should then determine if a linear model is the appropriate model for the analysis. Does the relationship appear to be linear or does it appear to be curved? Are there a few influential observations that need to be investigated?

You should not routinely remove all influential observations but, if the influential observation was caused by an error in collecting or recording the data it should be removed. If no error was made, the influential observation may indicate that the linear model is not appropriate. In this case, the model should be revised. There are many other regression models that can fit non-linear data that are described in textbooks on regression analysis. If you feel that the influential observation should be kept in the analysis, you should consult one of the textbooks on regression analysis or seek the advice of a statistician.

Exercise Set I

1. A exercise physiologist measured the heights and weights of a sample of $n = 25$ soccer players. She needed to publish some measure of association between the height and weight of the players. She made a scatterplot of the data and concluded that it was roughly bivariate normal. In your opinion, which measure of association should be used? Point out the advantages and disadvantages of each measure of association.

2. Sometimes researchers give a test to a group of individuals twice in order to determine how reliable the test is. To measure the test-retest reliability of the test, a correlation coefficient is sometimes computed between the first score and the second score. Suppose we give the same test to ten students and record their first and second scores on the test and plot the points on the scatterplot. In your opinion should the ordinary Pearson correlation be used or should the Spearman rank correlation be used? Explain.

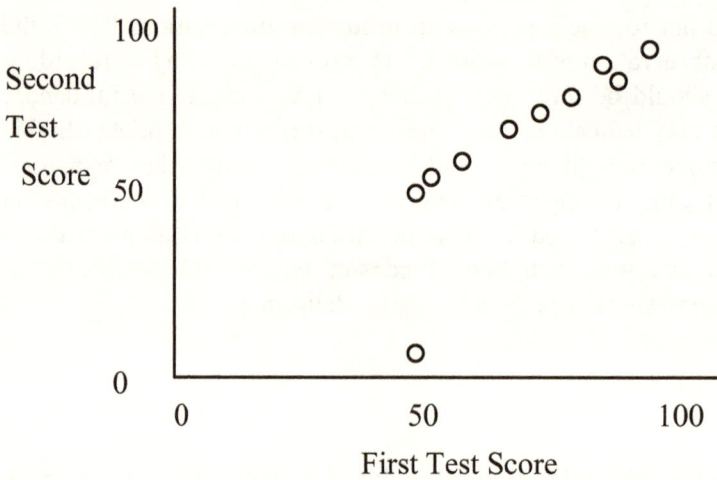

First Test Score

3. An experiment was designed to evaluate the effect of fertilizer on the amount of wheat that could be produced by a plot of land. The amount of fertilizer(X) and the crop yield (Y) were obtained for 12 plots. The points are identified by letters in this scatterplot:

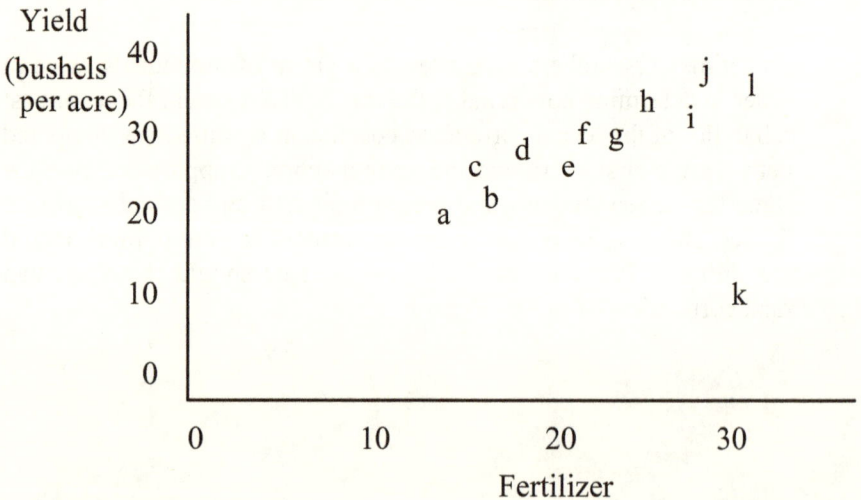

Fertilizer

a) Are there any influential points on the scatterplot. If there are, indicate why you believe they are influential.

b) If you wanted to obtain a good prediction equation what course of action would you take?

Chapter Review Exercises

1. The following data has been obtained on the size of homes and price of homes in one suburban location:

Observation	Size of Home (square feet)	Price (Dollars)
1	1200	$101,000
2	1100	$92,000
3	1400	$125,000
4	2200	$195,000
5	2400	$219,000
6	1400	$39,000
7	1800	$175,000

a) Make a scatterplot of the data.

b) If the primary objective of the analysis is to use the size to predict the price, would you use correlation or regression? Explain.

c) Are there any outliers in this data set? If you were to analyze this data what procedure would you follow?

2. Match the correlations with the scatterplot.

Correlations $r = -.5$ $r = 0$ $r = +.5$ $r = +.9$

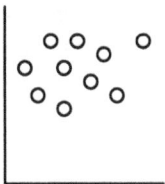

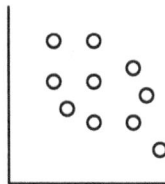

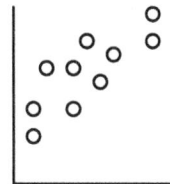

i ii iii iv

3. An educator wanted to determine the relationship between the time that eighth grade children spent watching television and their grade point average (GPA). The data are as follows:

Student	Time (X)	GPA (Y)
	(hours per week)	
1	32	1.5
2	2	3.5
3	0	3.8
4	17	3.0
5	12	2.7
6	25	2.0

The sample averages are $\bar{x} = 14.67$ and $\bar{y} = 2.750$.

The sample standard deviations are $s_x = 12.61$ and $s_y = .878$.

a) Standardize the x-values.

b) Standardize the y-values.

c) Compute the sample ordinary Pearson correlation coefficient.

d) If the educator had recorded the time in minutes, instead of hours, so that each time would be 60 times what is listed above, what would the correlation be?

4. Suppose that the educator in the previous exercise preferred to quote the Spearman rank correlation instead of the ordinary Pearson correlation. For this exercise let X=ranks of the viewing times and let Y=ranks of the GPA.

a) Compute the ranks of both variables.

b) After making the rank transformation, will $\bar{x} = \bar{y}$? Why or why not?

c) After making the rank transformation, will $s_x = s_y$? Why or why not?

d) Standardize the x-values and the y-values.

e) Compute the rank correlation and compare it to the ordinary correlation.

5. For a 10 day period the students in a class recorded the minimum temperature for the day (x-value) and the maximum temperature for the day (y-value) in degrees Fahrenheit. One student recorded the data in Fahrenheit while another student recorded the same data in degrees centigrade on the Celsius scale. The scatterplots using the two scales are as follows:

Fahrenheit scale

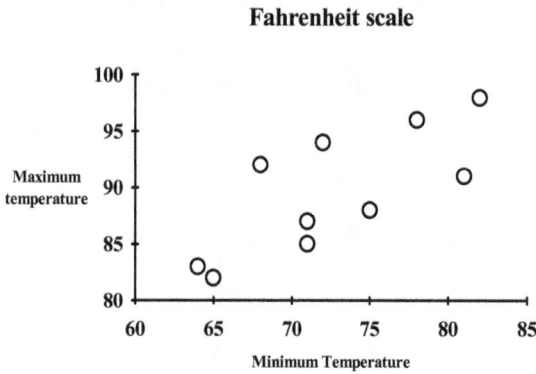

Celsius scale

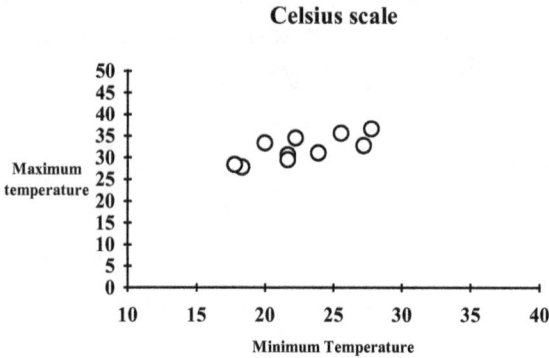

After seeing the scatterplots, a third student declared that the student who used the Celsius scale obtained a greater association. Is this true? Explain.

6. Fred was an exercise scientist who was interested in the relationship between the amount of arm exercise done by the middle aged males and their arm strength. He wanted to be able to predict the average arms strength for middle aged males who had performed a certain amount of arm exercises. He collected data on 47 subjects using

x_i=arms strength of the ith subject,

y_i=exercise duration of the ith subject.

a) Considering his objective, should he use correlation or regression? Explain.

b) He appears to be using an incorrect approach to the analysis of this data. How should the approach be modified before proceeding?

7. Choose the phrase that best describes the correlation.

a) A correlation of $r = .9$ indicates there was _____.

> A strong positive association.
>
> A weak positive association.
>
> A weak negative association.
>
> An error in the calculations.

b) A correlation of $r = -.5$ indicates that there was _____.

c) A correlation of $r = 1.1$ indicates that there was _____.

8. A civil engineer studied the relationship between temperature (X) and drying time (Y) for a certain type of concrete. He measured the temperature on the Fahrenheit scale and the time in hours. He obtained a correlation of $r = -.6$ with $\bar{x} = 65$ degrees and $s_x = 4$ degrees for the temperature and obtained $\bar{y} = 71$ hours and $s_y = 8$ hours for the drying time. Now suppose the calculations were repeated using the drying time in minutes, so that all previous y-values would be multiplied by 60.

a) Would the correlation change? Explain.

b) Would s_x change? If so, what would it change to?

c) Would s_y change? If so, what would it change to?

    d) Would b change? If so, what would it change to? Recall that
$$b = r\frac{s_y}{s_x}.$$

9. A researcher believed that apple trees that have many fruit have better quality fruit because the insects can only damage a certain number of fruit in an area. Consider the following data on the crop yield and the percentage of fruits that are wormy from six apple trees

Tree	Crop Yield	Percentage of Wormy Fruit
1	2200	48
2	1300	55
3	2800	37
4	2400	44
5	1800	8
6	1900	51

    a) Make a scatterplot. Identify any outliers.

    b) If we remove observation 5 from the data we have $\bar{x} = 2120$, $s_x = 563$ for the crop yield data and $\bar{y} = 47$, $s_y = 6.89$ for the wormy fruit data. Use these summary statistics for the remaining five trees to compute a correlation coefficient.

    c) Does the sample correlation describe the relationship that you observed in the scatterplot?

10. A music therapist noted that, among third graders who were enrolled in violin lessons, the correlation between their actual practice time and their grade point average was $r = .65$. This large correlation was cited as proof that playing the violin improves student performance. Is this conclusion justified by this data alone? Explain.

11. A music therapist wanted to determine if violin lessons could improve the performance on a language test. She obtained $n = 50$

children who obtained the consent of their parents to volunteer for the study. The children were randomly assigned, by a roll of a die, to one of six violin practice time groups having 10 minutes per day, 20 minutes per day, ..., 60 minutes per day. A language test was administered before the experiment and immediately after the one-month trial ended. The parents were instructed to have the children comply with the practice time that was assigned.

The researcher found a correlation of $r = .4$ between the practice time (X) and the improvement in language score (Y). The music therapist claimed that the violin practice caused the improvement in language scores.

a) Is the temporal relationship between the variable correct? Explain.

b) Could there be a common cause of the practice time and the improvement in language scores? If so, what is it?

c) Explain fully the advantage of using an experiment for doing research on this topic.

12. A physician took a random sample of $n = 7$ emphysema patients and recorded the number of years that the patients smoked (X) and their lung capacity (Y). The data is given in the table below. The mean values are $\bar{x} = 26.571$ and $\bar{y} = 53.571$ and the standard deviations are $s_x = 9.780$ and $s_y = 14.351$. The sample correlation is $r = -.841$.

Observations	Duration of Smoking (Years)	Lung Capacity (Percent)
1	15	75
2	25	40
3	22	65
4	38	45
5	42	35
6	20	60
7	24	55

a) Make a scatterplot of the data.

b) Compute the slope of the regression line.

c) Compute the estimated lung capacity for each observation and the residual for each observation.

13. An automobile manufacturer was attempting to improve the cold weather performance of their engines. They measured the air temperature (X) and the mileage (Y) of a vehicle operating at several temperatures and made a scatterplot of their data as follows:

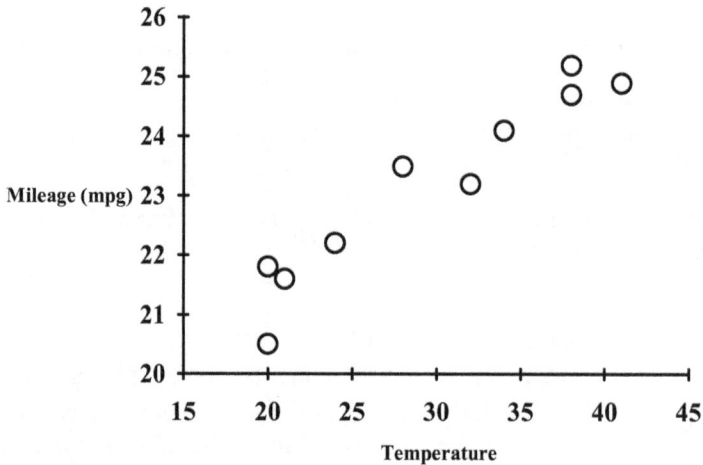

The means are $\bar{x} = 29.6$ and $\bar{y} = 23.17$ and the standard deviations are $s_x = 8.086$ and $s_y = 1.593$. The correlation is $r = .957$.

a) Compute the slope and intercept of the least squares line.

b) On average, what increase in mileage do you expect for an increase of one degree?

c) On average, what increase in mileage do you expect for an increase of ten degrees?

14. A real estate appraiser obtained the size and price of five homes that had sold recently on one location. The size is measured in square feet. The average size of the homes is $\bar{x} = 1660$ and the standard deviation is $s_x = 598$. The average price of the homes is $\bar{y} = 148400$ and the standard deviation is $s_y = 55864$. The correlation is $r = .9856$.

Observation	Size of Home	Price
	(square feet)	(Dollars)
1	1200	$101,000
2	1100	$99,000
3	1400	$125,000
4	2200	$212,000
5	2400	$205,000

a) Compute the slope and intercept.

b) Calculate the predicted price of these five homes based on the size of the homes.

c) Calculate the residuals. Check to be certain that the residuals sum to zero.

d) Calculate the root mean squared error.

e) Can the root mean squared error that was calculated in (d) be reduced by careful selection of a different slope and intercept.

15. A country was divided into crime reporting districts. Within each district the crime rate (number of serious crimes per 100000 population) and the police presence (number of police officers per 100000 population) was recorded. From the $n = 62$ crime reporting districts the correlation between the crime rate and the police presence was $r = -.22$. You may assume that these two variables follow a bivariate normal distribution. Using $\alpha = .05$ perform a one-tailed test of $H_o : \rho = 0$ versus $H_a : \rho < 0$.

Chapter 8. Categorical Data

8.1 Introduction

Chapters 4 through 7 concerned the analysis of continuous data. In this chapter we will turn our attention to categorical variables. You may recall, from Chapter 3, that categorical random variables usually have only a small number of possible outcomes. For example, if we let X be the random variable that equals the number of dots shown on the die, then there can be only six possible outcomes. Consequently, we say that X is a categorical, or discrete, random variable. In contrast, continuous random variables have an infinite number of possible outcomes.

In this chapter we will deal exclusively with categorical random variables. These categorical variables are commonly found in political science research, social science research, psychological research, and in many other areas. The most common type of categorical variable is the **binary random variable**, which is a random variable that takes on only one of two values. For example, if we flip a coin the outcome is either a head or a tail. These binary random variables were introduced in section 4.3, but we considered only a large sample confidence interval on the population proportion. In this chapter we will consider binary random variables in greater detail.

We will also consider categorical variables that have three or more outcomes, which can either be ordered or unordered. **Ordered categorical variables** have a natural ordering to the possible outcomes. For example, if we take a sample of students from a high school, each student can be classified into one of four grades (freshman, sophomore, junior, or senior) and the grades have a natural ordering from freshman to senior. Many categorical variables are ordered.

Unordered categorical variables do not have a natural ordering to the possible outcomes. An example of an unordered categorical variable is marital status (single, married, divorced, widowed). Most people would agree that being married is not superior to being single and that being single is not superior to being divorced. Although some people might believe there is some order to these categories, most people would find it difficult to order these four categories. Unordered categorical variables do not appear to be as common as ordered categorical variables, but they are common enough to be considered in this text. It is important to carefully distinguish between ordered and unordered variables because, as we shall see, ordered and unordered categorical variables require different statistical techniques.

A few variables have a large number of possible outcomes that are ordered. For example, we may examine the heights of $n = 250$ children of a certain age and obtain 15 distinct heights, when we measure height to the nearest inch. Strictly speaking, height is a categorical random variable having a large number of categories. However, we generally consider such a random variable as a continuous random variable because it takes on many distinct values and because a number (height) can be specified for each outcome that can be interpreted on an interval scale. Little is lost by considering such a variable as a continuous variable because the sampling distribution of the sample mean will have many possible values. Although there is no definite rule on this point, we will consider a variable to be categorical if the number of categories is less than 10, and will consider an interval scaled random variable to be continuous if has more than 10 categories. Because most random variable have either a few possible outcomes or have many possible outcomes, it is generally easy to tell if the variable should be considered as a continuous or a categorical variable.

Exercise Set A

1. An epidemiologist took a survey of $n = 200$ individuals and obtained the following results concerning their current smoking habits:

Heavy Smoker	Light Smoker	Non-smoker
37	25	138

   a) Should the response to the question on smoking status be considered a continuous or a categorical random variable?

   b) Do you believe that the response is an ordered or an unordered random variable?

2. A political scientist took a random sample of 600 voters in a state to survey opinions of voters concerning the job performance of their Governor. The results of the sample showed that 435 thought that the Governor was doing a good job and that 165 thought that the Governor was doing a bad job.

   a) Was this random variable categorical or continuous?

    b) Does it appear that the random variable is a binary random variable?

3. A wine connoisseur took a sample of $n = 35$ bottles of red wine and measured the alcohol content in each bottle. He was able to measure the alcohol content accurately to the tenth of a percent. He obtained 24 distinct values of the alcohol content ranging from 9.7% to 13.2%. Do you believe that you should analyze this data as categorical data or as continuous data?

8.2 Binomial Random Variables

The binomial experiment

    The most common categorical random variable is the **binomial random variable**. This variable is used for many types of experiments and surveys that have binary outcomes. These binary outcomes are traditionally referred to as successes (S) or as failures (F), but they can be any binary outcome. For example, a question on a survey may require a "yes" or a "no" response. The binomial random variable is the number of successes over the n trials of the experiment or it can be the number of "yes" responses over a sample of n individuals in a sample.

    Formally, a binomial random variable is the total number of successes in a **binomial experiment** which has the following characteristics:

    1) The experiment consists of n trials, and each trial can result in a success (S) or a failure (F).

    2) The trials are independent. This means that a success in one trial does not change the chance of getting a success on the next trial.

    3) The trials are identical so that the probability of success (p) does not change from trial to trial.

    To illustrate these points consider a binomial experiment consisting of $n = 5$ rolls of a die where the "success" on each roll is the event of rolling a "3". The binomial random variable (X) is the number of "3"s rolled with 5 rolls of a die. This is a binomial experiment because:

1) The experiment consists of $n = 5$ trials and a trial is a success if it is a "3".

2) The rolls are independent.

3) The probability of a success is $p = \frac{1}{6}$ for all rolls.

Examples of binomial experiments occur frequently in political polling. If we take a sample of $n = 600$ individuals, the number of "yes" responses may approximate a binomial random variable. In these surveys each individual is an independent trial. If the population is large compared to the sample the probability of a "yes" will be nearly constant for all samples so the total number of "yes" responses will be a binomial random variable.

Using a formula to compute binomial probabilities

If we have a binomial experiment consisting of n independent trials, with each trial having a probability of a success of p, then the chance of obtaining exactly x successes is

$$p(x) = \frac{n!}{x!(n-x)!} p^x (1-p)^{n-x},$$

where $n! = n(n-1)(n-2)\cdots(3)(2)(1)$. Now suppose we roll a die $n = 5$ times and want to compute the chance of obtaining exactly 2 "3"s. Since the chance of getting a "3" on a single roll is $p = \frac{1}{6}$, we substitute $n = 5$, $p = \frac{1}{6}$, and $x = 2$ into the formula to obtain

$$p(2) = \frac{5!}{2!(5-2)!}\left(\frac{1}{6}\right)^2\left(1-\frac{1}{6}\right)^{5-2} = \frac{5\times4\times3\times2\times1}{(2\times1)(3\times2\times1)}\left(\frac{1}{6}\right)^2\left(\frac{5}{6}\right)^3$$

$$= 10\left(\frac{1}{36}\right)\left(\frac{125}{216}\right) = .161$$

This means that we have a 16.1% chance of getting exactly 2 "3"s in 5 rolls.

If we wanted to compute the chance that we would obtain no more than 2 "3"s on five rolls of a die, we could write this as $P(X \le 2)$, where

X = number of "3"s that appear. Since the outcomes are mutually exclusive, we can write

$$P(X \le 2) = P(X = 0) + P(X = 1) + P(X = 2) = p(0) + p(1) + p(2).$$

We have already computed $p(2) = .161$. To compute $p(0)$ recall that $0! = 1$ and that $\left(\frac{1}{6}\right)^0 = 1$. Thus,

$$p(0) = \frac{5!}{0!(5-0)!}\left(\frac{1}{6}\right)^0\left(1-\frac{1}{6}\right)^{5-0} = \frac{5!}{5!}(1)\left(\frac{5}{6}\right)^5 = .402$$

and

$$p(1) = \frac{5!}{1!(5-1)!}\left(\frac{1}{6}\right)^1\left(1-\frac{1}{6}\right)^{5-1} = \frac{5!}{1!5!}\left(\frac{1}{6}\right)^1\left(\frac{5}{6}\right)^4 = .402$$

Thus, the chance of obtaining 2 or fewer "3"2 is

$$P(X \le 2) = p(0) + p(1) + p(2) = .402 + .402 + .161 = .965$$

Now suppose we wanted to calculate the chance of getting more than 2 "3"s in five rolls of a die. This is easy to calculate because

$$P(X > 2) = P(X \ge 3) = 1 - P(X \le 2) = 1 - .965 = .035$$

Using a Table of Binomial Probabilities

Although the formula for the binomial probability distribution can be used to calculate probabilities for any values of n and x, it can become tedious to calculate for large values of n. Fortunately, some spreadsheet software program and statistical programs have the binomial distribution function. If n and x are large, as they often are in political polling, the normal distribution can be used as a large sample approximation. For smaller values of n, some people prefer to use a table of the binomial distribution.

Table 5 in the Appendix has the binomial probabilities $p(x)$ tabulated for $n = 5, 10$, and 15 for $p = .1, .2, ...9$. More extensive tables are available, but are rarely needed because statistical software programs and spreadsheet software can also be used to calculate these probabilities.

To calculate binomial probabilities using these tables we need to determine the probabilities that are required, and then need to look up the appropriate probabilities in the table. For example, suppose we flip a fair coin 10 times and want to obtain the chance of obtaining 8 or more heads. For this example, $n = 10$ and $p = .5$. To compute this chance we express it as the sum of individuals probabilities that can be found in the table. The table entries give

$$P(X \geq 8) = p(8) + p(9) + p(10) = .044 + .010 + .001 = .055$$

The major disadvantage of the table is that, no matter how large a table we construct, real-world problems tend to fall outside the range of the table. For example, suppose we wanted to use the table to calculate probabilities for a random variable that is the number of "3"s rolled in 5 rolls of a die. Since $p = \frac{1}{6}$ does not appear in this table it cannot be used for that purpose; the calculations would need to be done by hand or by a computer program.

A large sample approximation to the binomial

If the random variable X is a binomial random variable then the expected value of X equals np. This makes sense because on each trial the chance that X will increase by one is p, so that with n trials we should have np successes. Furthermore, by the central limit theorem, the sampling distribution should be roughly normal if the sample size is large.

An example may illustrate these important points. Suppose we took many samples, each of size $n = 100$, from a population having $p = .6$. Since the sample size is large approximately 60% of the samples will be a success so that most of these samples will have around 60 successes and 40 failures. The sampling distribution of the number of successes (X) is shown in Figure 8.1

Figure 8.1 The binomial distribution for a large sample of
$n = 100$ **with a probability of** $p = .6$.

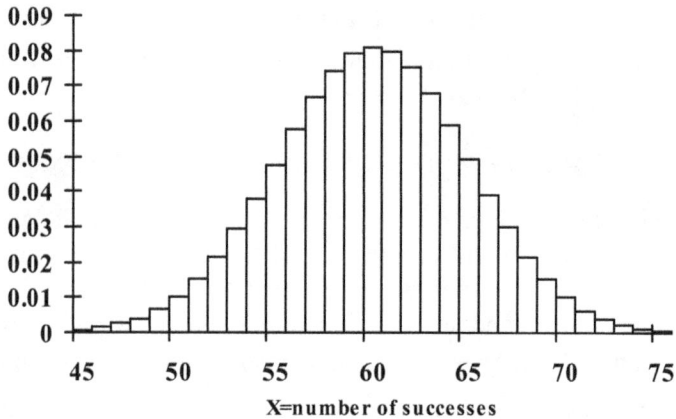

X=number of successes

Naturally, there will be some variability, from sample to sample, about the expected value of np. The variability of X can be expressed by the standard deviation of X which is $\sigma_x = \sqrt{np(1-p)}$. If $n = 100$ and $p = .6$ the standard deviation will be $\sigma_x = \sqrt{100(.4)(.4)} = 4.9$. Since for large samples the sampling distribution will be approximately normal, we can see from Figure 8.1 that approximately 68% of the repeated samples will fall within $\sigma_x = 4.9$ of the mean of 60.

This large sample approximation can be used in political polling because the sample size is large. We indicated in section 4.3 that the number of positive responses and the number of negative responses must both exceed 5 for the large sample approximation to be valid. The large sample approximation is often used to estimate probabilities of the form $P(a \le X \le b)$ by standardizing the variables by subtracting the mean np and by dividing by the standard deviation σ_x. Thus, to calculate the approximate value of $P(50 \le X \le 70)$ we standardize to obtain

$$P(50 \leq X \leq 70) = P\left(\frac{50 - np}{\sqrt{np(1-p)}} \leq \frac{X - np}{\sqrt{np(1-p)}} \leq \frac{70 - np}{\sqrt{np(1-p)}} \right)$$

$$= P\left(\frac{50 - 60}{4.9} \leq Z \leq \frac{70 - 60}{4.9} \right) = P(-2.04 \leq Z \leq 2.04)$$

$$= P(Z \geq -2.04) - P(Z \geq 2.04) = .9793 - .0207 = .9586$$

It should be remembered that this is not an exact answer; the result is an approximation based on the normal approximation.

Exercise Set B

1. A fair coin is flipped four times.

    a) What is the chance of getting exactly 2 heads?

    b) What is the chance of getting exactly 3 heads?

    c) What is the chance of getting exactly 4 heads?

    d) What is the chance of getting 2 or more heads?

2. A fair coin is flipped four times.

    a) Is the chance of getting exactly one tail equal to the chance of getting exactly one head? Explain.

    b) Is the chance of getting exactly one tail equal to the chance of getting exactly 3 heads? Explain.

    c) Use a result from the previous exercise to obtain the chance of obtaining exactly one head.

    d) Use a result from the previous exercise to obtain the chance of not getting any heads.

    e) Using the results from this exercise and the previous exercise, make a chart showing the probabilities of obtaining 0 heads, 1 head, ..., 4 heads.

3. A gambler shuffles a deck of cards and turns over the cards one at a time until he turns over 5 cards. Every time he turns over a card he

considers it to be an experiment with a red card counting as a success and a black card counting as a failure.

a) Is this a binomial experiment with $n = 5$? Why or why not?

b) What is the chance that the first card would be a red card?

c) If he turns over a red card on the first two cards what is the chance that the third card would be a red card?

4. A political scientist wondered if a small sample of $n = 10$ voters would often correctly indicate the support for a candidate. Suppose that 40% of the voters in the state favor candidate A and 60 % favor candidate B. If the political scientist took a sample of $n = 10$ voters from a population consisting of $N = 10,000,000$ voters, what is the approximate chance of selecting 6 or more voters who favor candidate A?

5. A child development specialist is planning to follow the first 15 children born in a certain hospital for 10 years. She expects that 7 or 8 of these children will be girls. For this exercise you may assume that the children are equally like to be males as females.

a) What is the chance that 7 or 8 will be girls?

b) What is the chance that there will be fewer than 7 girls?

c) What is the chance that there will be more than 8 girls?

6. Suppose that 40% of the voters in a state favor candidate A and that 60% favor candidate B. Suppose further than a political scientist takes a sample of $n = 1000$ voters.

a) What is the expected values of the number of voters who favor candidate A?

b) Compute the standard deviation of the number of voters who favor candidate A.

c) If you took repeated samples, each of size $n = 1000$, the number of voters who favor candidate A should fall between 380 and 420 for about _____% of the samples

8.3 The Chi-Squared Distribution and a Test for Goodness-of-fit.

In the last section we discussed the analysis of data from experiments that had only two possible outcomes. In this section we will investigate methods to analyze data from experiments that have more than two possible outcomes. We will denote the number of possible outcomes by r. These **multinomial experiments** have the same characteristics as the binomial experiments except that there are $r > 2$ possible outcomes.

An example of a multinomial experiment would be the experiment of rolling a die and counting the number of times that a "1" appears, the number of time at "2" appears, etc. For this experiment there are $r = 6$ possible outcomes and, if the die is fair, the probability of obtaining each outcome is $\frac{1}{6}$. Now suppose we are given a die that may have been altered in some way as to make these probabilities unequal. We would like to be able to perform some experiment to see if the die is fair or if it has been altered. If we roll the die 120 times we should obtain approximately 20 for each outcome. Intuitively, if all of the counts fall between 18 and 22 there is no reason to believe that it has been altered. However, if we obtain only 5 of a certain outcome when we expect to obtain 20, we would probably believe the die had been altered.

In order to proceed further we need some notation. Let the number of rolls that are "1" be indicated by n_1, the number of rolls that are "2" be indicated by n_2, etc. If we roll the die n times then $n_1 + n_2 + \ldots + n_6 = n$. Let the probabilities associated with each of these possible outcomes be indicated by $\{p_1, p_2, \ldots, p_6\}$. If the die is fair then $p_1 = p_2 = \ldots = p_6 = \frac{1}{6}$. We can calculated the expected number of "1"s by multiplying the number of rolls times the chance of that event occurring. Thus, the expected number of "1"s is $np_1 = \frac{1}{6}(120) = 20$, which we will denote by e_1. We can calculated the other expected counts in the same way to obtain $e_1 = e_2 = \ldots e_6 = 20$. In general, we will use n_i to indicate the actual cell count in the ith cell, p_i to indicate the chance of obtaining an outcome that would fall in the ith cell, and e_i to indicate the expected cell count in the ith cell. We will always compute the expected cell counts as $e_i = np_i$, where the p_i is determined under the null hypothesis.

Now suppose we roll a die 120 times and obtain the results that are shown in Table 8.1. Note that the actual number of "3"s and "4"s are somewhat more than we expected, as indicated by the differences $(n_i - e_i)$ shown in the fourth column. To test the null hypothesis that the die is fair we will use a test statistic that summarizes the relative magnitude of these difference. This **chi-squared test statistic** is given by

$$X^2 = \sum_{i=1}^{r} \frac{(n_i - e_i)^2}{e_i}.$$

Note that the summation is over all $r = 6$ cells in this table. The **contributions to the chi-squared test statistic** are given by $\frac{(n_i - e_i)^2}{n_i}$, which are given in the last column of Table 8.2. By summing these contributions   for   this   example   we   obtain $X^2 = 1.8 + .45 + ... + 2.45 = 13.1$.   We now need to determine if $X^2 = 13.1$ is so large that it is unlikely that it could have been due to chance.

Table 8.1 Results of an experiment consisting of 120 rolls of a die.

Outcome i	Actual n_i	Expected e_i	Difference $n_i - e_i$	Contributions $(n_i - e_i)^2/e_i$
1	14	20	-6	36/20=1.8
2	17	20	-3	9/20=.45
3	30	20	10	100/20=5.0
4	28	20	8	64/20=3.2
5	18	20	-2	4/20=0.2
6	13	20	-1	29/20=2.45

The chi-squared statistic was first proposed by Karl Pearson in 1900. He found that if the null hypothesis is true and if the expected cell counts are large then the distribution of the chi-squared statistic can be approximated by a chi-squared distribution with $v = r - 1$ degrees of freedom. That is, in this example, if the die were fair and we repeated the experiment many times, we would obtain a different X^2 value each time we rolled the die 120 times. If we made a sampling distribution from our repeated experiments we might obtain a distribution something like that shown in Figure 8.3, which is the chi-squared distribution with 5 degrees of freedom.

Figure 8.3 The sampling distribution of X^2 for repeated samples of $n = 120$ using a six-sided die.

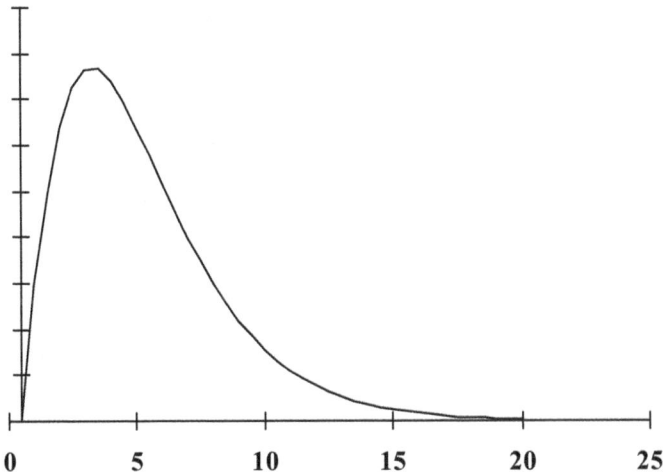

Note that this distribution does not include negative values. It cannot because the contributions to the chi-squared statistic cannot be negative. It has been shown that the exact shape of the sampling distribution depends on the number of rows in the table. That is, it is incorrect to refer to a chi-squared distribution because it is not one distribution. Rather, it is a family of distributions that have members which are specified by the number of degrees of freedom. In our example we use a chi-squared distribution of $v = 5$ degrees of freedom because there are $r = 6$ rows in this table. In general, the sampling distribution of X^2 for a table having r rows is the chi-squared distribution with $v = r - 1$ degrees of freedom, if the null hypothesis is true.   Table 7 in the appendix gives the upper-tail probabilities for the chi-squared distribution for many values of v. If we refer to the row for $v = 5$ degrees of freedom, we find that there is only a 5% chance of obtaining a chi-squared test statistic that exceeds 11.07. Since we obtained a values of $X^2 = 13.1$ the p-values must be less than 5%, and it appears from the table that $p \cong .02$. If we were performing a statistical test of the null hypothesis that the die is a fair die, we would reject the null hypothesis if we had used a significance level of $\alpha = .05$.

Note that in our example the expected cell counts were calculated based on the null hypothesis that the die was fair. The test we performed is called a **goodness-of-fit** test because we test how well the actual data fits our theory that the die is fair. The goodness-of-fit test can be used whenever we can specify the null hypothesis with enough precision that we can determine the probabilities $\{p_1, p_2, ..., p_r\}$.

The goodness-of-fit test can be used in genetic research to test a theory concerning the inheritance of certain traits. In a study involving tomato plants (Trans. Royal Can. Inst., 1931, pp. 1-19) investigators attempted to determine if the proportion of four types of plants agreed with what would be expected if Mendel's laws of inheritance applied. The null hypothesis is that the proportion of plant types is $\{\frac{9}{16}, \frac{3}{16}, \frac{3}{16}, \frac{1}{16}\}$, as indicated by Mendel's laws of inheritance. The actual counts of these plant types are listed in Table 8.2 along with the expected cell counts, which were based on the proportion of these plans that should be observed under Mendel's laws of inheritance. The alternative hypothesis is that the proportions of plant types differ in some way from that prescribed under the null hypothesis. To perform a goodness-of-fit test we first compute a chi-squared test statistic of

$$X^2 = \sum_{i=1}^{r} \frac{(n_i - e_i)^2}{e_i} = .43 + .66 + .27 + .11 = 1.47 .$$

Since there are $r = 4$ rows in the table, the number of degrees of freedom is $v = r - 1 = 4 - 1 = 3$, and it follows that, with $\alpha = .05$, the critical value is 7.81. Since $X^2 = 1.47$ does not exceed the critical value, we would not have sufficient evidence to reject the null hypothesis that Mendel's laws of inheritance apply to this experiment.

Table 8.2 Results of a genetics experiment on tomato plants.

Outcome i	Actual Count n_i	Expected Count e_i	Contributions $(n_i - e_i)^2 / e_i$
tall, cut leaf	926	906.2	.43
tall, potato leaf	288	302.1	.66
Dwarf, cut-leaf	293	302.1	.27
Dwarf, potato-leaf	104	100.7	.11

Exercise Set C

1. Suppose we have a coin that may not be fair. As a check on the fairness of the coin we flip the coin $n = 100$ times and obtain 55 heads and 45 tails.

   a) State the null hypothesis precisely by specifying the proportion of heads.

   b) Make a table showing the actual counts and the expected counts.

   c) Perform a goodness-of-fit test to determine if there is sufficient evidence to conclude that the coin is not fair. Use $\alpha = .05$.

2. Suppose that, of the people who are eligible for jury duty, 70% are over 40 years of age, 20% are between 30 and 40 years of age, and 10% are younger than 30 years of age. Suppose that a pool of 200 jurors was selected from this population and that a political scientist counted the number of jurors in the pool who fell into the three categories and found that 147 were over 40 years of age, 35 were between 30 and 40, and 18 were below 30. The political scientist, who suspected some age discrimination in the selection process, then used a goodness-of-fit test on this data.

   a) State the null hypothesis.

   b)   State the alternative hypothesis.

   c)   Make a table of the actual and expected cell counts.

   d)   Perform a goodness-of-fit test using $\alpha = .05$.

   e)   State your conclusion.

3.   The relationship between season and criminal behavior was investigated by D. Cheatwood in the paper "Is there a season for Homicide?" (Criminology (1988), vol. 26, pp. 287-306). The researcher classified the homicides by season and found, in Baltimore between 1974 and 1984, that 465 of the 1894 homicides occurred in the Winter, 453 occurred in the Spring, 496 occurred in the Summer, and 480 occurred in the Fall. Perform a Goodness-of-Fit test with $\alpha = .05$ by assuming that if homicides were unrelated to season then approximately $\frac{1}{4}$ of the total number of homicides would occur in each season.

4.   After playing a government lottery for several years, a player suspected that numbers in the range of 31 to 40 are selected as the first selection far more often than they should occur. He investigated this by observing the next 100 lotteries and observing the first number selected in each of those lotteries. In each of these lotteries six numbered balls were selected from 50 numbered balls.

Range of numbers	Actual counts
1-10	18
11-20	21
21-30	22
31-40	23
41-50	16

   a)   State the null hypothesis.

   b)   State the alternative hypothesis.

c) Make a table with the expected cell counts and the contributions to the chi-squared values.

d) Perform a goodness-of-fit test using $\alpha = .1$.

e) State your conclusion.

8.4 The Chi-squared Test for Independence for Two-dimensional Tables

In the previous section we analyzed one categorical variable to determine if the data were consistent with the theory that specified the probabilities, and hence the expected cell counts, in each cell of a one-dimensional table. In this section we will analyze two categorical variables to determine if they are independent. The analysis usually begins with an display of the count data in a two-dimensional table having r rows and c columns. These two-dimensional tables are sometimes called **contingency tables** or **cross-classification tables**. Within each cell we put the total number of observations that fall in that row and column category. For example, suppose a university has four colleges (Liberal Arts, Business, Education, and Engineering) and a researcher is interested in the proportion of males and females who are enrolled in these four programs. Table 8.3 gives the cell counts for this two-dimensional table having $r = 2$ rows and $c = 4$ columns.

Table 8.3 A two-dimensional table of gender and college.

	Liberal Arts	Business	Education	Engineering	Total	Percent
Females	318	78	167	18	581	52.4368
Males	252	120	82	73	527	47.5632
Total	570	198	249	91	1108	

The upper left cell entry indicates that 318 females were enrolled in the college of liberal arts. The entry in the second row under "Education" indicates that 82 males were enrolled in Education. Note that the row and column categories are not ordered. That is, we cannot assign a natural ordering to either the gender or the college. The test for independence that will be described in this section is suitable only for those table having unordered rows and columns.

In Table 8.3 we have also indicated the row and column totals. The total number of females is 581 which is 52.4368% of the 1108 students. The null hypothesis of independence states that the percentage of female students should be equal in the four colleges. Of course, in each college there may be a small departure from the overall proportion of 52.4368%, but if the null hypothesis is true and the number of students in the college is large the departure should be relatively small.

To perform a statistical test of the null hypothesis of independence, we need to calculate the number of males and females that we would expect to find in each college if the row and column classifications were independent. In the college of liberal arts there were 570 students and, if gender is independent of college then 52.4368% of these 570 students should be females. Thus, we would expect .524368(570)=298.89 female students in the college of Liberal Arts. This expected cell count is placed in the appropriate cell in Table 8.4 and the other expected cell counts were computed in a similar manner and placed in the other cells of that table.

Table 8.4 Expected cell counts under the null hypothesis of independence.

	Liberal Arts	Business	Education	Engineering	Total	Percent
Females	298.89	103.82	130.57	47.72	581	52.4368
Males	271.11	94.18	118.43	43.28	527	47.5632
Total	570	198	249	91	1108	

The test statistic for the null hypothesis of independence will depend on the difference between the observed cell counts and the expected cell count. We will denote the observed count in the ith row and the jth column as n_{ij} and the expected cell count as e_{ij}. If these difference between the observed and the expected cell counts $\left(n_{ij} - e_{ij}\right)$ are large then we will reject the null hypothesis. The test statistic for the test of independence is

$$X^2 = \sum_{i=1}^{r} \sum_{j=1}^{c} \frac{\left(n_{ij} - e_{ij}\right)^2}{e_{ij}},$$

which is similar to the goodness-of-fit test statistic except that we sum the contributions to the chi-squared value over all cells in the table.

For our example we computed the contributions

$$\frac{\left(n_{ij} - e_{ij}\right)^2}{e_{ij}}$$

to the chi-squared value and entered these contribution into Table 8.5. Thus, the chi-squared test statistic is the sum of these contributions over the r rows and the c columns, which is

$$X^2 = 1.22 + 6.42 + \ldots + 20.40 = 76.36.$$

The sampling distribution for this statistic is a chi-squared distribution with $v = (r-1) \times (c-1)$ degrees of freedom, which is used to calculate a p-value. In our example, $r = 2$ and $c = 4$ so that $v = (2-1) \times (4-1) = 3$. From table 7 in the Appendix we see that $p < .001$ for $X^2 = 76.36$ with $v = 3$. Thus, for any reasonable level of α, we would reject the null hypothesis.

Table 8.5 Contributions to the chi-squared statistic.

	Liberal Arts	Business	Education	Engineering
Females	1.22	6.42	10.17	18.51
Males	1.35	7.08	11.21	20.40

It should be remembered that the chi-squared test is a large sample approximation. This approximation is reasonably accurate if at least 80% of the expected cell counts exceed 5 and all of the expected cell counts exceed 1. Most statistical software can compute the chi-squared test statistic and the p-value based on the chi-squared distribution with $v = (r-1) \times (c-1)$ degrees of freedom. Furthermore, several software programs indicate if the cell counts are large enough to use the chi-squared approximation. If the cell counts are not large enough to use the chi-squared approximation then an exact method is recommended. The exact test will be discussed in the next section for tables having 2 rows and 2 columns.

The procedure for performing a chi-squared test for independence can be summarized as follows:

- The test is suitable for unordered row and column categories. If either the rows or the columns are ordered then other methods may be more powerful.
- For the chi-squared approximation to be valid at least 80% of the cells should have expected frequencies that exceed 5 and all cells should have expected frequencies that exceed 1.
- For a table having r rows and c columns compute the chi-squared test statistic using

- $$X^2 = \sum_{i=1}^{r} \sum_{j=1}^{c} \frac{(n_{ij} - e_{ij})^2}{e_{ij}},$$

  where n_{ij} is the actual count and e_{ij} is the expected count in the ith row and jth column.
- Use a chi-squared distribution with $v = (r-1) \times (c-1)$ degrees of freedom to compute the p-value.

Sometimes researchers take r samples from r populations and display the results in a two-dimensional table having r rows, with the results from the ith sample displayed in the ith row. In this situation the objective is to use the sample results to test the null hypothesis that the responses in the populations are identical. Such a test is called a test of **homogeneity**. Although the test differs conceptually from the chi-squared test of independence, the chi-squared test statistic is used in the same manner to test homogeneity as it is used to test independence. Thus, for many practical purposes there is no need to make a distinction between a test of homogeneity and the test of independence.

Exercise Set D

1. A criminologist took a sample of 300 prisoners from a large population of prisoners in a state. The objective was to determine if there was a relationship between gender and crime. The results were tabulated in the following table:

	Crime		
	Drug	Violent	Property
Females	26	8	47
Males	117	131	185

a) Calculate the expected cell counts under the null hypothesis.

b) For which crimes do females have smaller actual counts that expected counts?

c) Perform a test of association using $\alpha = .05$.

2. Graubard and Korn (Biometrics, 1987, pp.471-476) reported data on congenital sex organ malformation and maternal alcohol consumption. The following table, which was obtained by collapsing over some of the alcohol consumption categories, shows the relationship between alcohol consumption and malformations:

	Alcohol Consumption	
Malformation	<1 drink per day	1 or more drinks per day
Absent	31530	951
Present	86	7

a) Compute the expected cell frequencies.

b) How many cells have expected frequencies below 5?

c) Based on the guidelines given in this section, is the chi-squared test appropriate?

d) If possible, perform a test of independence using $\alpha = .05$. State your conclusion.

3. Suppose a survey was performed in a community to determine the lifetime prevalence of certain psychiatric disorders. In order to compare the prevalence for males to the prevalence for females the following data was obtained on alcohol abuse:

	Alcohol Abuse	No Alcohol Abuse
Males	192	291
Females	47	505

a) Compute the chi-squared test statistic.

b) Test the null hypothesis using $\alpha = .05$.

c) Roughly estimate the p-value for this test. State your conclusion.

4. A nutritionist attempted to describe the eating behavior of middle-aged men as a function of their marital status. She obtained marital status and eating behavior, classified into three categories, for a simple random sample of $n = 272$ males between the ages of 40 and 60. The results are as follows:

	Single	Married	Divorced	Total
Eat-at-home vegetarian	5	12	2	19
Eat-at-home non-vegetarian	38	76	28	142
Eat-at-Restaurant	47	28	36	111
Total	90	116	66	272

a) Compute the expected cell counts under the null hypothesis.

b) Are any of the expected cell counts less than 5.

c) Is the chi-squared test for independence appropriate for this data set?

d) Perform the chi-squared test for independence using $\alpha = .05$ and state your conclusion.

*8.5 An Exact Test of Independence

In the last section we used a chi-squared test statistic to test the null hypothesis of independence. We noted that the test was a good approximation provided the expected cell counts were sufficiently large. But what should we do if the expected cell counts are too small for the chi-squared approximation? Fortunately, an exact test has been developed by R. A. Fisher which allows us to test the null hypothesis of independence even if the expected cell counts are small.

The exact test does require a fair amount of calculation. To simplify the calculation, some additional notation will be required. Consider Table 8.6 which has the same data that we analyzed in the last section. As before, the actual cell count in the ith row and jth column is n_{ij}. We will use $\{n_{+1}, n_{+2}, n_{+3}, n_{+4}\}$ for the column totals. The "+" in these column total subscripts indicates that the cell counts should be added over the r rows. We will use $\{n_{1+}, n_{2+}\}$ for the row totals. The "+" in these row total subscripts indicates that the cell counts should be added over the c columns. For example, $n_{2+} = n_{21} + n_{22} + n_{23} + n_{24}$. The sum of all cells in the table is n, which equals the number of observations.

Table 8.6 The notation for a two-dimensional table of gender and college.

	Liberal Arts	Business	Education	Engineering	Total
Females	$n_{11} = 318$	$n_{12} = 78$	$n_{13} = 167$	$n_{14} = 18$	$n_{1+} = 581$
Males	$n_{21} = 252$	$n_{22} = 120$	$n_{23} = 82$	$n_{24} = 73$	$n_{+2} = 527$
Total	$n_{+1} = 570$	$n_{+2} = 198$	$n_{+3} = 249$	$n_{+4} = 91$	$n = 1108$

Although it is possible to perform exact tests for large tables, our discussion of exact tests will concentrate on tables having 2 rows and 2 columns, which is the most common table size. The cell counts and row and column totals are shown in Table 8.7 .

Table 8.7 Notation for exact tests.

	Column 1	Column 2	Total
Row 1	n_{11}	n_{12}	n_{1+}
Row 2	n_{21}	n_{22}	n_{2+}
Total	n_{+1}	n_{+2}	n

If the row and column totals equal the values in the observed table and if the row and column categories are independent, it can be shown that the chance of obtaining this table is

$$f(n_{11}, n_{12}, n_{21}, n_{22}) = \frac{n_{1+}! \, n_{2+}! \, n_{+1}! \, n_{+2}!}{n! \, n_{11}! \, n_{12}! \, n_{21}! \, n_{22}!}$$

The exact test is best explained with an example. Suppose we have randomly assigned 40 depressed subjects to either a standard drug therapy or to an experimental drug therapy group. After 4 weeks the subjects were interviewed to determine the proportion that have an adverse event (a side effect) from taking the drug. One adverse event of interest was insomnia, which was tabulated in Table 8.8.

Table 8.8 A two-dimensional table of drug treatment and insomnia.

	Insomnia	No Insomnia	Total
Experimental Drug	4	16	20
Standard Drug	1	19	20
Total	5	35	40

Note that the expected cell count for the two cells in the first column are $e_{11} = e_{21} = \left(\frac{20}{20}\right)5 = 2.5$. Since half of the cells have expected counts less than 5 the chi-squared approximation may not be accurate, which implies that an exact test is necessary. We begin the exact test by computing the chance of observing the table shown in Table 8.8 for the row and column totals specified in that table, assuming that insomnia is independent of drug. The chance is

$$f(4,16,1,19) = \frac{20!\,20!\,5!\,35!}{40!\,4!\,16!\,1!\,19!}$$

$$= \frac{\left(2.433 \times 10^{18}\right)\left(2.433 \times 10^{18}\right)(120)\left(1.033 \times 10^{40}\right)}{\left(8.159 \times 10^{47}\right)(24)\left(2.092 \times 10^{13}\right)(1)\left(1.216 \times 10^{17}\right)}$$

$$= .147$$

This is the chance of observing 4 of the 5 reports of insomnia in the experimental treatment group. It is possible that we would have observed the following table, which has the same row and column totals as the table we observed.

Table 8.9 Another two-dimensional table of drug treatment and insomnia having the same row and column totals as Table 8.8.

	Insomnia	No Insomnia	Total
Experimental Drug	5	15	20
Standard Drug	0	20	20
Total	5	35	40

The chance of observing this table, under the null hypothesis of independence, is

$$f(5,15,0,20) = \frac{20!\,20!\,5!\,35!}{40!\,5!\,15!\,0!\,20!} = 0.024$$

Thus, the chance of observing either 4 or 5 reports of insomnia is
$$P(4 \text{ or } 5 \text{ reports of insomnia}) = .147 + .024 = .171.$$

This is the chance of observing a table as extreme or more extreme than that observed if the row and column categories are independent. This is the exact p-value for the test of independence for a table having the row and column totals that we observed. If we used $\alpha = .05$ for the test we would not reject the null hypothesis. However, if we had observed the cell counts in Table 8.9 we would reject the null hypothesis.

Exercise Set E

1. A new treatment for cancer was compared to an older treatment in a clinical trial involving 38 patients. The patients were assigned at random to be given the new treatment or the old treatment. The primary outcome measure was the survival of the patient for five or more years. The results are given in the following table:

	survived < 5 years	survived > 5 years
New treatment	1	17
Old treatment	8	12

   a) Use the exact test to test the null hypothesis of independence against the alternative that the new treatment improves survival.

   b) Would you reject the null hypothesis using $\alpha = .05$?

   c) State your conclusion.

2. A forester planted 22 white pine trees in a park. Approximately 2 years after he planted the trees he wanted to determine how many were still alive. He felt that the survival of the trees was related to the slope of the ground (highly sloped or flat) and made a two-dimensional classification table of tree survival and ground slope.

	Flat ground	Sloped ground
Did not survive	2	1
Survived	13	6

a) Use the exact test to test the null hypothesis of independence against the alternative that the slope of the ground is related to survival of the tree.

b) Would you reject the null hypothesis using $\alpha = .1$?

c) Do you believe that a strong conclusion can be made with this data? If you feel that you cannot make a strong conclusion, explain why you cannot.

*8.6 A Test for Trend

The last two sections were devoted to the analysis of data in tables that had unordered categories. Now we turn our attention to the analysis of ordered categorical data. We will concentrate on the most common type of ordered categorical data classified in tables having $r = 2$ rows and c columns, where the columns are ordered categories.

Suppose we have obtained data from a sample of healthy adults aged 65-70 and from a sample of adults aged 65-70 who have a type of cancer. Suppose we obtain data on smoking history and construct a two-dimensional table, as shown in Table 8.10, to display the results. Note that this table has $c = 3$ columns that are ordered. If we ignore the ordering and perform a test of independence, as described in section 8.4, we will obtain a test statistic of $X^2 = 4.81$, which will gives a p-value of $p \cong .09$ for this $v = 2$ degree of freedom test. If we had used $\alpha = .05$ as our significance level we would fail to reject the null hypothesis of independence. However, the test of independence ignores the natural ordering of the smoking categories. Perhaps a test that uses the order of the column categories would produce a more powerful test.

Table 8.10 A table of cancer and smoking history.

	Non-smoker	Light smoker	Heavy smoker	Total
Cancer present	30	26	25	81
Cancer absent	114	62	49	225
Total	144	88	74	306

One method of using the order of the smoking history is to assign scores to the three categories. It is traditional to assign the integer scores $\{1,2,3,\ldots,c\}$ to these columns but any other reasonable set of scores could be assigned. For this example the average number of cigarettes per day could be assigned if these numbers were available. The scores for the columns will be denoted by $\{x_1, x_2, \ldots, x_c\}$. For this example we will not use the average number of cigarettes per day; we will use the integer scores $\{x_1 = 1, x_2 = 2, x_3 = 3\}$.

The test for trend was proposed by Mantel in 1963 and is related to a test proposed by Cochran in 1954 and to a test proposed by Mantel and Haenszel in 1959. Consequently, this test is sometimes called the Cochran-Mantel-Haenszel (CMH) test. The CMH test is based on a one degree of freedom chi-squared test statistic which will be denoted by X^2_{CMH}. This statistic can be written as

$$X^2_{CMH} = (n-1)r^2,$$

where r is the Pearson correlation between the column scores and any scores assigned to the $r = 2$ rows. However, this is not a convenient formula for hand calculation. We can express the CMH test statistic for tables having $r = 2$ rows and c columns in our notation as,

$$X^2_{CMH} = \frac{\left(\sum_{j=1}^{c} n_{1j} x_j - \frac{n_{1+}}{n} \sum_{j=1}^{c} n_{+j} x_j \right)^2}{\frac{n_{1+} n_{2+}}{n^2(n-1)} \left[n \sum_{j=1}^{c} n_{+j} x_j^2 - \left(\sum_{j=1}^{c} n_{+j} x_j \right)^2 \right]}.$$

For our example we would like to test the null hypothesis of no trend in the proportion of subject who have cancer over the smoking categories. The alternative hypothesis is that there is a trend in the proportion of people in the samples who have cancer over the smoking categories. For this data in Table 8.10 we compute the test statistics as

$$X^2_{CMH} = \frac{\left[30\cdot1+26\cdot2+25\cdot3-\frac{81}{306}(144\cdot1+88\cdot2+74\cdot3)\right]^2}{\frac{81\cdot225}{306^2\cdot305}\left[306(144\cdot1+88\cdot4+74\cdot9)-(144\cdot1+88\cdot2+74\cdot3)^2\right]}$$

$$= \frac{\left[157-\frac{81}{306}(542)\right]^2}{.0006382\left[306(1162)-542^2\right]} = \frac{183.04}{39.44} = 4.641$$

Since this is a one degree of freedom test we find $p \cong .03$ from the chi-squared table and we would reject the null hypothesis of no trend in proportion if we had used $\alpha = .05$.

It is important to note that if we had used $\alpha = .05$ with a chi-squared test of independence we would have obtained $p \cong .09$, which would not allow us to reject the null hypothesis. The test of trend is more powerful than the test of independence because there are fewer alternatives considered by the test of trend. The test of independence will reject the null hypothesis if the proportion of cancer subjects vary in any way from a constant proportion. The test of trend is sensitive to trends in those proportions. In this situation the test for trend is appropriate because ordered column categories imply that the non-smoking category should have the lowest proportion of cancer subjects and the heavy smoking category should have the highest proportion of cancer subjects.

Exercise Set F

1. A researcher wanted to determine the relationship, if any, between place of residence and academic performance in college. He selected 100 students at random from a large population of students. He obtained the grade in a required history course and determined if they were living near campus in a dormitory or apartment of if they were living off campus. The results were:

	Grade					
	A	B	C	D	F	Total
On Campus	8	27	9	3	7	54
Off Campus	10	19	7	5	5	46
Total	18	46	16	8	12	100

a) Are the column categories ordered?

b) Use the integer scores $\{A = 4, B = 3,..., F = 0\}$ to compute a test statistic for the CMH test.

c) Using $\alpha = .05$ perform a CMH test.

d) State your conclusion.

2. Graubard and Korn (Biometrics, 1987, pp. 471-476) reported data on congenital sex organ malformations and maternal alcohol consumption. After we combine the categories for those mothers who had fewer than one drink per day we obtain the following data:

Malformation	Drinks/Day			
	<1	1-2	3-5	>5
Absent	31530	788	126	37
Present	86	5	1	1

In this analysis we will use scores that represent the "average" drinks per day. We will use $x_1 = 0, x_2 = 1.5, x_3 = 4$, and $x_4 = 7$ as column scores. Compute the CMH test statistic and estimate the p-value. Is there a relationship between maternal alcohol consumption and malformations? Explain.

3. The relationship between hemostatic abnormalities in patients with head injuries and survival was investigated by Olson and other (Neurosurgery, 1989, pp. 825-832). There were 41 patients who had a Glascow Coma Score of 6, 7, or 8 and who had their fibrin degradation products (FDP) measured. The degree of abnormality is measured on a scale of 1 to 3.

Outcome	FDP score		
	1	2	3
Discharged	12	7	2
Dead	2	2	16

a) Use the FDP scores to compute the CMH test statistic and use it to estimate the p-value.

b) Does it appear that the FDP score is related to survival?

4. A researcher, in an effort to investigate the relationship between drivers education and subsequent injuries in motor vehicle accidents, took a sample of 2000 students in a state that requires drivers education and another sample of 3000 students in another state where drivers education is not required. For these purposes of this exercise we will assume that the drivers in these two states were equivalent in all other respects. The researcher followed the students for 5 years and recorded the outcome of the most serious accident that these drivers had during those 5 years. The results were tabulated as follows:

		Injury			
Driver's Education	None	Slight	Severe	Died	Total
Yes	1433	518	43	6	2000
No	2165	767	58	10	3000
Total	3638	1285	101	16	5000

a) As a practical matter, is the difference between death and serious injury approximately equal to the difference between slight injury and no injury? Explain.

b) Which set of scores do you believe is most appropriate for this data set?

$$\text{Set A } \{x_1 = 1, x_2 = 2, x_3 = 3, x_4 = 4\}$$

$$\text{Set B } \{x_1 = 1, x_2 = 2, x_3 = 4, x_4 = 10\}$$

c) Using the set of scores that you choose in part (b), compute the CMH test statistic.

d) Based on this data and on the scores assigned to the data, does driver's education appear to be effective?

*8.7 Odds Ratios and Measures of Association

The odds ratio

We have described several methods for performing tests with categorical data but have not yet described a way to measure the strength of the relationship between the row variable and the column variable. This section will concern some commonly used methods to estimate the strength of the association between the two variables. It should be pointed out that this topic is closely related to correlation. Indeed, if scores can be assigned to the row and column categories then the Pearson correlation coefficient can be used as a measure of association.

We will first consider the table having $r = 2$ rows and $c = 2$ columns which we will call the 2×2 table. For reference, the notation for the 2×2 table is shown in Table 8.11.

Table 8.11 Notation for 2×2 tables.

	Column 1	Column 2
Row 1	n_{11}	n_{12}
Row 2	n_{21}	n_{22}

The most common measure of association for the 2×2 table is the odds ratio which is defined as

$$\text{Odds ratio} = \frac{n_{11}n_{22}}{n_{21}n_{12}} = \frac{n_{11}/n_{12}}{n_{21}/n_{22}}.$$

This ratio is sometimes referred to as the **cross-product** ratio because it is the ratio of the products that are on the diagonals. To develop some feel for the odds ratio consider the following 2×2 table of smoking and coffee.

Table 8.12 A two-dimensional table for smoking and coffee.

	Coffee Drinkers	Not Coffee Drinkers
Smokers	87	13
Non-smokers	174	26

Note that there are twice as many non-smokers as smokers among the coffee drinkers and twice as many non-smokers as smokers among non-coffee drinkers. Thus, there is no relationship between coffee and smoking in this hypothetical data set; there are simply twice as many non-smokers as there are smokers. The odds ratio linking smoking to coffee drinking is

$$\text{Odds Ratio} = \frac{87 \cdot 26}{174 \cdot 13} = 1.0$$

The odds ratio of 1.0 also indicates that there is no association between smoking and coffee consumption.

If smokers tended to drink coffee, the table might look something like Table 8.13, which will have an odds ratio of $\text{Odds ratio} = \dfrac{91 \cdot 40}{160 \cdot 9} = 2.53$.

Table 8.13 A table showing an association between smoking and coffee.

	Coffee Drinkers	Not Coffee Drinkers
Smokers	91	9
Non-smokers	160	40

The usual procedure is to perform of test of significance of the null hypothesis of independence, which is equivalent to a test of the null hypothesis that the odds ratio is equal to one. If we reject the null hypothesis of independence we then use the odds ratio to measure the association between the variables.

Measures of Association

The odds ratio is the most popular measure of association for 2×2 tables. For larger tables having ordered row and column classifications there are several measures of association that have been proposed. The two measures that we will consider are based on concordant and discordant pairs of observations, which require some explanation.

Concordant and discordant pairs are best explained with an example. Consider Table 8.14 which is a tabulation of the frequency of eating outside the home and body weight for $n = 200$ individuals. We will assume that eating out and body weight are both ordered categories. We now consider all pairs of observations. That is, we compare the 1st observation to the 2nd, the first to the 3rd, ..., the 199th to the 200th observation. There are many such pairs. Now suppose we compare one individuals who seldom eats out and had a normal weight to another individual who often eats out and is overweight. This pair is shown in Figure 8.15(a) as a concordant pair because the second individual ate out more often and had a greater body mass.

Table 8.14 A two-dimensional table of frequency of eating outside the home to body mass for $n = 200$ individuals.

	Body Mass			
	Normal	Overweight	Obese	Total
Seldom eat out	36	14	2	52
Occasionally eat out	54	57	3	114
Often eat out	15	10	9	34
Total	105	81	14	200

Now consider a second pair of individuals. The first member of the pair is an overweight individual who occasionally eats out and the second member of the pair is an normal weight pair who often eats out. This pair, which is shown in Figure 8.15(b) is discordant because the second member of the pair, who ate out more often, had a lower body mass.

Figure 8.15 Concordant and discordant pairs.

a) A concordant pair

Body Mass

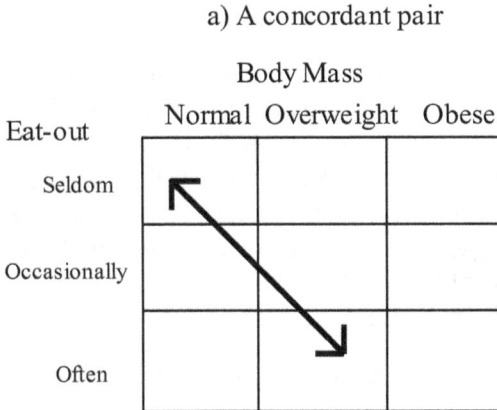

b) A discordant pair

Body Mass

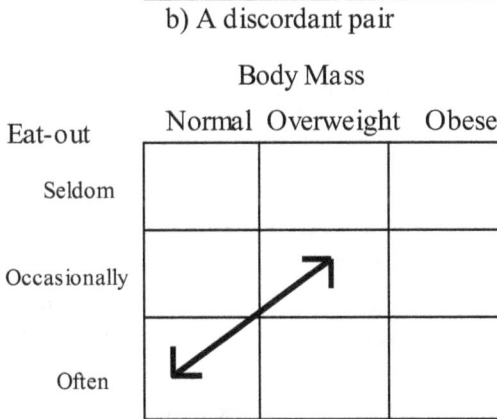

To compute the total number of concordant pairs we refer to the frequency counts in Table 8.14. The 36 normal weight individuals who seldom eat out, when paired with the 57 overweight people who occasionally eat out, result in $36 \times 57 = 2052$ concordant pairs. These pairs were concordant because the members of the pairs that had a greater body mass ate out more frequently. Those 36 normal weight individuals who seldom ate out, when paired with the 3 obese individuals who occasionally ate out, result in $36 \times 3 = 108$ concordant pairs. By continuing in this fashion throughout the table, we find the total number of concordant pairs (C) to be

$$C = 36 \cdot 57 + 36 \cdot 3 + 36 \cdot 10 + 36 \cdot 9 + 14 \cdot 3$$

$$+ 14 \cdot 9 + 54 \cdot 10 + 54 \cdot 9 + 57 \cdot 9 = 4551$$

We calculate discordant pairs in a similar fashion. The 14 overweight individuals who seldom ate out are discordant with the 54 normal weight individuals who ate out occasionally, which implies that there are $15 \cdot 54 = 810$ discordant pairs from these cells. These pairs were discordant because the members of the pair that had a greater body mass ate out less often that the other member of the pair. The 14 overweight individuals who seldom ate out are discordant with the 15 normal weight individuals who often eat out, which results in $14 \times 15 = 210$ discordant pairs. The total number of discordant pairs (D) is

$$D = 14 \cdot 54 + 14 \cdot 15 + 2 \cdot 54 + 2 \cdot 57 + 2 \cdot 15$$

$$+ 2 \cdot 10 + 57 \cdot 15 + 3 \cdot 15 + 3 \cdot 10 = 2168$$

Note that there are more concordant pairs than discordant pairs, which implies that there is some association between body mass and frequency of eating outside the home. Several measures of association have been proposed for tables having ordered row and column categories that are based on the differences between the concordant pairs and the discordant pairs $(C - D) = 4551 - 2168 = 2383$. If the number of concordant pairs is approximately equal to the number of discordant pairs, which implies that $C - D$ is close to zero, then there is little association between the row and column classifications.

We will now define the **gamma statistic**, which is one of two measures of association that we will describe in this section. The gamma statistic compares the difference $C - D$ to the total number of concordant and discordant pairs, which is $C + D = 4551 + 2168 = 6719$ for this data. The gamma statistic is defined as

$$\gamma = \frac{C - D}{C + D}.$$

For the data in Table 8.14 the gamma statistic is

$$\gamma = \frac{C-D}{C+D} = \frac{2383}{6719} = .355 \, .$$

If there is no relationship between the row and column classification, the difference $C-D$ will be very small compared to $C+D$ so the gamma statistic will be close to zero. If there is a very strong relationship between the row and column classifications then the difference $C-D$ will nearly equal $C+D$ so the gamma statistic will be near +1. If there is a negative relationship between the row and column categories then most of the pairs will be discordant and it follows that $C-D$ will be the opposite sign of $C+D$ so that the gamma statistic will be near -1. Thus, the gamma statistic always falls between -1 and +1 and a value near zero indicate that there is little, if any, association between the variables.

Note that we computed concordant pairs and discordant pairs for pairs that fell in different rows and columns of the table. But how do we classify pairs whose members are both overweight? We say the members are tied on body weight and that they cannot be classified as concordant or discordant. Hence, $C+D$ is the total number of pairs that are concordant or discordant but are not tied on either the row category or on the column category. Thus, if we write gamma as

$$\gamma = \frac{C-D}{C+D} = \frac{C}{C+D} - \frac{D}{C+D} \, ,$$

it can be seen that the gamma statistic is the proportion of untied pairs that are concordant minus the proportion of untied pairs that are discordant.

Another measure of association that is closely related to the gamma statistic is Kendall's tau-b statistic which we will write as τ_b. The numerator of τ_b is identical to the numerator of γ, but the denominator is slightly more complex. The measure of association τ_b is defined as

$$\tau_b = \frac{C-D}{\frac{1}{2}\sqrt{\left(n^2 - \sum_{i=1}^{r} n_{i+}^2\right)\left(n^2 - \sum_{j=1}^{c} n_{+j}^2\right)}}.$$

Using the data from the $n = 200$ individuals in Table 8.14, we compute τ_b to be

$$\tau_b = \frac{4551 - 2168}{\frac{1}{2}\sqrt{[200^2 - (52^2 + 114^2 + 34^2)][200^2 - (105^2 + 81^2 + 14^2)]}}$$

$$= \frac{2383}{11339} = .210$$

It should be pointed out that both γ and τ_b can be computed for data that are continuous. In fact, Kendall's τ_b is an excellent measure of correlation between two continuous variables because it is not sensitive to the presence of an outlier. It may be used instead of Pearson's ordinary correlation (r) or Spearman's rank-order correlation (r_s) to describe the association between two continuous or categorical variables. However, it should be noted that γ and τ_b both require ordered categories to compute the number of concordant and discordant pairs.

Exercise Set G

1. Consider the following 2×2 table:

	low	high
low	5	0
high	0	8

a) Compute the total number of concordant pairs.

b) Compute the total number of discordant pairs.

c) Compute γ.

2. Consider the following 2×2 table:

	low	high
low	0	8
high	13	0

a) Compute the total number of concordant pairs.

b) Compute the total number of discordant pairs.

c) Compute γ.

3. A sociologist wanted to investigate the relationship between educational attainment and job satisfaction. He selected 25 subjects at random and classified their educational attainment into three categories. The were also asked if they were unsatisfied, satisfied, or very satisfied with their current job. The results are:

Education	Job Satisfaction			Total
	Unsatisfied	Satisfied	Very Satisfied	
H. S. drop out	5	1	0	6
Completed H. S.	2	6	3	11
Completed College	2	2	4	8
Total	9	9	7	25

a) Are the rows and column classifications ordered?

b) Compute the total number of concordant pairs.

c) Compute the total number of discordant pairs.

d) Compute the measure of association γ.

e) Does γ indicate that there is a positive association between education and job satisfaction? Explain.

4. A criminologist obtained information about early drug use (before 15 years of age) and criminal behavior for 400 males between 24 and 30 years of age. The results were summarized in the following table:

	Criminal Behavior		
Drug Use	Yes	No	Total
Often	43	10	53
Sometimes	29	34	63
None	7	277	284

a) Compute the total number of concordant pairs and the total number of discordant pairs.

b) Compute γ.

c) Does there appear to be a positive association between drug use and criminal behavior?

d) Would it be possible to compute the ordinary correlation coefficient for this data? Why or why not?

5. Use the data in the previous exercise to compute Kendall's τ_b.

6. Without doing any calculations, roughly estimate the measure of association γ for the data in the following tables. For each table select either

$$\gamma = -.8, \quad \gamma = 0, \text{ or } \gamma = +.8:$$

a)

	low	medium	high
low	3	2	5
medium	4	5	11
high	6	7	14

b)

	low	medium	high
low	1	2	1
medium	3	5	2
high	1	2	8

c)

	low	medium	high
low	0	1	7
medium	2	5	3
high	6	2	1

*8.8 A Test for Matched-Pairs and Repeated Measures Categorical Data

We have seen data from matched pairs and repeated measures designs before in Chapter 5. In that chapter we analyzed continuous data produced by these designs. In this section we will analyze categorical data from paired comparisons and repeated measures designs.

Matched pair and repeated measure categorical data is always displayed in square tables because the first and second measurements are always on the same scale so they always have the same number of categories. The most common study design has only two possible responses so that the resulting table is a (2×2) table. We will restrict our discussion of repeated measures and matched pair data to the analysis of (2×2) tables. Consider the repeated measures data for the president's job approval rating. In this study design there were 145 respondents who expressed their approval or their disapproval of presidential job performance on two occasions that were separated by two months. This data is tabulated in Table 8.16.

Table 8.16 Comparison of the job approval rating of a President on two occasions.

		2nd interview	
		Approve	Disapprove
1st interview	Approve	82	12
	Disapprove	7	44

The analysis of repeated measures data differs greatly from the analysis we presented so far in this chapter. The analysis is different because with repeated measures data we are not too interested in a test of independence because we assume that most of the people who approved of the president's performance on the 1st interview would approve of the president's performance on the second interview and that most of the people who disapproved of the president's performance on the first interview would disapprove of the president's performance on the second interview. Therefore we are not surprised that $\frac{82+44}{145} = 87\%$ of those interviewed responded in the same way on both occasions. Clearly, the responses are dependent which implies there is no need to perform a chi-squared test of independence.

With repeated measures data we are more interested in the changes in approval over time. The 82 subjects in the upper left cell of the table and the 44 subjects in the lower right cell of the table give us no information about the change in approval. The important information is obtained from the people who changed their minds. Our test will be based on the 7 individuals who had a negative response on the first survey and a positive response on the second survey and on the 12 individuals who had a positive response on the first survey and a negative response on the second survey.

The sign test

The sign test is used to find out if there is a difference in the proportion of positive responses between the first and second interviews. If there really has been no change in the President's job approval then about half of the people who changed their minds about the President should fall in the lower left cell and about half should fall in the upper right cell. That is, if the null hypothesis is true, for each of the 19 respondents who changed their minds, there was a $p = .5$ chance that the individual would fall into the lower left cell. This is equivalent to flipping a coin 19 times and counting the number of heads in the lower left cell. We can perform a one-tailed test of significance by computing the chance of obtaining a count in that cell as large or larger than that obtained in the study. The sampling distribution of the count in the lower left cell must be the same as the sampling distribution of the number of heads in a coin flipping experiment, which is a binomial distribution with $n = 19$ and $p = .5$. A test that uses the binomial distribution with $p = .5$ is called the **sign test**, which was discussed briefly in section 5.1.

Our general formula for the binomial distribution can be used to compute the chance of obtaining exactly x successes in n trials. The general formula is

$$p(x) = \frac{n!}{x!(n-x)!} p^x (1-p)^{n-x},$$

where p is the probability of a success. For the sign test we always use $p = (1-p) = .5$ so the formula simplifies to

$$p(x) = \frac{n!}{x!(n-x)!} (.5)^n.$$

In our example there are $n = 19$ off diagonal responses. The chance of obtaining exactly 7 respondents in the lower left cell is

$$p(7) = \frac{19!}{7!(19-7)!} (.5)^{19} = 50388\left(1.90734 \times 10^{-6}\right) = .096.$$

To compute a one-tailed p-value we need to compute

$$\text{p - value} = p(7) + p(6) + \ldots + p(0)$$

The calculations can be greatly simplified by using Table 6 which gives individual probabilities for the sign test, which is identical to the individual probabilities for the binomial distribution with $p = .5$. From this table we can obtain the probabilities to easily compute the p-value. For our example the p-value, which is the probability of obtaining 7 or fewer in the lower left cell of the table, is

$$p\text{-}value = p(7) + p(6) + ... + p(0)$$

$$= .096 + .052 + .022 + .007 + .002 = .179$$

The sign test gives the exact p-value. It can easily be computed if $n_{12} + n_{21} \leq 20$ using the probabilities in Table 6. If a two-tailed p-value is required it can be obtained by doubling the one-tailed p-value. In this example the two-tailed p-value would be $p = 2(.179) = .358$.

McNemar's Test

This sign test gives exact probabilities for the test and is the preferred test method for data sets having $n_{12} + n_{21} \leq 20$. For studies having a large number of off diagonal values, the computations are tedious and the tables are of little value. Fortunately, a large sample approximate test, known as **McNemar's test**, is available to test the null hypothesis that the probability of an off diagonal response falling the lower cell is $p = .5$. McNemar's test statistic is

$$X^2 = \frac{\left(\left|n_{12} - n_{21}\right| - 1\right)^2}{n_{12} + n_{21}}.$$

The McNemar X^2 test statistic has a sampling distribution that approximates a chi-square distribution with one degree of freedom if the probability of an off diagonal response falling in the lower left cell is $p = .5$. Thus, we reject the null hypothesis if X^2 exceeds the critical value of a chi-squared statistic with $v = 1$ degrees of freedom and a level of significance of α, or if the p-value is less than α. This is a two-tailed test because the alternative hypothesis is $p \neq .5$.

For example, suppose we had 255 volunteers who participated in an experiment to compare the effectiveness of two treatments. The 255 volunteers were first given one drug for 4 weeks to reduce their chronic pain. After a 2 week period where no drugs were given the same volunteers were given a second treatment with a different drug. The researchers recorded the improvement in pain after both treatments. The results are tabulated in Table 8.17.

Table 8.17 Comparison of the effectiveness of two drug treatments.

		2nd Drug Treatment	
		Improvement	No Improvement
1st Drug	Improvement	155	21
Treatment	No Improvement	38	41

For this data McNemar's test statistic is

$$X^2 = \frac{(|21-38|-1)^2}{21+38} = \frac{16^2}{59} = 4.34.$$

With $v = 1$ degrees of freedom we obtain a p-value of $p \cong .04$. Thus, if we used $\alpha = .05$ we would reject the null hypothesis and would conclude that the second drug is more effective than the first drug.

Exercise Set H

1. An environmental researcher attempted to relate the reduction in air pollution in a city to decreases in respiratory symptoms experienced by inhabitants of the city. The researcher reported respiratory symptoms on a sample of 305 randomly selected individuals before air pollution was reduced and six months later after the air pollution had been cut by 20%. The researcher obtained the following data:

		Symptoms 2nd Observation		
		Yes	No	Total
Symptoms				
1st Observation	Yes	77	11	88
	No	7	210	217
	Total	84	221	305

a) Should a sign test or a McNemar's test be used?

b) Estimate the p-value based on the test statistic you chose in (a).

c) State your conclusion.

2. A physical therapist enrolled 200 patients who were suffering from chronic wrist pain in a clinical trial. Over a period of several weeks the volunteers used two pain relievers (Drug A and Drug B) and they indicted the effectiveness of these drugs in reducing pain.   There results were tabulated as follows:

		Drug B Relieved Symptoms		
Drug A		Yes	No	Total
Relieved Symptoms	Yes	77	32	109
	No	9	82	91
	Total	86	114	200

a) Should a sign test or a McNemar's test be used?

b) Estimate the p-value based on the test statistic you chose in (a).

c) State your conclusion.

8.9 A Guide to Categorical Data

A large number of techniques have been presented in this chapter. It is important to use the technique that is most appropriate for the data that is available and that meets your objective. It is best to carefully consider the design of the study and the data itself to decide if a categorical procedure is appropriate. This is not always easy if there are a large number of categories that are on an interval scale. For example, if someone has data on age and income and they have categorized it into 10 age categories and 15 incomes categories, it may not be obvious if you should you use a categorical or a continuous method to describe the relationship between the two variables. For this data a researcher could use a Pearson's correlation coefficient (r), which is normally used with continuous data, or the researcher could use γ or Kendall's τ_b which are usually used on categorical data.

Now suppose you have determined that there are a small number of row and column categories and that you need to use a categorical method. The first step is to determine if the row and column categories are naturally ordered or if they can be put in a reasonable order. If both the row and column categories cannot be ordered and if you want to test the independence of the row and column classifications, then a chi-squared test of independence is appropriate, provided the expected cell counts are large enough for the chi-squared approximation to be valid.

Very often you will find that the column categories can be ordered. If there are only 2 rows then the Cochran-Mantel-Haenszel (CMH) test for trend is appropriate. If the CMH test is appropriate it should be used instead of the chi-squared test for independence because it provided a more powerful test of the null hypothesis. The big difference between the chi-squared test for independence and the CMH test for trend is that the CMH test has a more specific alternative which allows it to be more sensitive to trends in proportions. If there are 2 columns and more than 2 rows that are ordered you should switch the rows and column and compute the CMH test. More advanced texts describe extensions to the CMH test for the rare cases where there are more than 2 ordered columns and more than two unordered rows.

You should be alert to recognize matched-pairs data or repeated measures data. This data will look like unmatched data but will always have the same classifications for the rows as for the columns. The major difference with this data is that the null hypothesis of independence is usually not of interest. Usually, interest centers on the off diagonal elements which provide the information for a sign test or for McNemar's test.

Sometimes you may be more interested in describing the relationship between the row and column variables. For 2×2 tables the odds ratio is often used to describe the relationship between the variables. For larger tables that have ordered rows and column categories, the γ statistic and the Kendall τ_b statistic are recommended as measures of association.

Chapter Review Exercises

1. The probability of rolling a "3" or a "4" with a single roll of a fair die is $\frac{1}{3}$. If you roll a die 6 times, what is the chance of getting a "3" or a "4" exactly twice?

2. Suppose a supplier of automotive parts has a poor quality control program which results in 10% of the parts being defective. Suppose the manufacturer that obtains the parts from the supplier selects $n = 15$ parts for inspection. If a defective part is found the entire lot may be rejected.

   a) What is the chance that the manufacturer will find no defective parts?

   b) What is the chance that the manufacturer will find at least one defective part?

3. A student, who knows nothing about chemistry, must get 5 out of 10 multiple chance questions correct on an exam in order to pass the course. We will assume that he will make a random guess on each question and that the chance that he will be successful on any given question is $p = \frac{1}{5}$.

   a) Is this a binomial experiment?

   b) What is his chance of receiving a passing grade?

4. An administrator in a large city wanted to determine the proportion of housing units that had a functioning smoke detector. She planned to take a random sample of 500 housing units. If 45% of the homes have a functioning smoke detector, what is the chance that the sample proportion would be greater than 50%? That is, what is the chance that more than 250 of 500 housing units that were sampled would have working smoke detectors?

5. Some people believe that the number of serious crimes increases around the time of the full moon. To test this theory a researcher classified 682 serious crimes as occurring in the seven day period

around the full moon. He noted that 165 serious crimes occurred in that period and that 517 occurred outside that period. Since a full moon occurs approximately once every 28 days, about $\frac{1}{4}$ of the crime should occur in the 7 day period around the full moon, if the full moon is unrelated to crime.

a) Compute the expected number of serious crimes in each of the two time periods if the full moon is unrelated to crime.

b) Compute the goodness-of-fit test statistic.

c) Estimate the p-value. State your conclusion using a significance level of $\alpha = .05$.

6. An educator analyzed the results of an achievement test given to eighth grade students. The educator classified the $n = 100$ students into one of four quartiles based on their achievement. The test publisher indicted the cutpoints for the quartiles by providing the 25th, 50th, and 75th percentiles. The actual and expected counts were :

	Actual	Expected
1st (lowest) quartile	2	25
2nd quartile	12	25
3th quartile	39	25
4th (highest) quartile	49	25

a) Compute the goodness-of-fit test statistic.

b) Estimate the p-value. State your conclusion using a significance level of $\alpha = .05$.

7. A dietician took a survey of adults in a city to determine the relationship of reduced calorie soft drinks (diet drinks) to body mass. She included into her survey only people who regularly consume soft drinks. The results were :

Soft Drink	Overweight	Not overweight	Total
Non-diet	6	27	33
Diet	22	32	54
Total	28	59	87

a) Using $\alpha = .05$ perform a test for independence.
b) State the conclusion that you would draw from the test.
c) In your opinion, what is the relationship between these two variables? Does consumption of diet soft drinks contribute to weight problems? Explain.

8. A neurologist designed an experiment to determine what physical problems are aggravated by the use of laptop computers. Over the course of several months he randomly assigned 1000 volunteers to use a desktop or a laptop computer for 75 minutes of typing. At the end of the typing session the volunteers were asked to indicate their most serious problem. The results were tabulated as follows:

	Most Serious Problem		
Type	None	Neck or Head Pain	Wrist or Hand Pain
Desktop	435	32	33
Laptop	382	47	71

a) Are the columns ordered or unordered?
b) Should a test of independence or a test for trend be used?
c) Perform the appropriate test and state your conclusions. Use $\alpha = .05$.

9. A University recently took a survey to determine if their graduates were finding employment in their field of study. They classified the

students according to the school they attended and by whether they were able to find full time work in their major or minor field of study. The results were classified as:

Job in field

College	Yes	No	Total
Liberal Arts	81	237	318
Business	63	18	81
Engineering	47	4	51
Total	191	259	450

a) Are the cell frequencies large enough to justify the use of a chi-squared test of independence? Note that there were only 4 Engineering graduates who failed to get a job in their field.

b) Compute the chi-squared test statistic for the chi-squared test of independence.

c) Does the ability to obtain a job in the field of study appear to be related to the college? Explain.

10. A safety expert investigated all motor vehicle accidents in a state that involved a passenger car. Trucks, vans, mini-vans, and were excluded from the study. The passenger cars were classified as light (<2700 lbs.), average (between 2700 lbs. and 3500 lbs.), and heavy (>3500 lbs.). The number of driver's fatalities and the weights of the automobiles were recorded as follows:

Passenger Car Weight

Outcome	Light	Average	Heavy	Total
Not Fatal	1204	3080	976	5260
Fatal	17	36	5	58
Total	1221	3116	981	5318

a) Are the column categories ordered or unordered?

b) In your opinion, should a test of independence or a test of trend be used?

c) Compute the appropriate test statistic and estimate the p-value. If you decide to do a test of trend use the integer scores $\{x_1 = 1, x_2 = 2, x_3 = 3\}$.

d) Using $\alpha = .10$ perform the test and state your conclusion.

11. A school district was criticized for allowing students to participate in sports even though they had low grades. In order to determine the relationship between grades and sports they classified students as having low grades if their grade average was below a "C" and made a two-dimensional table of grades and participation in sports. They obtained the following table:

Participants	Grades		Total
	Satisfactory	Low	
Yes	74	21	95
No	212	103	315
Total	286	124	410

a) Should an exact test be used or should a chi-squared test be used to test independence?

b) Perform the test, estimate the p-value, and state your conclusions.

c) Compute the odds ratios relating sports participation to grades. Does there appear to be a positive association between participation in sports and grades.

12. A polling organization wanted to determine the support for a candidate (Candidate A) for a political office and wanted to determine if there was more support for that candidate from people who describe themselves as liberals. They asked 1270 probable voters their preference and asked them to describe themselves as liberal,

moderate, or conservative. The polling organization obtained the following data:

	Liberal	Moderate	Conservative	Total
Candidate A	143	457	110	710
Candidate B	113	356	91	560
Total	256	813	201	1270

a) Are the column categories ordered or unordered?

b) In your opinion, should a test of independence or a test of trend be used?

c) Compute the appropriate test statistic and estimate the p-value. If you decide to do a test of trend use the integer scores $\{x_1 = 1, x_2 = 2, x_3 = 3\}$.

d) Using $\alpha = .05$ perform the test and state your conclusion.

13. A sociologist wanted to determine the relationship, if any, between income and attitude toward "professional" wrestling. She obtained information from 20 individuals and classified their income as below or above the median. She determined if they believed that much of the wrestling performance was "fake" and made the following table:

	Opinion of wrestling		
Income	Fake	Not Fake	Total
Below Median	7	3	10
Above Median	9	1	10
Total	16	4	20

a) Should a chi-squared test or an exact test be used to test independence?

b) Estimate the p-value for the appropriate test.

c) State your conclusion.

14. Refer to the motor vehicle accident data given in Exercise 10. Suppose we want to describe the relationship between the weights of passenger cars and the number of fatalities.

   a) Compute the number of concordant pairs.

   b) Compute the number of discordant pairs.

   c) Calculate the γ statistic.

   d) Interpret the γ statistic. Does this measure of association indicate a positive or a negative relationship between weight and fatalities?

15. A neurologist enrolled 35 volunteers in an experiment which was designed to determine the effect of caffeine on certain coordination skills. She required each volunteer to ingest 560 mg. of caffeine, which is the amount in four cups of coffee, over a 4 hour period. A block stacking test was administered immediately before and immediately after the 4 hour period. The results were recorded as follows:

		Symptoms Post-test		
		Pass	Fail	Total
Symptoms Pre-test	Pass	12	9	21
	Fail	6	8	14
	Total	18	17	35

   a) Should a sign test or a McNemar's test be used?

   b) Estimate the p-value based on the test statistic you chose in (a).

   c) State your conclusion.

16. Army volunteers in a certain country take a physical exam before they are accepted into the military, which includes a hearing exam. In the examination the right and left ears are evaluated separately, and the hearing in the ears is judged to be acceptable or deficient. Data on a total of 957 Army volunteers were recorded as follows:

		Left ear		
		Acceptable	Deficient	Total
Right Ear	Acceptable	906	7	913
	Deficient	24	20	44
	Total	930	27	957

a) Should a sign test or a McNemar's test be used?

b) Estimate the p-value based on the test statistic you chose in (a).

c) State your conclusion.

Chapter 9. Inferences for Three or More Samples

9.1 The F Test

Introduction

In Chapters 4 and 5 we worked with confidence intervals and tests based on data obtained from a single sample and in Chapter 6 we analyzed data from two samples. In this chapter we will develop methods for analyzing data from three or more samples.

These methods are very popular in experimental work where there is a need to compare three or more drugs or treatments. For example, a scientist may investigate the effectiveness of four drugs in one experiment by randomly assigning 40 rats to one of four drug groups. These methods are also popular in observational studies because they can be used to compare subpopulations. For example, a sociologist may identify five occupational groups and then may use the methods in this chapter to compare the average income for individuals who are classified into these groups.

Notation

The most popular test for comparing three or more groups is the F test, which uses the F statistic. Before we state the test statistic we must introduce some notation. Table 9.1 gives the notation for an experiment that has $G = 3$ groups. In the first group there are $n_1 = 4$ observations that are indicated by $\{x_{1,1}, x_{1,2}, x_{1,3}, x_{1,4}\}$. Note that the first subscript indicates the group and the second subscript indicates the observation number within the group. The average, over all observations, in the first group is

$$\bar{x}_1 = \frac{\sum_{j=1}^{n_1} x_{1,j}}{n_1}.$$

In the example given in Table 9.1, $\bar{x}_1 = 76$.

In this chapter we will use $x_{i,j}$ to indicate the jth observation in the ith group. Using this notation, the mean value in the ith group is

$$\bar{x}_i = \frac{\sum_{j=1}^{n_i} x_{i,j}}{n_i}.$$

The mean values for the three groups are given at the bottom of the table. The total number of observations, over all G groups, is $n = n_1 + n_2 + \ldots + n_G$. We can write the overall mean as

$$\bar{x} = \frac{\sum_{i=1}^{G} \sum_{j=1}^{n_i} x_{i,j}}{n},$$

which is simply the sum of the observations divided by the total number of observations. Note that $\bar{x}$ is not the average of the three group means, it is the average of the 14 observations. For the data in Table 9.1 the overall mean is $\bar{x} = 81$.

Table 9.1 Scores on an exam for 14 students who were assigned to one of three teaching methods.

Method A	Method B	Method C
$x_{1,1} = 72$	$x_{2,1} = 81$	$x_{3,1} = 94$
$x_{1,2} = 78$	$x_{2,2} = 68$	$x_{3,2} = 84$
$x_{1,3} = 69$	$x_{2,3} = 76$	$x_{3,3} = 97$
$x_{1,4} = 85$	$x_{2,4} = 79$	$x_{3,4} = 89$
	$x_{2,5} = 66$	$x_{3,5} = 96$
$\bar{x}_1 = 76$	$\bar{x}_2 = 74$	$\bar{x}_3 = 92$

As you might expect, the analysis will depend heavily on the values of $\bar{x}_i$, which are the treatment group means. If the treatment group means are far apart then it is reasonable to conclude that the treatment groups are really different.   If the treatment group means are close together then there is no reason to believe that the treatment groups are really different.   An overall measure of the differences between these treatment group means is called the **Sum of Squares for Treatments**, which is defined as

$$SS_{treatments} = \sum_{i=1}^{G} n_i (\bar{x}_i - \bar{x})^2 .$$

Note that $SS_{treatments}$ will be small if the means are close together because each $(\bar{x}_i - \bar{x})$ will be close to zero.   However, not all treatment means need to be different for $SS_{treatments}$ to be large.   If one of the treatment groups has a mean value that is apart from the other treatment group means then $(\bar{x}_i - \bar{x})$ will be large for that group which will cause $SS_{treatments}$ to be large.   Using the sample means in Table 9.1 and using $\bar{x} = 81$, we find

$$SS_{treatments} = 4(76-81)^2 + 5(74-81)^2 + 5(92-81)^2$$

$$= 100 + 245 + 605 = 950$$

Although $\bar{x}$ is not a simple average of the treatment group means, it can be shown that $\bar{x}$ is a weighted average of the treatment group means. This implies that if we knew $\bar{x}$ and the group means for $G-1$ of the groups, then we could calculate the last treatment mean.   Since, for a given $\bar{x}$, the $SS_{treatments}$ depends on only $G-1$ treatment means, we say that there are $G-1$ degrees of freedom associated with $SS_{treatments}$, which we will write as $DF_{treatments} = G-1$.   Our overall measure of the variability between groups is given by the **Mean Square for Treatments**, which is defined as

$$MS_{treatments} = \frac{SS_{treatments}}{DF_{treatments}} = \frac{\sum_{i=1}^{G} n_i (\bar{x}_i - \bar{x})^2}{G-1}$$

In   our   example, $DF_{treatments} = G-1 = 2$,   which   implies   that $MS_{treatments} = \frac{950}{2} = 475$.

The formula shows that the value of $MS_{treatments}$ is a function of the sample means in the G groups. But these sample means will reflect the population mean values. If the data comes from an experiment having G treatments and these treatments are not effective then the population mean values should be equal. In this chapter the null hypothesis is always a statement that the population means are equal. Consequently, we will usually write the null hypothesis as $H_o : \mu_1 = \mu_2 = ... = \mu_G$. We will use μ_i to denote the population mean in the ith group. The alternative hypothesis is that at least one of the mean values differs from one of the other mean values. We will usually write the alternative hypothesis as $H_a : \mu_i \neq \mu_j$, for some i,j .

If the population means are quite variable then it is likely that the sample means will be quite variable which will produce a large $MS_{treatments}$. On the other hand, if the null hypothesis is true the value of $MS_{treatments}$ should be small. Of course, even if the population means are equal, there may still be some small differences in the sample means which could cause $MS_{treatments}$ to exceed zero. To proceed further, we need to know how large the $MS_{treatments}$ could be under the null hypothesis. In order to assess how large $MS_{treatments}$ could be if the null hypothesis is true, we need to take the variability of the observations into account. We begin by calculating the **Sum of Squares for Errors**, which is defined as

$$SS_{errors} = \sum_{i=1}^{G} \sum_{j=1}^{n_i} (x_{i,j} - \bar{x}_i)^2 .$$

Note that SS_{errors} measures the variability about the treatment groups means. If the values in each treatment group are closely grouped around the mean then the SS_{errors} will be small, even if the treatment group means differ greatly. The data in Table 9.1 give

$$SS_{errors} = \left\{ (-4)^2 + 2^2 + (-7)^2 + 9^2 \right\}$$
$$+ \left\{ 7^2 + (-6)^2 + 2^2 + 5^2 + (-8)^2 \right\}$$
$$+ \left\{ 2^2 + (-8)^2 + 5^2 + (-3)^2 + 4^2 \right\}$$
$$= 150 + 178 + 118 = 446$$

We can also write SS_{errors} as a function of the within-group variances. Recall that the formula for the variance for the observations in the ith group is

$$s_i^2 = \frac{\displaystyle\sum_{j=1}^{n_i}(x_{i,j} - \bar{x}_i)^2}{n_i - 1},$$

which implies that $\displaystyle\sum_{j=1}^{n_i}(x_{i,j} - \bar{x}_i)^2 = (n_i - 1)s_i^2$. We can substitute this expression into the formula for SS_{errors} to obtain

$$SS_{errors} = \sum_{i=1}^{G}(n_i - 1)s_i^2 .$$

This equation is often used for calculation when the sample means and sample standard deviations are available.

We also associate degrees of freedom with SS_{errors}. Recall that the variability in the ith sample is measured by the variance in the ith group, which has $n_i - 1$ degrees of freedom. It can be shown that the degrees of freedom associated with the SS_{errors} is the sum of the degrees of freedom associated with the variances, which is

$$DF_{errors} = \sum_{i=1}^{G}(n_i - 1) = n - G .$$

In our example $DF_{errors} = 14 - 3 = 11$.

In order to estimate the overall variability in the experiment we compute the **mean squared error** (MSE) which is defined as

$$MSE = \frac{SS_{errors}}{DF_{errors}} = \frac{\sum\limits_{i=1}^{G}\sum\limits_{j=1}^{n_i}\left(x_{i,j} - \bar{x}_i\right)^2}{n-G}.$$

In our example $MSE = \frac{446}{11} = 40.55$. This is a pooled estimate of the variance of the observations within the treatment groups. A large value of MSE indicates that there is a great deal of variability within the groups. A small value of the MSE indicates that there is little variability within the groups. Note that the MSE is not related to the variability that is observed between groups, which is measured by $MS_{treatments}$.

The F Statistic

As we indicated previously, the $MS_{treatments}$ is sensitive to the differences between the group means. However, it is also related to the within-group variability. In order to obtain a statistic that depends solely on the population means we divide the $MS_{treatments}$ by the MSE to obtain the F statistic, which is used to test the null hypothesis $H_o: \mu_1 = \mu_2 = \ldots = \mu_G$. This test statistic is usually written as

$$F = \frac{MS_{treatments}}{MSE}$$

Since the $MS_{treatments}$ is sensitive to differences in the population means and since the MSE is not sensitive to these differences, the F statistic will be sensitive to differences between the groups. Thus, we will reject the null hypothesis for large values of F. In the next section we will determine how large F needs to be in order to reject the null hypothesis.

Another practical reason for dividing the $MS_{treatments}$ by MSE is that the resulting F statistic becomes insensitive to changes in scale. That is, if we have observations that are recorded in inches and we change our scale to record the observations in centimeters then, because $1\,inch = 2.54\,centimeters$, the $MS_{treatments}$ will increase by a factor of 2.54 and the MSE will increase by a factor of 2.54, so that F will not change. This insensitivity to changes in scale is very desirable for a test statistic.

Exercise Set A

1. In order to reduce the cost of maintaining the highways a Highway Department wanted to determine the most durable paint to use on the highways. Four paints were used in an experiment designed to determine if any of the paints lasted longer than any other paint. The four paints were each used on 6 stretches of road that were selected at random. The number of months that the paints were clearly visible were recorded as follows:

	Paint			
	A	B	C	D
	49	24	52	49
	46	28	47	53
	48	25	48	47
	51	29	50	52
	47	27	51	51
	48	26	48	54
Mean	48.16	26.50	49.33	51.00
Standard Dev.	1.72	1.87	1.97	2.61

a) Identify $\bar{x}_2$ and s_2.

b) Compute the sum of squares for treatments, the degrees of freedom for treatments, and the mean square for treatments. Label each properly.

c) Compute the sum of squares for errors, the degrees of freedom for errors, and the mean square error. Label each properly.

d) Compute F.

e) Are the measurements for each paint fairly consistent or do they vary a great deal.

f) Why is F so large?

2. In order to obtain approval to market a new drug, pharmaceutical companies usually need to show that the new drug is effective and

relatively safe. In order to determine the effectiveness of a new pain reliever, a company obtained 22 volunteers who suffered from occasional headaches. These volunteers were assigned at random to one of three treatment groups, and the time between taking the drug and pain relief was recorded. Summary statistics for the volunteers in the three groups were recorded as follows:

	Number in Group	Mean	Standard Deviation
Drug A	8	8.80	1.78
Drug B	7	9.43	1.47
Drug C	7	9.13	2.01

a) Compute the sum of squares for treatments, the degrees of freedom for treatments, and the mean square for treatments. Label each properly.

b) Compute the variances for each treatment group.

c) Compute the sum of squares for errors, the degrees of freedom for errors, and the mean square error. Label each properly.

d) Compute F.

3. Consider the following two data sets. In each of the data sets there are 5 observations in groups 1 and 2 and 7 observations in group 3. The observations in the first group are indicated by circles, the observations in the second group are indicated by squares, and the observations in the third group are indicated by triangles.

Data Set A

Data Set B

a) Is the MSE in data set A equal to the MSE in data set B? No calculations should be necessary.

b) Is the $MS_{treatments}$ in data set A equal to the MSE in data set B? No calculations should be necessary.

c) Which data set has the larger F statistic? Explain why.

4. Consider the following two data sets. In each of the data sets there are 5 observations in groups 1 and 2 and 7 observations in group 3. The observations in the first group are indicated by circles, the observations in the second group are indicated by squares, and the observations in the third group are indicated by triangles.

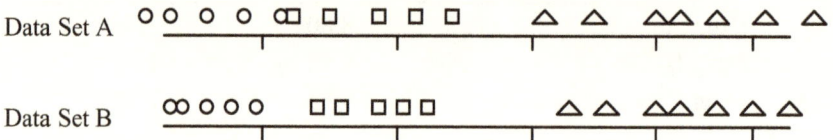

Data Set A

Data Set B

a) Is the MSE in data set A equal to the MSE in data set B? No calculations should be necessary.

b) Is the $MS_{treatments}$ in data set A equal to the MSE in data set B? No calculations should be necessary.

c) Which data set has the larger F statistic? Explain why.

9.2 The F test

Assumptions for the F test

We need to develop the sampling distribution of F under the null hypothesis, but this can only be determined by making several assumptions about the observations. The three assumptions, which are often called the **inference assumptions**, are :

- The observations are normally distributed about the population mean value for the sample. This assumption is illustrated in Figure 9.1 below.

- The variability in any population equals the variability in any other population. Of course, there may be some differences in the sample variability, but the assumption is that there is no difference in the population variances between groups. This point is also illustrated in Figure 9.1 by the equal width of the normal distribution.

- The observations are independent. That is, the value of one observation in one sample must not depend on any other observations in that sample or in any other sample.

Figure 9.1 Three normal populations having equal variances.

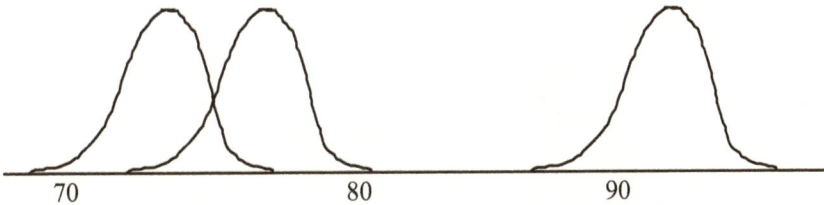

70   80   90

The Sampling Distribution of F

If the three inference assumptions are met it is possible to calculate the sampling distribution of the F statistic. We will assume throughout this section that the inference assumptions are satisfied. It can be shown that if the null hypothesis is true then the F statistic is distributed as a F distribution with degrees of freedom equal to $DF_{treatments} = G - 1$ and $DF_{errors} = n - G$. The F distribution is not a single distribution, it is a family of distributions indexed by the degrees of freedom in the numerator ($DF_{treatments} = G - 1$) and by the degrees of freedom in the denominator ($DF_{errors} = n - G$). Tables of the F distribution are given in the Appendix.

An example may clarify the use of the F distribution. Suppose we need to analyze the results of an experiment that had $G = 4$ treatment groups with 8 animals in each group. If we obtained an F statistic of

$F = 1.7$ we would like to know if the null hypothesis should be rejected. To determine this we need the sampling distribution of F under the null hypothesis with $DF_{treatments} = 4 - 1 = 3$ and $DF_{errors} = 32 - 4 = 28$. The F distribution with 3 and 28 degrees of freedom is illustrated in Figure 9.2. It can be seen that the chance of getting an F statistic as large or larger than 1.7 is sizeable. Therefore, we have no reason to reject the null hypothesis.

Figure 9.2 The sampling distribution of the F statistic with 3 and 28 degrees of freedom. The p-value for $F = 1.7$ is indicated in the shaded region.

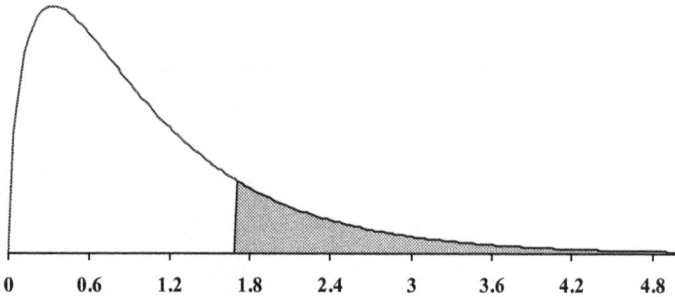

| 0 | 0.6 | 1.2 | 1.8 | 2.4 | 3 | 3.6 | 4.2 | 4.8 |

Tables of the F Distribution

In most applications of the F test, it is necessary to compute the p-value, which is the probability of obtaining an F statistic that is greater than the observed F statistic under the null hypothesis. Once the F test statistic is calculated the p-value can be estimated by using one of the four F tables in the Appendix. Each F table was computed for a specified value of $DF_{treatments}$. Table 8a is for $DF_{treatments} = 2$, Table 8b is for $DF_{treatments} = 3$, etc. The degrees of freedom in the denominator are listed in the rows and the p-values are listed in the columns. If $F = 1.7$ and $DF_{treatments} = 3$ and $DF_{errors} = 28$ we find an approximate p-value of $p \cong .2$. The tables is the Appendix will be sufficient for $G = 3, 4, 5,$ and 6. Larger tables are available, but these should be sufficient for the exercises in this text.

The F test

We have defined the F statistic and have described its sampling distribution under the null hypothesis. We now need to put it all together to perform an F test of the null hypothesis $H_o : \mu_1 = \mu_2 = \ldots = \mu_G$ against the alternative hypothesis $H_o : \mu_i \neq \mu_j$, for some i,j. The steps in an F-test can be summarized as follows:

- Compute the mean values $\bar{x}_i$ for $i=1,\ldots,G$.
- Compute $SS_{treatments}$, $DF_{treatments}$, and $MS_{treatment}$.
- Compute SS_{errors}, DF_{errors}, and MSE.
- Compute the test statistic F.
- Use the appropriate table in the Appendix to estimate the p-value.
- If the p-value is less than α reject the null hypothesis.
- State your conclusion.

For the data given in Table 9.1 we have already computed many of these quantities, including $MS_{treatment} = 475$ and $MS_{errors} = 40.55$. Thus, $F = \frac{475}{40.55} = 11.71$. To test the null hypothesis, we would then use Table 8a with $DF_{errors} = 11$ to estimate a p-value of approximately $p \cong .002$. Thus, for $\alpha = .05$ we will reject the null hypothesis and conclude that at least one of the teaching methods has a different mean value than one of the others. Note that we do not, based solely on this F test, know which population mean values may differ. We will discuss how to compare the population mean values in section 9.5.

The ANOVA Table

It is traditional to arrange the $SS_{treatments}$, SS_{errors}, and other statistics in a table form. Computer output is also often arranged in this tabular form, which is called an **analysis of variance table**, or an **ANOVA** table. The table usually includes the **total sum of squares**, which is defined as

$$SS_{total} = \sum_{i=1}^{G}\sum_{j=1}^{n_i}\left(x_{i,j} - \bar{x}\right)^2 .$$

Note that, for SS_{total}, the deviations are about the overall mean $\bar{x}$. For SS_{errors} the deviations are about the group means $\bar{x}_i$.

There is usually no need to calculate SS_{total} directly because it can be shown that $SS_{treatments} + SS_{errors} = SS_{total}$. Since there are n observations over the G treatment groups and since we are computing the deviations about the overall mean, the degrees of freedom associated with SS_{total} is $DF_{total} = n-1$. We have now defined the statistics that are tabulated in an ANOVA table. The general form of the ANOVA table is shown in Table 9.2

Table 9.2 The General Form of the ANOVA Table

Source	D.F.	Sum of Squares	Mean Square	F
Treatments	$G-1$	$SS_{treatments}$	$MS_{treatments}$	$\dfrac{MS_{treatments}}{MSE}$
Error	$n-G$	SS_{error}	MSE	
Total	$n-1$	SS_{total}		

Table 9.3 gives the ANOVA table for the data in Table 9.1. We have computed the table entries, except for SS_{total}, which can easily be obtained by adding the other two sums of squares. Note that the mean square associated with the total is not needed, so it is usually not computed. Computer outputs sometimes use "Group" or "Model" as the source instead of "Treatments" and often use the abbreviations DF for degrees of freedom and SS for sums of squares. The p-value is often listed in the last column of the ANOVA table.

Table 9.3 The ANOVA Table for the data in table 9.1

Source	D.F.	Sum of Squares	Mean Square	F
Treatments	2	950	475	11.71
Error	11	446	40.55	
Total	13	1440		

Exercise Set B

1. Estimate the p-value corresponding to the following F statistics:

   a) $F = 4.62, G = 5, n = 35$.

   b) $F = 1.90, G = 3, n = 25$.

   c) $F = 2.49, G = 6, n = 115$.

2. A researcher used $n = 25$ animals in $G = 5$ treatment groups in an experiment and obtained $F = 3.49$. If she performed another experiment using $n = 25$ animals in $G = 5$ treatment groups and obtained $F = 4.82$, would the p-value be smaller or larger than in the previous experiment?

3. A physical therapist enrolled 18 patients who were suffering from lower back pain in an experiment. Patients were assigned at random to either a control group, which received no treatment, or to a drug treatment group, or to an acupuncture treatment group. After treating these patients for 6 weeks, the therapist had another physical therapist, who did not know which treatments were assigned to the patients, to assign a mobility score to each patient. These mobility scores were recorded on a 0 to 100 scale with a score of 100 indicating the maximum amount of pain-free mobility. Thus, larger scores are better. The data are:

	Mobility Scores	
No Treatment	Drug Treatment	Acupuncture
85	67	80
89	87	93
70	85	92
83	74	69
68	78	73
91	89	91
$\bar{x}_1 = 81$	$\bar{x}_2 = 80$	$\bar{x}_3 = 83$
$s_1 = 9.74$	$s_2 = 8.53$	$s_3 = 10.49$

a) Compute $SS_{treatments}$ and SS_{errors}.

b) Make an ANOVA table.

c) Calculate the F test statistic and roughly estimate the p-value.

d) Has the physical therapist shown that the acupuncture treatment is more effective than the other treatments? Explain.

4. A dietitian conducted a large study to determine which weight loss program would have the best long-term benefits. She randomly assigned 105 volunteers to one of three treatment groups. Volunteers who were assigned to the first treatment group attempted to reduce weight by dietary restrictions only. Volunteers assigned to the second treatment group attempted to reduce their weight by an exercise program, but there were no dietary restrictions. Volunteers assigned to the third treatment group used dietary restriction and an exercise program. The number of pounds that were lost over a one-year period were:

Diet Only	Exercise Only	Diet and Exercise
$n_1 = 35$	$n_2 = 35$	$n_3 = 35$
$\bar{x}_1 = 1.82$	$\bar{x}_2 = 3.27$	$\bar{x}_3 = 5.71$
$s_1 = 6.32$	$s_2 = 5.66$	$s_3 = 6.82$

a) Compute $SS_{treatments}$ and SS_{errors}.

b) Make an ANOVA table.

c) Calculate the F test statistic and roughly estimate the p-value.

d) Would you reject the null hypothesis using $\alpha = .05$? State your conclusion.

9.3 The Kruskal-Wallis test

The Kruskal-Wallis test can generally be used in the same situations that the F test can be used. The major difference between these two tests is that the Kruskal-Wallis test is based on the ranks of the data, rather than on the data itself. Another difference is that the Kruskal-Wallis test does not use the assumption of normality in the justification of the sampling distribution of the test statistic. The Kruskal-Wallis test was derived under the assumption that the observations are independent and that the distributions have the same variance and the same overall shape. That is, we assume that the distributions are the same except for some change in location but we do not assume that the distributions are normal. Since the distributions have the same shape the null hypothesis is that the distributions are identical and the alternative hypothesis is that at least one of the distributions is shifted relative to another distribution. Figure 9.3 illustrates skewed distributions that are shifted relative to one another. The Kruskal-Wallis test is suitable for such skewed distributions. A complete comparison of the two tests will be presented in the next section.

Figure 9.3 Three skewed distributions that are shifted relative to one another.

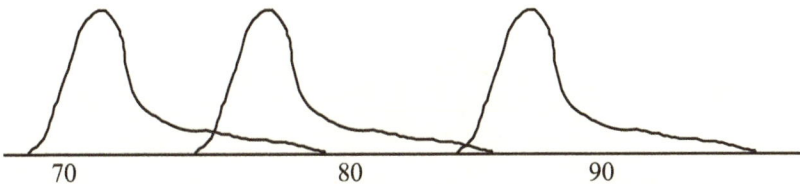

The Kruskal-Wallis test is based on the ranks of the observations that are obtained by combining the n observations in the G treatment groups. We will let $R_{i,j}$ be the rank of $X_{i,j}$ among the n observations. Thus, the smallest observation among the n observations will have a rank of 1 and the largest will have a rank of n. After we determine the ranks for the observations we can ignore the observations because the Kruskal-Wallis test is based solely on the ranks.

We begin the calculations by computing the average rank for each treatment group. For the ith treatment group the average rank is

$$\overline{R}_i = \frac{\sum_{j=1}^{n_i} R_{i,j}}{n_i}.$$

If the observations in one group tend to be greater than the observations in the other groups then the ranks will tend to be greater than the ranks in the other groups and the average rank in that group will be larger than the average ranks in the other groups. Conversely, if there is little difference between the populations then the average ranks in the groups will be approximately equal to the overall average rank, which equals $\overline{\overline{R}} = \frac{n+1}{2}$.

A test statistic that is sensitive to the differences in average rank between the groups is the Kruskal-Wallis test statistic, which is defined as

$$K = \frac{\sum_{i=1}^{G} n_i \left(\overline{R}_i - \overline{\overline{R}}\right)^2}{\left[\sum_{i=1}^{G}\sum_{j=1}^{n_i} \left(R_{i,j} - \overline{\overline{R}}\right)^2\right]\Big/(n-1)}.$$

If there are no ties in the data the denominator in this formula can be simplified to obtain

$$K = \frac{\sum_{i=1}^{G} n_i \left(\overline{R}_i - \overline{\overline{R}}\right)^2}{n(n+1)/12}.$$

Since ties are often present we will usually use the first formula.

We now need to determine how large K must be to reject the null hypothesis. To determine this we need the sampling distribution of the test statistic under the null hypothesis, which states that the distributions are identical. The exact sampling distribution is complicated, but for data sets that have group sizes that exceed 5, the sampling distribution of K approximates the chi-squared distribution with $G-1$ degrees of freedom. In this text we will ignore the exact distribution and use only the large sample approximation. This is not a severe limitation because, as we shall see in the next section, the Kruskal-Wallis test is recommended only for large samples.

Table 9.4 gives the ranks for the data given in Table 9.1. Note that the ranks are obtained by ranking over the 14 observations in the three treatment groups and that the average rank in the third group (Method C) is much larger than the average of the ranks in the other two groups.

Table 9.4 Ranks for the data in Table 9.1

Method A	Method B	Method C
$R_{1,1} = 4$	$R_{2,1} = 8$	$R_{3,1} = 12$
$R_{1,2} = 6$	$R_{2,2} = 2$	$R_{3,2} = 9$
$R_{1,3} = 3$	$R_{2,3} = 5$	$R_{3,3} = 14$
$R_{1,4} = 10$	$R_{2,4} = 7$	$R_{3,4} = 11$
	$R_{2,5} = 1$	$R_{3,5} = 13$
$\overline{R}_1 = 5.75$	$\overline{R}_2 = 4.60$	$\overline{R}_3 = 11.80$

For this data we compute the average rank $\overline{\overline{R}} = \frac{14+1}{2} = 7.5$ and determine the Kruskal-Wallis test statistic to be

$$K = \frac{4(5.75 - 7.5)^2 + 5(4.6 - 7.5)^2 + 5(11.8 - 7.5)^2}{17.5}$$

$$= \frac{12.25 + 42.05 + 92.45}{17.5} = \frac{146.75}{17.5} = 8.39$$

For an approximate p-value we refer to the chi-squared distribution with $G-1=2$ degrees of freedom, which gives an approximate p-value of

$p \cong .015$ for the Kruskal-Wallis test statistic of $K = 8.39$. The exact p-value, based on a table (see "A Nonparametric Introduction to Statistics" by Kraft and Van Eeden, 1968) is $p = .005$.

There were no ties in the hypothetical data given in Table 9.1. However, ties are often present in data sets. We deal with these ties by assigning midranks to the tied observations. For example, if we had obtained $x_{1,2} = 76$ instead of $x_{1,2} = 78$ then it would tie with $x_{2,3} = 76$ and we would assign the midranks of 5.5 to those two observations. We would then calculate $\overline{R}_1, \overline{R}_2$, and $\overline{R}_3$ based on these midranks and compute the test statistic

$$K = \frac{4(5.625 - 7.5)^2 + 5(4.7 - 7.5)^2 + 5(11.8 - 7.5)^2}{17.46}$$

$$= \frac{14.06 + 39.2 + 92.45}{17.46} = \frac{145.71}{17.46} = 8.35$$

We would then use the chi-squared distribution with 2 degrees of freedom to estimate the approximate p-value. The presence of ties makes it difficult to find an exact p-value for small samples because tables are not readily available for tied data.

Exercise Set C

1. The results of an experiment that was designed to evaluate the effectiveness of three weight loss programs were given in exercise 4 of the previous section. If possible, compute the Kruskal-Wallis test statistic using the summary statistics stated in that exercise. If it is not possible, state why the Kruskal-Wallis test cannot be computed.

2. Consider the following two data sets. In each of the data sets there were 5 observations in groups 1 and 2 and 7 observations in group 3. The observations in the first group were indicated by circles, the observations in the second group were indicated by squares, and the observations in the third group were indicated by triangles.

Data Set A

Data Set B

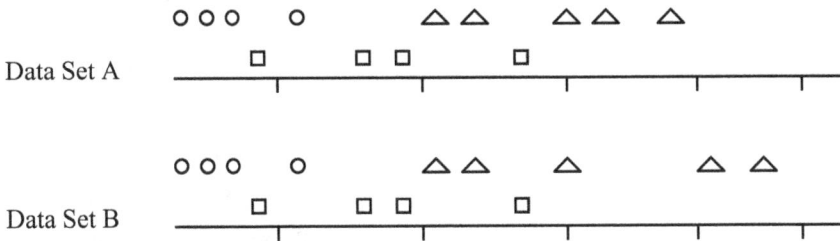

Is the Kruskal-Wallis statistic for the data in Data Set A greater than the Kruskal-Wallis statistic for the data in Data Set B ? Why or why not? No calculations should be necessary.

3. A golfer wanted to determine if there was any difference in the average driving distance between three golf balls. Seven balls of each type were obtained and a professional golfer was hired to drive the 21 golf balls. The professional did not know which balls she was hitting. The distances in yards for the 3 balls were recorded as follows:

Distance in Yards

Inexpensive	Moderately Priced	Expensive
182	208	201
193	215	218
202	221	213
187	212	209
194	217	204
186	205	207
172	210	211

a) Are there any ties in this data?

b) Compute the average ranks for each group.

c) Compute the Kruskal-Wallis test statistic.

d) Is the large sample approximation appropriate for computing the p-value? Why or why not?

e) Estimate the p-value.

f) Using $\alpha = .05$, would you reject the null hypothesis?

g) What is your conclusion concerning the driving distances of these golf balls?

4. A physician obtained informed consent from 24 female volunteers who agreed to participate in an experiment designed to evaluate the effectiveness of four programs to improve the iron status of women. The volunteers, who had been diagnosed as having an iron deficiency, were randomly assigned to one of four treatment groups and their serum levels (in micrograms per 100 milliliters) were obtained after 20 days of treatment. The results were recorded as follows:

| | Program | | |
A	B	C	D
102	101	93	114
105	89	91	110
111	88	83	117
98	93	97	93
113	97	94	115
107	102	87	116

a) Compute the average rank for each group.

b) Compute the Kruskal-Wallis test statistic.

c) Estimate the p-value. Was the large sample approximation appropriate?

d) Using $\alpha = .05$, would you reject the null hypothesis?

e) State your conclusion.

9.4 A Guide to the Selection of Tests for Three or More Samples.

When data is available from three or more samples the researcher must choose which test is most appropriate. First, it should be remembered that the methods presented in this chapter are suitable only for data obtained from independent samples. For these independent samples, each experimental unit falls into one, and only one, treatment group. If the experimenter uses three or more treatments but each experimental unit receives more than one treatment, then the observations are not independent. An example of one of these designs is the randomized complete block design in which each experimental unit is assigned to all treatments. For example, to evaluate the effectiveness of three drugs, the drugs may be given to all subjects who volunteered for an experiment. Since individuals tend to vary in their responsiveness to drugs, the observations are not independent. You need to fully understand exactly how the experiment was conducted to determine if the observations are independent or dependent. If they are independent then either the F test or the Kruskal-Wallis test can be used; if they are dependent other procedures must be used.

Once we have determined that the samples are independent we need to use the F test, which is the most powerful test if the observations are normally distributed, or the Kruskal-Wallis test, which may be more powerful than the F test if the observations are not normally distributed. This is not an easy decision because, in most circumstances, the distribution of the observations is not known. However, we may be able to decide which test to use based on the number of observations in the samples and on the general shape of the distribution. For example, for a study having 25 observations in each of 4 treatment groups, we may know that the measurements cannot be negative and that the observations are often, but not always, small positive values. This suggest that the distribution is skewed to the right. The objective would be to pick the most powerful test for studies having 25 observations in 4 groups with skewed distributions.

Fortunately, the choice of a test can often be based on the sample size. Research on the relative performance of the F test and the Kruskal-Wallis test (see "An adaptive test for the one-way layout" by O'Gorman, T. W. in *The Canadian Journal of Statistics*, 1997, pp. 269-279), shows that for studies having 10 or fewer observations per group the F test was more powerful than the Kruskal-Wallis for several distributions. However, for larger studies having more than 10 observations per group the Kruskal-Wallis test was the most powerful test for many distributions and the power of the F test never greatly exceeded that of the Kruskal-Wallis test. Based on these results the recommended strategy would be:

> If the size of all treatment groups exceeds 10 use the Kruskal-Wallis test; otherwise, the F test is recommended.

This strategy is useful for many real world situations, but there are some exceptions. If we knew that the distributions were approximately normal then we would use the F test, regardless of the sample size. On the other hand, if the measurements were so crude that we had little confidence in the values but we did have confidence in the ranks then we would use the Kruskal-Wallis test, regardless of the sample size. This might occur if a physician recorded his or her perception of the pain a patient experienced on a scale of 0 to 100. In this example the pain measures may not be too useful but the ranks based on those pain measures could be used in a Kruskal-Wallis test.

Exercise Set D

1. In exercise 3 of Exercise Set C, three types of golf balls were analyzed. Would you recommend that an F test or a Kruskal-Wallis test be used? Justify your choice.

2. In exercise 4 of Exercise Set D, a dietitian conducted an experiment to determine the most effective weight loss program. Would you recommend that an F test or a Kruskal-Wallis test be used? Justify your choice.

9.5 Individual and Multiple Comparisons

If we have data from three or more samples we will usually begin the analysis with an F test or a Kruskal-Wallis test. If we reject the null hypothesis that $\mu_1 = \mu_2 = \ldots = \mu_G$, we believe the alternative hypothesis is true, which states that $\mu_i \neq \mu_j$, for some i, j. However, if we reject the null hypothesis we know that at least one mean differs from one other mean, but the F test statistic and the Kruskal-Wallis test statistics do not give a clue about which population group means might be different. Consequently, we usually want to look more closely at the results from the experiment to decide which population means are different.

Typically, if we had $G = 3$ treatment groups and we performed an F test or a Kruskal-Wallis test and rejected the null hypothesis, we would want to follow up the overall test with comparisons of treatments. That is, we want to know if $\mu_1 = \mu_2$ or if $\mu_1 = \mu_3$ or if $\mu_2 = \mu_3$. For example, if the first group received the standard treatment and the other two groups received experimental treatments we would probably want to compare the experimental treatments to the standard treatment. Consequently, we would want to test $H_o : \mu_2 = \mu_1$ against the alternative $H_a : \mu_2 \neq \mu_1$ and to test $H_o : \mu_3 = \mu_1$ against the alternative $H_a : \mu_3 \neq \mu_1$. Comparisons of this type are very commonly done in applied research.

In general, if we want to test $H_o : \mu_i = \mu_j$ against the alternative $H_a : \mu_i \neq \mu_j$, we rely primary on the differences between the group means $\bar{x}_i$ and $\bar{x}_j$. To determine how large the difference $(\bar{x}_i - \bar{x}_j)$ must be before we reject the null hypothesis, we compare the difference to the standard error of the difference which is $\sqrt{MSE\left(\frac{1}{n_i} + \frac{1}{n_j}\right)}$. It can be shown, under $H_o : \mu_i = \mu_j$, that the **individual comparison test statistic**

$$T_{ij} = \frac{\bar{x}_i - \bar{x}_j}{\sqrt{MSE\left(\dfrac{1}{n_i} + \dfrac{1}{n_j}\right)}}$$

has a t distribution with $n - G$ degrees of freedom. Thus, if we wanted to perform a test of $H_o : \mu_i = \mu_j$ at the α level of significance against the two sided alternative $H_a : \mu_i \neq \mu_j$, we would reject the null hypothesis if $\left| T_{ij} \right| > t_{\alpha, n-G}$.

At first glance, the test statistic T_{ij} may appear to be the same test statistic as that used in the two-sample test, but there are important differences between the tests. The individual comparison test statistic (T_{ij}) uses the MSE, which is obtained from the observations in the G groups, as an estimate of the variance, whereas the two-sample t test statistic used only data from the two samples. Also, the individual comparison test uses a t distribution with $n - G$ degrees of freedom to compute the p-value while the two-sample test uses a t distribution with $n_1 + n_2 - 2$ degrees of freedom.

To illustrate this method consider again the data in Table 9.1 which had a mean squared error of $MSE = 40.55$. If we wanted to compare Method A to Method C and we had no prior knowledge of the direction of the difference between these two methods, we would test $H_o : \mu_1 = \mu_3$ at the α level of significance against the two sided alternative $H_a : \mu_1 \neq \mu_3$. The test statistic for this comparison would be

$$T_{13} = \frac{\bar{x}_1 - \bar{x}_3}{\sqrt{MSE\left(\frac{1}{n_1} + \frac{1}{n_3}\right)}} = \frac{76 - 92}{\sqrt{40.55\left(\frac{1}{5} + \frac{1}{4}\right)}} = \frac{-16}{4.27} = -3.75 .$$

To obtain the two-tailed p-value we look up $|T_{13}|$ in the t table with $n - G = 14 - 3 = 11$ degrees of freedom to find an approximate p-value of $p = .005$. Thus, we would reject the null hypothesis using $\alpha = .05$.

We could use the same approach to test the null hypothesis $H_o : \mu_1 = \mu_2$ against the two-sided alternative $H_a : \mu_1 \neq \mu_2$. We compute the test statistic $T_{12} = \frac{76-74}{4.27} = .47$ and refer to the t table with 11 degrees of freedom to find that $p > .4$. Our conclusion is that Method C is superior to Method A but we have insufficient evidence to conclude that Method B is superior or inferior to Method A.

If we had used the Kruskal-Wallis test instead of the F test, we would usually use the Wilcoxon rank sum test to do the individual comparisons between treatment groups. Let W_{ij} be the Wilcoxon rank sum statistic for comparing the ith treatment group and the jth treatment group. If the minimum group size exceeds 10 and if $|W_{ij}| > z_{\alpha/2}$ we would reject the null hypothesis that the population distribution for the ith group is identical to that for the jth group. Note that the ranks used to compute W_{ij} are based solely on the observations in the ith and jth treatment groups.

Multiple Comparisons

Researchers often need to make several individual comparisons after they perform an F test or a Kruskal-Wallis test. Suppose we have $G = 6$ groups in an experiment designed to test the effectiveness of 6 drugs in lowering blood pressure. If we wanted to compare the first drug to the other five drugs we would need to do 5 individual comparisons. If we wanted to compare every drug to every other drug there would be 15 individual comparisons. The problem with doing so many comparisons is that it is quite possible that one or more of the comparisons would be "significant " if each comparison uses the $\alpha = .05$ level of significance, even if there actually are no differences between the population means. That is, if there is no real difference between any of these drugs, there is a sizeable chance that one or more of these individual tests would be rejected if we use $\alpha = .05$ for each comparison. This is the multiple comparison problem.

The problem with multiple comparisons is quite serious because researchers may erroneously believe that there is a difference between two treatment when there really is no difference between them. To deal with this problem we must carefully distinguish between the significance level that we use in an individual comparison, which we will call the **comparisonwise error rate**, and the error rate for the experiment as a whole, which we will call the **experimentwise error rate**. The experimentwise error rate is the chance that one or more of the comparisons would be in error, over all the comparisons that are actually being performed. To make a distinction between these two error rates we will use α for the experimentwise error rate and will use α' for the comparisonwise error rate.

There are many multiple comparisons procedures available in statistical software packages. Rather than explore the extensive literature on these methods, we will concentrate on the **Bonferroni** multiple comparisons procedure, which is a simple procedure that is most suitable when a few comparisons are to be made. The idea behind this procedure is straightforward. If we want to have an experimentwise error rate of $\alpha = .05$, we should reduce the significance level for each individual comparison. It can be shown that if we set the comparisonwise error rate to

$$\alpha' = \frac{\alpha}{\text{Number of comparisons}},$$

then the experimentwise error rate will be no larger than α. For example, if we have $G = 6$ treatment groups and we want to compare the first group to each of the 5 other groups, we will test $H_o : \mu_1 = \mu_i$ against the two-sided alternative $H_a : \mu_1 \neq \mu_i$ for $i = 2,...,6$. If we want the experimentwise error rate to be $\alpha = .05$ after doing these 5 comparisons, then we need to set the comparisonwise error rate to

$$\alpha' = \frac{\alpha}{\text{Number of comparisons}} = \frac{.05}{5} = .01.$$

Thus, using the Bonferroni multiple comparison procedure we need to use a significance level of $\alpha' = .01$ for each of the five comparisons. If two-tailed tests were used we would use the critical value $t_{\alpha'/2, n-G}$. This will require a larger value of the test statistic $|T_{1i}|$ to reject the null hypothesis, which implies that we must see a larger difference between the treatment group means that would be required if we had used $\alpha = .05$ for each comparison. Thus, by using a multiple comparison procedure we are assured that the experimentwise error rate does not exceed α, but we pay a price for that assurance in the decreased sensitivity of the individual comparisons.

The Bonferroni approach can be used with the individual comparison t-test statistics T_{ij} or with individual comparison Wilcoxon test statistics W_{ij}. Usually, to compare the ith population with the jth population, we would compute the T_{ij} test statistic if we had used an F test, and would use the W_{ij} test statistic if we had used the Kruskal-Wallis test. Whatever approach is used, the significance level is computed using the comparisonwise error rate of $\alpha' = \frac{\alpha}{\text{number of comparisons}}$. For example, if $G = 6$ and we wanted to compare the first treatment to all the other treatments there would be 5 comparisons, which implies that $\alpha' = .05/5 = .01$. If there were $n = 36$ observations then the critical value for the individual t tests would be $t_{\alpha'/2, n-G} = t_{.005,30} = 2.75$. If individual Wilcoxon tests were used then the critical value would be $z_{\alpha'/2} = z_{.005} = 2.57$.

The Bonferroni multiple comparison procedure is a conservative approach in the sense that the true experimentwise error rate cannot exceed α. Unlike some other procedures, the Bonferroni approach can be used when the sample sizes in the treatment groups are unequal. However, the Bonferroni approach cannot be honestly used when the researcher uses the data to determine which comparisons should be performed. For example, if a researcher has $G = 6$ treatment groups and the researcher looks at the treatment means $\{\bar{x}_1, \bar{x}_2, \ldots, \bar{x}_6\}$ after the data is collected and determines that only two comparisons should be done, then the researcher may have consciously or unconsciously selected those comparisons that have large differences between the means.   The Bonferroni approach should not be used for this situation because the experimentwise error rate may greatly exceed α when the comparisons are selected after then data is analyzed. The Bonferroni multiple comparison procedure should only be used when the selection of the comparisons is not determined by the results of the experiment.

Exercise Set E

1. An experimenter used $G = 3$ groups with $n = 90$ subjects in an experiment and found the individual test statistic for comparing the population means of the first and third group to be $T_{13} = 1.93$. Another 30 subjects were added to a fourth experimental group so that $G = 4$. Although no observations had changed in the first three groups, the revised individual comparison statistic became $T_{13} = 1.85$. Explain how T_{13} could change. Which value(s) in T_{13} must have changed?

2. An experimenter used $G = 5$ treatment groups in a study and wanted to compare the 2nd, 3rd, 4th, and 5th groups to the first group. That is, the experimenter wanted to test the null hypotheses $H_o : \mu_1 = \mu_i$, for $i = 2, 3, 4,$ and 5. If the experimenter wanted to maintain an experimentwise error rate of $\alpha = .05$, describe how these individual comparisons would be made.

3. Suppose an experimenter used 5 individual comparisons in an experiment with a comparisonwise error rate of $\alpha' = .05$ for each of these tests. In your opinion what would the experimentwise error rate be closest to?

$$\alpha = .20 \qquad\qquad \alpha = .05 \qquad\qquad \alpha = .01$$

4. Exercise 4 in Exercise Set B contains summary statistics for the weight loss experienced by volunteers in a weight loss program. Use the Bonferroni multiple comparisons procedure to compare the "Exercise Only" group to the "Diet Only" group and to compare the "Diet and Exercise" group to the "Diet Only" group. Use the appropriate critical values necessary to maintain an experimentwise error rate of $\alpha = .05$.

Chapter Review Exercises

1. A researcher at a health service at a large university noticed that many of the students did not appear to be getting enough sleep. She selected a sample of $n = 20$ unmarried Juniors on a Wednesday in the 5th week of the semester. In an effort to determine which students got insufficient sleep, she obtained information on the amount of sleep they got the night before and on their living arrangements. The results for the three types of living arrangements were:

Dorm	Apartment	At Home
7.5	3.0	7.0
2.0	4.5	4.0
6.0	6.0	8.5
4.5	3.5	6.5
7.0	7.0	7.5
3.5	4.0	7.0
5.0	0.0	
$\bar{x}_1 = 5.07$	$\bar{x}_2 = 4.00$	$\bar{x}_3 = 6.75$
$s_1 = 1.95$	$s_2 = 2.25$	$s_3 = 1.51$

a) Compute $SS_{treatments}$, SS_{errors}, and SS_{total}.

b) Compute $DF_{treatments}$ and DF_{errors}.

c) Compute F.

d) Use the results to fill in the ANOVA table.

e) Estimate the p-value associated with the F test statistic.

2. Suppose that, in the previous exercise, the data had been recorded in minutes, instead of hours.

a) How would the $SS_{treatments}$ change?

b) How would the SS_{errors} change?

c) Would F change? Why or why not?

3. A petroleum company wanted to evaluate the effectiveness of three types of motor oil on gas mileage. Mechanics at a highway patrol fleet maintenance shop cooperated with the petroleum company in an experiment involving $n = 43$ cars. The summary statistics for the three groups are:

Type A	Type B	Type C
$n_1 = 15$	$n_2 = 15$	$n_3 = 13$
$\bar{x}_1 = 18.7$	$\bar{x}_2 = 19.2$	$\bar{x}_3 = 16.2$
$s_1 = 1.4$	$s_2 = 1.9$	$s_3 = 1.7$

a) Fill in the ANOVA table.

b) Estimate the p-value.

c) State your conclusion.

4. A dietitian wanted to compare the coffee consumption of Business students, Liberal Arts students, and Engineering students. She randomly selected students at the university and determined their average daily coffee consumption to be:

Business	Liberal Arts	Engineering
3.0	1.0	0.0
2.0	2.0	1.5
2.5	3.5	2.0
1.5	2.0	3.0
4.5	8.0	1.0
0.0	1.5	
	2.5	
	6.0	
	5.0	

a) Compute the average rank for each group.

b) Compute the Kruskal-Wallis test.

c) Estimate the p-value and state your conclusion.

5. An exercise scientist wanted to determine the relationship between academic achievement and exercise. She obtained the number of hours spent per week exercising for 385 students who had been classified into 4 categories (below average, average, above average, superior). The summary statistics for these four groups are:

Below Average	Average	Above Average	Superior
$n_1 = 88$	$n_2 = 143$	$n_3 = 97$	$n_4 = 57$
$\bar{x}_1 = 2.45$	$\bar{x}_2 = 2.53$	$\bar{x}_3 = 4.21$	$\bar{x}_4 = 4.14$
$s_1 = 2.61$	$s_2 = 2.41$	$s_3 = 2.96$	$s_4 = 2.78$

a) Compute $SS_{treatments}$, $DF_{treatments}$, and $MS_{treatments}$.

b) Compute SS_{errors}, DF_{errors}, and MSE.

c) Compute the F test statistic and roughly estimate the p-value.

d) Using a significance level of $\alpha = .05$, would you reject the null hypothesis?

6. Suppose that, in the previous exercise, the exercise scientist had decided, before the data was collected, that she would compare the below average students to the average students, the above average students to the average students, and the above average students to the average students. Suppose further that she wanted to maintain an overall experimentwise error rate of $\alpha = .05$.

   a) What is the approximate critical value that should be used for these three tests?

   b) Perform the tests and state the conclusion of each.

7. In exercise 5 the exercise scientist used an F test. If you had the raw data would you recommend an F test or a Kruskal-Wallis test? Is it possible to compute the Kruskal-Wallis test using the summary data provided in exercise 5? Explain.

8. A physician was interested in the possible relationship between weight and inactivity in adolescents. He classified adolescents into three weight categories (Underweight, Normal weight, and Overweight) based on their weight and height. He also obtained the number of hours of television viewing, not including time spent playing video games. The data was recorded as follows:

Underweight	Normal weight	Overweight
8	11	21
65	17	77
2	6	19
16	25	42
18	36	22
7	57	23
19	21	31
5	15	36
$\bar{x}_1 = 17.5$	$\bar{x}_2 = 23.5$	$\bar{x}_3 = 33.875$
$s_1 = 20.21$	$s_2 = 16.32$	$s_3 = 19.22$

a) Compute $SS_{treatments}$, $DF_{treatments}$, and $MS_{treatments}$.

b) Compute SS_{errors}, DF_{errors}, and MSE.

c) Compute the F test statistic and roughly estimate the p-value.

d) Using a significance level of $\alpha = .05$, would you reject the null hypothesis?

9. For the data given in the previous exercise, compute the Kruskal-Wallis test statistic and perform the Kruskal-Wallis test using $\alpha = .05$.   The average of the ranks for the three groups are $\overline{R}_1 = 8.0625$, $\overline{R}_2 = 12.125$, and $\overline{R}_3 = 17.3125$.   State your conclusion.

10. Compare the conclusion you reached in exercise 8 to the conclusion you reached in exercise 9. For many data sets of this size, the p-values obtained from the F test and Kruskal-Wallis test would not be dramatically different. Why do you suppose the differences are large for this data set? [Hint: Considering the large differences in the means, what caused F to be small?]

Tables

Table 1. Upper-tail Probabilities for the Standard Normal Distribution. For negative values of z.

z	0.00	0.01	0.02	0.03	0.04
-3.0	0.9987	0.9987	0.9987	0.9988	0.9988
-2.9	0.9981	0.9982	0.9982	0.9983	0.9984
-2.8	0.9974	0.9975	0.9976	0.9977	0.9977
-2.7	0.9965	0.9966	0.9967	0.9968	0.9969
-2.6	0.9953	0.9955	0.9956	0.9957	0.9959
-2.5	0.9938	0.9940	0.9941	0.9943	0.9945
-2.4	0.9918	0.9920	0.9922	0.9925	0.9927
-2.3	0.9893	0.9896	0.9898	0.9901	0.9904
-2.2	0.9861	0.9864	0.9868	0.9871	0.9875
-2.1	0.9821	0.9826	0.9830	0.9834	0.9838
-2.0	0.9772	0.9778	0.9783	0.9788	0.9793
-1.9	0.9713	0.9719	0.9726	0.9732	0.9738
-1.8	0.9641	0.9649	0.9656	0.9664	0.9671
-1.7	0.9554	0.9564	0.9573	0.9582	0.9591
-1.6	0.9452	0.9463	0.9474	0.9484	0.9495
-1.5	0.9332	0.9345	0.9357	0.9370	0.9382
-1.4	0.9192	0.9207	0.9222	0.9236	0.9251
-1.3	0.9032	0.9049	0.9066	0.9082	0.9099
-1.2	0.8849	0.8869	0.8888	0.8907	0.8925
-1.1	0.8643	0.8665	0.8686	0.8708	0.8729
-1.0	0.8413	0.8438	0.8461	0.8485	0.8508
-0.9	0.8159	0.8186	0.8212	0.8238	0.8264
-0.8	0.7881	0.7910	0.7939	0.7967	0.7995
-0.7	0.7580	0.7611	0.7642	0.7673	0.7704
-0.6	0.7257	0.7291	0.7324	0.7357	0.7389
-0.5	0.6915	0.6950	0.6985	0.7019	0.7054
-0.4	0.6554	0.6591	0.6628	0.6664	0.6700
-0.3	0.6179	0.6217	0.6255	0.6293	0.6331
-0.2	0.5793	0.5832	0.5871	0.5910	0.5948
-0.1	0.5398	0.5438	0.5478	0.5517	0.5557
0.0	0.5000	0.5040	0.5080	0.5120	0.5160

Table 1. Upper-tail Probabilities for the Standard Normal Distribution
(continued). For negative values of z.

z	0.05	0.06	0.07	0.08	0.09
-3.0	0.9989	0.9989	0.9989	0.9990	0.9990
-2.9	0.9984	0.9985	0.9985	0.9986	0.9986
-2.8	0.9978	0.9979	0.9979	0.9980	0.9981
-2.7	0.9970	0.9971	0.9972	0.9973	0.9974
-2.6	0.9960	0.9961	0.9962	0.9963	0.9964
-2.5	0.9946	0.9948	0.9949	0.9951	0.9952
-2.4	0.9929	0.9931	0.9932	0.9934	0.9936
-2.3	0.9906	0.9909	0.9911	0.9913	0.9916
-2.2	0.9878	0.9881	0.9884	0.9887	0.9890
-2.1	0.9842	0.9846	0.9850	0.9854	0.9857
-2.0	0.9798	0.9803	0.9808	0.9812	0.9817
-1.9	0.9744	0.9750	0.9756	0.9761	0.9767
-1.8	0.9678	0.9686	0.9693	0.9699	0.9706
-1.7	0.9599	0.9608	0.9616	0.9625	0.9633
-1.6	0.9505	0.9515	0.9525	0.9535	0.9545
-1.5	0.9394	0.9406	0.9418	0.9429	0.9441
-1.4	0.9265	0.9279	0.9292	0.9306	0.9319
-1.3	0.9115	0.9131	0.9147	0.9162	0.9177
-1.2	0.8944	0.8962	0.8980	0.8997	0.9015
-1.1	0.8749	0.8770	0.8790	0.8810	0.8830
-1.0	0.8531	0.8554	0.8577	0.8599	0.8621
-0.9	0.8289	0.8315	0.8340	0.8365	0.8389
-0.8	0.8023	0.8051	0.8078	0.8106	0.8133
-0.7	0.7734	0.7764	0.7794	0.7823	0.7852
-0.6	0.7422	0.7454	0.7486	0.7517	0.7549
-0.5	0.7088	0.7123	0.7157	0.7190	0.7224
-0.4	0.6736	0.6772	0.6808	0.6844	0.6879
-0.3	0.6368	0.6406	0.6443	0.6480	0.6517
-0.2	0.5987	0.6026	0.6064	0.6103	0.6141
-0.1	0.5596	0.5636	0.5675	0.5714	0.5753
0.0	0.5199	0.5239	0.5279	0.5319	0.5359

Table 1. Upper-tail Probabilities for the Standard Normal Distribution.
(continued) For positive values of z.

z	0.00	0.01	0.02	0.03	0.04
0.0	0.5000	0.4960	0.4920	0.4880	0.4840
0.1	0.4602	0.4562	0.4522	0.4483	0.4443
0.2	0.4207	0.4168	0.4129	0.4090	0.4052
0.3	0.3821	0.3783	0.3745	0.3707	0.3669
0.4	0.3446	0.3409	0.3372	0.3336	0.3300
0.5	0.3085	0.3050	0.3015	0.2981	0.2946
0.6	0.2743	0.2709	0.2676	0.2643	0.2611
0.7	0.2420	0.2389	0.2358	0.2327	0.2296
0.8	0.2119	0.2090	0.2061	0.2033	0.2005
0.9	0.1841	0.1814	0.1788	0.1762	0.1736
1.0	0.1587	0.1562	0.1539	0.1515	0.1492
1.1	0.1357	0.1335	0.1314	0.1292	0.1271
1.2	0.1151	0.1131	0.1112	0.1093	0.1075
1.3	0.0968	0.0951	0.0934	0.0918	0.0901
1.4	0.0808	0.0793	0.0778	0.0764	0.0749
1.5	0.0668	0.0655	0.0643	0.0630	0.0618
1.6	0.0548	0.0537	0.0526	0.0516	0.0505
1.7	0.0446	0.0436	0.0427	0.0418	0.0409
1.8	0.0359	0.0351	0.0344	0.0336	0.0329
1.9	0.0287	0.0281	0.0274	0.0268	0.0262
2.0	0.0228	0.0222	0.0217	0.0212	0.0207
2.1	0.0179	0.0174	0.0170	0.0166	0.0162
2.2	0.0139	0.0136	0.0132	0.0129	0.0125
2.3	0.0107	0.0104	0.0102	0.0099	0.0096
2.4	0.0082	0.0080	0.0078	0.0075	0.0073
2.5	0.0062	0.0060	0.0059	0.0057	0.0055
2.6	0.0047	0.0045	0.0044	0.0043	0.0041
2.7	0.0035	0.0034	0.0033	0.0032	0.0031
2.8	0.0026	0.0025	0.0024	0.0023	0.0023
2.9	0.0019	0.0018	0.0018	0.0017	0.0016
3.0	0.0013	0.0013	0.0013	0.0012	0.0012

Table 1. Upper-tail Probabilities for the Standard Normal Distribution.

(continued) For positive values of z.

z	0.05	0.06	0.07	0.08	0.09
0.0	0.4801	0.4761	0.4721	0.4681	0.4641
0.1	0.4404	0.4364	0.4325	0.4286	0.4247
0.2	0.4013	0.3974	0.3936	0.3897	0.3859
0.3	0.3632	0.3594	0.3557	0.3520	0.3483
0.4	0.3264	0.3228	0.3192	0.3156	0.3121
0.5	0.2912	0.2877	0.2843	0.2810	0.2776
0.6	0.2578	0.2546	0.2514	0.2483	0.2451
0.7	0.2266	0.2236	0.2206	0.2177	0.2148
0.8	0.1977	0.1949	0.1922	0.1894	0.1867
0.9	0.1711	0.1685	0.1660	0.1635	0.1611
1.0	0.1469	0.1446	0.1423	0.1401	0.1379
1.1	0.1251	0.1230	0.1210	0.1190	0.1170
1.2	0.1056	0.1038	0.1020	0.1003	0.0985
1.3	0.0885	0.0869	0.0853	0.0838	0.0823
1.4	0.0735	0.0721	0.0708	0.0694	0.0681
1.5	0.0606	0.0594	0.0582	0.0571	0.0559
1.6	0.0495	0.0485	0.0475	0.0465	0.0455
1.7	0.0401	0.0392	0.0384	0.0375	0.0367
1.8	0.0322	0.0314	0.0307	0.0301	0.0294
1.9	0.0256	0.0250	0.0244	0.0239	0.0233
2.0	0.0202	0.0197	0.0192	0.0188	0.0183
2.1	0.0158	0.0154	0.0150	0.0146	0.0143
2.2	0.0122	0.0119	0.0116	0.0113	0.0110
2.3	0.0094	0.0091	0.0089	0.0087	0.0084
2.4	0.0071	0.0069	0.0068	0.0066	0.0064
2.5	0.0054	0.0052	0.0051	0.0049	0.0048
2.6	0.0040	0.0039	0.0038	0.0037	0.0036
2.7	0.0030	0.0029	0.0028	0.0027	0.0026
2.8	0.0022	0.0021	0.0021	0.0020	0.0019
2.9	0.0016	0.0015	0.0015	0.0014	0.0014
3.0	0.0011	0.0011	0.0011	0.0010	0.0010

Table 2. t distribution critical values.

one-tail p	0.2	0.1	0.075	0.05	0.04
two-tailed p	0.4	0.2	0.15	0.1	0.08
d.f.					
1	1.38	3.08	4.17	6.31	7.92
2	1.06	1.89	2.28	2.92	3.32
3	0.98	1.64	1.92	2.35	2.61
4	0.94	1.53	1.78	2.13	2.33
5	0.92	1.48	1.70	2.02	2.19
6	0.91	1.44	1.65	1.94	2.10
7	0.90	1.41	1.62	1.89	2.05
8	0.89	1.40	1.59	1.86	2.00
9	0.88	1.38	1.57	1.83	1.97
10	0.88	1.37	1.56	1.81	1.95
11	0.88	1.36	1.55	1.80	1.93
12	0.87	1.36	1.54	1.78	1.91
13	0.87	1.35	1.53	1.77	1.90
14	0.87	1.35	1.52	1.76	1.89
15	0.87	1.34	1.52	1.75	1.88
16	0.86	1.34	1.51	1.75	1.87
17	0.86	1.33	1.51	1.74	1.86
18	0.86	1.33	1.50	1.73	1.86
19	0.86	1.33	1.50	1.73	1.85
20	0.86	1.33	1.50	1.72	1.84
21	0.86	1.32	1.49	1.72	1.84
22	0.86	1.32	1.49	1.72	1.84
23	0.86	1.32	1.49	1.71	1.83
24	0.86	1.32	1.49	1.71	1.83
25	0.86	1.32	1.49	1.71	1.82
26	0.86	1.31	1.48	1.71	1.82
27	0.86	1.31	1.48	1.70	1.82
28	0.85	1.31	1.48	1.70	1.82
29	0.85	1.31	1.48	1.70	1.81
30	0.85	1.31	1.48	1.70	1.81
35	0.85	1.31	1.47	1.69	1.80
40	0.85	1.30	1.47	1.68	1.80
60	0.85	1.30	1.46	1.67	1.78
80	0.85	1.29	1.45	1.66	1.77
1000	0.84	1.28	1.44	1.65	1.75

Table 2. t distribution critical values. (Continued)

one-tail p	0.03	0.025	0.02	0.01	0.005	0.001
two-tailed p	0.06	0.05	0.04	0.02	0.01	0.002
d.f.						
1	10.58	12.71	15.89	31.82	63.66	318.29
2	3.90	4.30	4.85	6.96	9.92	22.33
3	2.95	3.18	3.48	4.54	5.84	10.21
4	2.60	2.78	3.00	3.75	4.60	7.17
5	2.42	2.57	2.76	3.36	4.03	5.89
6	2.31	2.45	2.61	3.14	3.71	5.21
7	2.24	2.36	2.52	3.00	3.50	4.79
8	2.19	2.31	2.45	2.90	3.36	4.50
9	2.15	2.26	2.40	2.82	3.25	4.30
10	2.12	2.23	2.36	2.76	3.17	4.14
11	2.10	2.20	2.33	2.72	3.11	4.02
12	2.08	2.18	2.30	2.68	3.05	3.93
13	2.06	2.16	2.28	2.65	3.01	3.85
14	2.05	2.14	2.26	2.62	2.98	3.79
15	2.03	2.13	2.25	2.60	2.95	3.73
16	2.02	2.12	2.24	2.58	2.92	3.69
17	2.02	2.11	2.22	2.57	2.90	3.65
18	2.01	2.10	2.21	2.55	2.88	3.61
19	2.00	2.09	2.20	2.54	2.86	3.58
20	1.99	2.09	2.20	2.53	2.85	3.55
21	1.99	2.08	2.19	2.52	2.83	3.53
22	1.98	2.07	2.18	2.51	2.82	3.50
23	1.98	2.07	2.18	2.50	2.81	3.48
24	1.97	2.06	2.17	2.49	2.80	3.47
25	1.97	2.06	2.17	2.49	2.79	3.45
26	1.97	2.06	2.16	2.48	2.78	3.43
27	1.96	2.05	2.16	2.47	2.77	3.42
28	1.96	2.05	2.15	2.47	2.76	3.41
29	1.96	2.05	2.15	2.46	2.76	3.40
30	1.95	2.04	2.15	2.46	2.75	3.39
35	1.94	2.03	2.13	2.44	2.72	3.34
40	1.94	2.02	2.12	2.42	2.70	3.31
60	1.92	2.00	2.10	2.39	2.66	3.23
80	1.91	1.99	2.09	2.37	2.64	3.20
1000	1.88	1.96	2.06	2.33	2.58	3.10

Table 3. Signed-rank table for $n \leq 14$.

Sample size = n

S	1	2	3	4	5	6	7	8	9	10
0	1.000	1.000	1.000	1.000	1.000	1.000	1.000	1.000	1.000	1.000
1	0.500	0.750	0.875	0.938	0.969	0.984	0.992	0.996	0.998	0.999
2		0.500	0.750	0.875	0.938	0.969	0.984	0.992	0.996	0.998
3		0.250	0.625	0.813	0.906	0.953	0.977	0.988	0.994	0.997
4			0.375	0.688	0.844	0.922	0.961	0.980	0.990	0.995
5			0.250	0.563	0.781	0.891	0.945	0.973	0.986	0.993
6			0.125	0.438	0.688	0.844	0.922	0.961	0.980	0.990
7				0.313	0.594	0.781	0.891	0.945	0.973	0.986
8				0.188	0.500	0.719	0.852	0.926	0.963	0.981
9				0.125	0.406	0.656	0.813	0.902	0.951	0.976
10				0.063	0.313	0.578	0.766	0.875	0.936	0.968
11					0.219	0.500	0.711	0.844	0.918	0.958
12					0.156	0.422	0.656	0.809	0.898	0.947
13					0.094	0.344	0.594	0.770	0.875	0.935
14					0.063	0.281	0.531	0.727	0.850	0.920
15					0.031	0.219	0.469	0.680	0.820	0.903
16						0.156	0.406	0.629	0.787	0.884
17						0.109	0.344	0.578	0.752	0.862
18						0.078	0.289	0.527	0.715	0.839
19						0.047	0.234	0.473	0.674	0.813
20						0.031	0.188	0.422	0.633	0.784
21						0.016	0.148	0.371	0.590	0.754
22							0.109	0.320	0.545	0.722
23							0.078	0.273	0.500	0.688
24							0.055	0.230	0.455	0.652
25							0.039	0.191	0.410	0.615
26							0.023	0.156	0.367	0.577
27							0.016	0.125	0.326	0.539

Table 3. Signed-rank table for $n \leq 14$. (Continued)

Sample size = n

S	1	2	3	4	5	6	7	8	9	10
28							0.008	0.098	0.285	0.500
29								0.074	0.248	0.461
30								0.055	0.213	0.423
31								0.039	0.180	0.385
32								0.027	0.150	0.348
33								0.020	0.125	0.313
34								0.012	0.102	0.278
35								0.008	0.082	0.246
36								0.004	0.064	0.216
37									0.049	0.188
38									0.037	0.161
39									0.027	0.138
40									0.020	0.116
41									0.014	0.097
42									0.010	0.080
43									0.006	0.065
44									0.004	0.053
45									0.002	0.042
46										0.032
47										0.024
48										0.019
49										0.014
50										0.010
51										0.007
52										0.005
53										0.003
54										0.002
55										0.001

Table 3. Signed-rank table for $n \leq 14$. (Continued)

Sample size=n

S	11	12	13	14
40	0.289	0.485	0.658	0.787
41	0.260	0.455	0.632	0.768
42	0.232	0.425	0.607	0.749
43	0.207	0.396	0.580	0.729
44	0.183	0.367	0.554	0.708
45	0.160	0.339	0.527	0.687
46	0.139	0.311	0.500	0.665
47	0.120	0.285	0.473	0.643
48	0.103	0.259	0.446	0.620
49	0.087	0.235	0.420	0.596
50	0.074	0.212	0.393	0.572
51	0.062	0.190	0.368	0.548
52	0.051	0.170	0.342	0.524
53	0.042	0.151	0.318	0.500
54	0.034	0.133	0.294	0.476
55	0.027	0.117	0.271	0.452
56	0.021	0.102	0.249	0.428
57	0.016	0.088	0.227	0.404
58	0.012	0.076	0.207	0.380
59	0.009	0.065	0.188	0.357
60	0.007	0.055	0.170	0.335
61	0.005	0.046	0.153	0.313
62	0.003	0.039	0.137	0.292
63	0.002	0.032	0.122	0.271
64	0.001	0.026	0.108	0.251
65	0.001	0.021	0.095	0.232
66	0.000	0.017	0.084	0.213
67		0.013	0.073	0.195

Table 3. Signed-rank table for $n \leq 14$. (Continued)

Sample size=n

S	11	12	13	14
68		0.010	0.064	0.179
69		0.008	0.055	0.163
70		0.006	0.047	0.148
71		0.005	0.040	0.134
72		0.003	0.034	0.121
73		0.002	0.029	0.108
74		0.002	0.024	0.097
75		0.001	0.020	0.086
76		0.001	0.016	0.077
77		0.000	0.013	0.068
78		0.000	0.011	0.059
79			0.009	0.052
80			0.007	0.045
81			0.005	0.039
82			0.004	0.034
83			0.003	0.029
84			0.002	0.025
85			0.002	0.021
86			0.001	0.018
87			0.001	0.015
88			0.001	0.012
89			0.000	0.010
90			0.000	0.008
91			0.000	0.007
92				0.005
93				0.004
94				0.003
95				0.003

Table 4. Rank-sum table for $n_1, n_2 \leq 10$.

n1= 2 n2= 2

W	p
5	0.667
6	0.333
7	0.167

n1= 2 n2= 3

W	p
9	0.600
10	0.400
11	0.200
12	0.100

n1= 2 n2= 4

W	p
14	0.600
15	0.400
16	0.267
17	0.133
18	0.067

n1= 2 n2= 5

W	p
20	0.571
21	0.429
22	0.286
23	0.190
24	0.095
25	0.048

n1= 2 n2= 6

W	p
27	0.571
28	0.429
29	0.321
30	0.214
31	0.143
32	0.071
33	0.036

n1= 2 n2= 7

W	p
35	0.556
36	0.444
37	0.333
38	0.250
39	0.167
40	0.111
41	0.056
42	0.028

n1= 2 n2= 8

W	p
44	0.556
45	0.444
46	0.356
47	0.267
48	0.200
49	0.133
50	0.089
51	0.044
52	0.022

n1= 2 n2= 9

W	p
54	0.545
55	0.455
56	0.364
57	0.291
58	0.218
59	0.164
60	0.109
61	0.073
62	0.036
63	0.018

n1= 2 n2=10

W	p
65	0.545
66	0.455
67	0.379
68	0.303
69	0.242
70	0.182
71	0.136
72	0.091
73	0.061
74	0.030
75	0.015

n1= 3 n2= 2

W	p
6	0.600
7	0.400
8	0.200
9	0.100

n1= 3 n2= 3

W	p
11	0.500
12	0.350
13	0.200
14	0.100
15	0.050

n1= 3 n2= 4

W	p
16	0.571
17	0.429
18	0.314
19	0.200
20	0.114
21	0.057
22	0.029

n1= 3 n2= 5

W	p
23	0.500
24	0.393
25	0.286
26	0.196
27	0.125
28	0.071
29	0.036
30	0.018

n1= 3 n2= 6

W	p
30	0.548
31	0.452
32	0.357
33	0.274
34	0.190
35	0.131
36	0.083
37	0.048
38	0.024
39	0.012

n1= 3 n2= 7

W	p
39	0.500
40	0.417
41	0.333
42	0.258
43	0.192
44	0.133
45	0.092
46	0.058
47	0.033
48	0.017
49	0.008

n1= 3 n2= 8

W	p
48	0.539
49	0.461
50	0.388
51	0.315
52	0.248
53	0.188
54	0.139
55	0.097
56	0.067
57	0.042
58	0.024
59	0.012
60	0.006

n1= 3 n2= 9

W	p
59	0.500
60	0.432
61	0.364
62	0.300
63	0.241
64	0.186
65	0.141
66	0.105
67	0.073
68	0.050
69	0.032
70	0.018
71	0.009
72	0.005

n1= 3 n2=10

W	p
70	0.531
71	0.469
72	0.406
73	0.346
74	0.287
75	0.234
76	0.185
77	0.143
78	0.108
79	0.080
80	0.056
81	0.038
82	0.024
83	0.014
84	0.007
85	0.003

n1= 4 n2= 2

W	p
7	0.600
8	0.400
9	0.267
10	0.133
11	0.067

n1= 4 n2= 3

W	p

W	p
12	0.571
13	0.429
14	0.314
15	0.200
16	0.114
17	0.057
18	0.029

n1= 4 n2= 4

W	p
18	0.557
19	0.443
20	0.343
21	0.243
22	0.171
23	0.100
24	0.057
25	0.029
26	0.014

n1= 4 n2= 5

W	p
25	0.548
26	0.452
27	0.365
28	0.278
29	0.206
30	0.143
31	0.095
32	0.056
33	0.032
34	0.016
35	0.008

n1= 4 n2= 6

W	p
33	0.543
34	0.457
35	0.381
36	0.305
37	0.238
38	0.176
39	0.129
40	0.086
41	0.057
42	0.033
43	0.019
44	0.010
45	0.005

n1= 4 n2= 7

W	p
42	0.536
43	0.464
44	0.394
45	0.324
46	0.264
47	0.206
48	0.158
49	0.115
50	0.082
51	0.055
52	0.036
53	0.021
54	0.012
55	0.006
56	0.003

n1= 4 n2= 8

W	p
52	0.533
53	0.467
54	0.404
55	0.341
56	0.285
57	0.230
58	0.184
59	0.141
60	0.107
61	0.077
62	0.055
63	0.036
64	0.024
65	0.014
66	0.008
67	0.004
68	0.002

n1= 4 n2= 9

W	p
63	0.530
64	0.470
65	0.413
66	0.355
67	0.302
68	0.252
69	0.207
70	0.165
71	0.130
72	0.099

W	p
73	0.074
74	0.053
75	0.038
76	0.025
77	0.017
78	0.010
79	0.006
80	0.003
81	0.001

n1= 4 n2=10

W	p
75	0.527
76	0.473
77	0.420
78	0.367
79	0.318
80	0.270
81	0.227
82	0.187
83	0.152
84	0.120
85	0.094
86	0.071
87	0.053
88	0.038
89	0.027
90	0.018
91	0.012
92	0.007
93	0.004
94	0.002
95	0.001

Table 4. Rank-sum table for $n_1, n_2 \leq 10$. (continued)

n1= 5 n2= 2			n1= 5 n2= 8		n1= 5 n2=10	

n1= 5 n2= 2

W	p
8	0.571
9	0.429
10	0.286
11	0.190
12	0.095
13	0.048

n1= 5 n2= 3

W	p
14	0.500
15	0.393
16	0.286
17	0.196
18	0.125
19	0.071
20	0.036
21	0.018

n1= 5 n2= 4

W	p
20	0.548
21	0.452
22	0.365
23	0.278
24	0.206
25	0.143
26	0.095
27	0.056
28	0.032
29	0.016
30	0.008

n1= 5 n2= 5

W	p
28	0.500
29	0.421
30	0.345
31	0.274
32	0.210
33	0.155
34	0.111
35	0.075
36	0.048
37	0.028
38	0.016
39	0.008
40	0.004

n1= 5 n2= 6

W	p
36	0.535
37	0.465
38	0.396
39	0.331
40	0.268
41	0.214
42	0.165
43	0.123
44	0.089
45	0.063
46	0.041
47	0.026
48	0.015
49	0.009
50	0.004
51	0.002

n1= 5 n2= 7

W	p
46	0.500
47	0.438
48	0.378
49	0.319
50	0.265
51	0.216
52	0.172
53	0.134
54	0.101
55	0.074
56	0.053
57	0.037
58	0.024
59	0.015
60	0.009
61	0.005
62	0.003
63	0.001

n1= 5 n2= 8

W	p
56	0.528
57	0.472
58	0.416
59	0.362
60	0.311
61	0.262
62	0.218
63	0.177
64	0.142
65	0.111
66	0.085
67	0.064
68	0.047
69	0.033
70	0.023
71	0.015
72	0.009
73	0.005
74	0.003
75	0.002
76	0.001

n1= 5 n2= 9

W	p
68	0.500
69	0.449
70	0.399
71	0.350
72	0.303
73	0.259
74	0.219
75	0.182
76	0.149
77	0.120
78	0.095
79	0.073
80	0.056
81	0.041
82	0.030
83	0.021
84	0.014
85	0.009
86	0.006
87	0.003
88	0.002
89	0.001
90	0.000

n1= 5 n2=10

W	p
80	0.523
81	0.477
82	0.430
83	0.384
84	0.339
85	0.297
86	0.257
87	0.220
88	0.185
89	0.155
90	0.127
91	0.103
92	0.082
93	0.065
94	0.050
95	0.038
96	0.028
97	0.020
98	0.014
99	0.010
100	0.006
101	0.004
102	0.002
103	0.001
104	0.001
105	0.000

n1= 6 n2= 2

W	p
9	0.571
10	0.429
11	0.321
12	0.214
13	0.143
14	0.071
15	0.036

n1= 6 n2= 3

W	p
15	0.548
16	0.452
17	0.357
18	0.274
19	0.190
20	0.131
21	0.083
22	0.048
23	0.024
24	0.012

n1= 6 n2= 4

W	p
22	0.543
23	0.457
24	0.381
25	0.305
26	0.238
27	0.176
28	0.129
29	0.086
30	0.057
31	0.033
32	0.019
33	0.010
34	0.005

n1= 6 n2= 5

W	p
30	0.535
31	0.465
32	0.396
33	0.331
34	0.268
35	0.214
36	0.165
37	0.123
38	0.089
39	0.063
40	0.041
41	0.026
42	0.015
43	0.009
44	0.004
45	0.002

n1= 6 n2= 6

W	p

W	p
39	0.531
40	0.469
41	0.409
42	0.350
43	0.294
44	0.242
45	0.197
46	0.155
47	0.120
48	0.090
49	0.066
50	0.047
51	0.032
52	0.021
53	0.013
54	0.008
55	0.004
56	0.002
57	0.001

n1= 6 n2= 7

W	p
49	0.527
50	0.473
51	0.418
52	0.365
53	0.314
54	0.267
55	0.223
56	0.183
57	0.147
58	0.117
59	0.090
60	0.069
61	0.051
62	0.037
63	0.026
64	0.017
65	0.011
66	0.007
67	0.004
68	0.002
69	0.001
70	0.001

n1= 6 n2= 8

W	p
60	0.525
61	0.475
62	0.426
63	0.377
64	0.331
65	0.286

W	p
66	0.245
67	0.207
68	0.172
69	0.141
70	0.114
71	0.091
72	0.071
73	0.054
74	0.041
75	0.030
76	0.021
77	0.015
78	0.010
79	0.006
80	0.004
81	0.002
82	0.001
83	0.001
84	0.000

n1= 6 n2= 9

W	p
72	0.523
73	0.477
74	0.432
75	0.388
76	0.344
77	0.303
78	0.264
79	0.228
80	0.194
81	0.164
82	0.136
83	0.112
84	0.091
85	0.072
86	0.057
87	0.044
88	0.033
89	0.025
90	0.018
91	0.013
92	0.009
93	0.006
94	0.004
95	0.002
96	0.001
97	0.001
98	0.000
99	0.000

n1= 6 n2=10

W	p

W	p
85	0.521
86	0.479
87	0.437
88	0.396
89	0.356
90	0.318
91	0.281
92	0.246
93	0.214
94	0.184
95	0.157
96	0.132
97	0.110
98	0.090
99	0.074
100	0.059
101	0.047
102	0.036
103	0.028
104	0.021
105	0.016
106	0.011
107	0.008
108	0.005
109	0.004
110	0.002
111	0.001
112	0.001
113	0.000
114	0.000
115	0.000

450   Appendix

Table 4. Rank-sum table for $n_1, n_2 \leq 10$. (continued)

37	0.006
38	0.003

n1= 7 n2= 2

W	p
10	0.556
11	0.444
12	0.333
13	0.250
14	0.167
15	0.111
16	0.056
17	0.028

n1= 7 n2= 3

W	p
17	0.500
18	0.417
19	0.333
20	0.258
21	0.192
22	0.133
23	0.092
24	0.058
25	0.033
26	0.017
27	0.008

n1= 7 n2= 4

W	p
24	0.536
25	0.464
26	0.394
27	0.324
28	0.264
29	0.206
30	0.158
31	0.115
32	0.082
33	0.055
34	0.036
35	0.021
36	0.012

n1= 7 n2= 5

W	p
33	0.500
34	0.438
35	0.378
36	0.319
37	0.265
38	0.216
39	0.172
40	0.134
41	0.101
42	0.074
43	0.053
44	0.037
45	0.024
46	0.015
47	0.009
48	0.005
49	0.003
50	0.001

n1= 7 n2= 6

W	p
42	0.527
43	0.473
44	0.418
45	0.365
46	0.314
47	0.267
48	0.223
49	0.183
50	0.147
51	0.117
52	0.090
53	0.069
54	0.051
55	0.037
56	0.026
57	0.017
58	0.011
59	0.007
60	0.004
61	0.002
62	0.001
63	0.001

n1= 7 n2= 7

W	p
53	0.500
54	0.451
55	0.402
56	0.355
57	0.310
58	0.267
59	0.228
60	0.191
61	0.159
62	0.130
63	0.104
64	0.082
65	0.064
66	0.049
67	0.036
68	0.027
69	0.019
70	0.013
71	0.009
72	0.006
73	0.003
74	0.002
75	0.001
76	0.001
77	0.000

n1= 7 n2= 8

W	p
64	0.522
65	0.478
66	0.433
67	0.389
68	0.347
69	0.306
70	0.268
71	0.232
72	0.198
73	0.168
74	0.140
75	0.116
76	0.095
77	0.076
78	0.060
79	0.047
80	0.036
81	0.027
82	0.020
83	0.014
84	0.010
85	0.007
86	0.005
87	0.003
88	0.002
89	0.001
90	0.001
91	0.000
92	0.000

n1= 7 n2= 9

W	p
77	0.500
78	0.459
79	0.419
80	0.379
81	0.340
82	0.303
83	0.268
84	0.235
85	0.204
86	0.176
87	0.150
88	0.126
89	0.105
90	0.087
91	0.071
92	0.057
93	0.045
94	0.036
95	0.027
96	0.021
97	0.016

W	p
98	0.011
99	0.008
100	0.006
101	0.004
102	0.003
103	0.002
104	0.001
105	0.001
106	0.000
107	0.000
108	0.000

n1= 7 n2=10

W	p
90	0.519
91	0.481
92	0.443
93	0.406
94	0.370
95	0.335
96	0.300
97	0.268
98	0.237
99	0.209
100	0.182
101	0.157
102	0.135
103	0.115
104	0.097
105	0.081
106	0.067
107	0.054
108	0.044
109	0.035
110	0.028
111	0.022
112	0.017
113	0.012
114	0.009
115	0.007
116	0.005
117	0.003
118	0.002
119	0.002
120	0.001
121	0.001
122	0.000
123	0.000
124	0.000
125	0.000

n1= 8 n2= 2

W	p
11	0.556
12	0.444
13	0.356
14	0.267
15	0.200
16	0.133
17	0.089
18	0.044
19	0.022

n1= 8 n2= 3

W	p
18	0.539
19	0.461
20	0.388
21	0.315
22	0.248
23	0.188
24	0.139
25	0.097
26	0.067
27	0.042
28	0.024
29	0.012
30	0.006

n1= 8 n2= 4

W	p
26	0.533
27	0.467
28	0.404
29	0.341
30	0.285
31	0.230
32	0.184
33	0.141
34	0.107
35	0.077
36	0.055
37	0.036
38	0.024
39	0.014
40	0.008
41	0.004
42	0.002

n1= 8 n2= 5

W	p
35	0.528
36	0.472
37	0.416
38	0.362
39	0.311
40	0.262
41	0.218
42	0.177
43	0.142
44	0.111
45	0.085
46	0.064
47	0.047
48	0.033
49	0.023
50	0.015
51	0.009
52	0.005
53	0.003
54	0.002
55	0.001

n1= 8 n2= 6

W	p
45	0.525
46	0.475
47	0.426
48	0.377
49	0.331
50	0.286
51	0.245
52	0.207
53	0.172
54	0.141
55	0.114
56	0.091
57	0.071
58	0.054
59	0.041
60	0.030
61	0.021
62	0.015
63	0.010
64	0.006
65	0.004
66	0.002
67	0.001
68	0.001
69	0.000

n1= 8 n2= 7

W	p
56	0.522
57	0.478
58	0.433
59	0.389
60	0.347
61	0.306
62	0.268
63	0.232
64	0.198
65	0.168
66	0.140
67	0.116
68	0.095
69	0.076
70	0.060
71	0.047
72	0.036
73	0.027
74	0.020
75	0.014
76	0.010
77	0.007
78	0.005
79	0.003
80	0.002
81	0.001
82	0.001
83	0.000
84	0.000

452   Appendix

Table 4. Rank-sum table for $n_1, n_2 \leq 10$. (continued)

n1= 8 n2= 8			
W	p		
68	0.520		
69	0.480		
70	0.439		
71	0.399		
72	0.360		
73	0.323		
74	0.287		
75	0.253		
76	0.221		
77	0.191		
78	0.164		
79	0.139		
80	0.117		
81	0.097		
82	0.080		
83	0.065		
84	0.052		
85	0.041		
86	0.032		
87	0.025		
88	0.019		
89	0.014		
90	0.010		
91	0.007		
92	0.005		
93	0.003		
94	0.002		
95	0.001		
96	0.001		
97	0.001		
98	0.000		
99	0.000		
100	0.000		

n1= 8 n2= 9

W	p
81	0.519
82	0.481
83	0.444
84	0.407
85	0.371
86	0.336
87	0.303
88	0.271
89	0.240
90	0.212
91	0.185
92	0.161
93	0.138
94	0.118
95	0.100
96	0.084
97	0.069
98	0.057
99	0.046
100	0.037
101	0.030
102	0.023
103	0.018
104	0.014
105	0.010
106	0.008
107	0.006
108	0.004
109	0.003
110	0.002
111	0.001
112	0.001
113	0.000
114	0.000
115	0.000
116	0.000
117	0.000

n1= 8 n2=10

W	p
95	0.517
96	0.483
97	0.448
98	0.414
99	0.381
100	0.348
101	0.317
102	0.286
103	0.257
104	0.230
105	0.204
106	0.180
107	0.158
108	0.137
109	0.118
110	0.102
111	0.086
112	0.073
113	0.061
114	0.051
115	0.042
116	0.034
117	0.027
118	0.022
119	0.017
120	0.013
121	0.010
122	0.008
123	0.006
124	0.004
125	0.003
126	0.002
127	0.002
128	0.001
129	0.001
130	0.000
131	0.000
132	0.000
133	0.000
134	0.000
135	0.000

n1= 9 n2= 2

W	p
12	0.545
13	0.455
14	0.364
15	0.291
16	0.218
17	0.164
18	0.109
19	0.073
20	0.036
21	0.018

n1= 9 n2= 3

W	p
20	0.500
21	0.432
22	0.364
23	0.300
24	0.241
25	0.186
26	0.141
27	0.105
28	0.073
29	0.050
30	0.032
31	0.018
32	0.009
33	0.005

n1= 9 n2= 4

W	p
28	0.530
29	0.470
30	0.413
31	0.355
32	0.302
33	0.252
34	0.207
35	0.165
36	0.130
37	0.099
38	0.074
39	0.053
40	0.038
41	0.025
42	0.017
43	0.010
44	0.006
45	0.003
46	0.001

n1= 9 n2= 5

W	p
38	0.500
39	0.449
40	0.399
41	0.350
42	0.303
43	0.259
44	0.219
45	0.182
46	0.149
47	0.120
48	0.095
49	0.073
50	0.056
51	0.041
52	0.030
53	0.021
54	0.014
55	0.009
56	0.006
57	0.003
58	0.002

W	p
59	0.001
60	0.000

n1= 9 n2= 6

W	p
48	0.523
49	0.477
50	0.432
51	0.388
52	0.344
53	0.303
54	0.264
55	0.228
56	0.194
57	0.164
58	0.136
59	0.112
60	0.091
61	0.072
62	0.057
63	0.044
64	0.033
65	0.025
66	0.018
67	0.013
68	0.009
69	0.006
70	0.004
71	0.002
72	0.001
73	0.001
74	0.000
75	0.000

n1= 9 n2= 7

W	p
60	0.500
61	0.459
62	0.419
63	0.379
64	0.340
65	0.303
66	0.268
67	0.235
68	0.204
69	0.176
70	0.150
71	0.126
72	0.105
73	0.087
74	0.071
75	0.057
76	0.045

W	p
77	0.036
78	0.027
79	0.021
80	0.016
81	0.011
82	0.008
83	0.006
84	0.004
85	0.003
86	0.002
87	0.001
88	0.001
89	0.000
90	0.000
91	0.000

n1= 9 n2= 8

W	p
72	0.519
73	0.481
74	0.444
75	0.407
76	0.371
77	0.336
78	0.303
79	0.271
80	0.240
81	0.212
82	0.185
83	0.161
84	0.138
85	0.118
86	0.100
87	0.084
88	0.069
89	0.057
90	0.046
91	0.037
92	0.030
93	0.023
94	0.018
95	0.014
96	0.010
97	0.008
98	0.006
99	0.004
100	0.003
101	0.002
102	0.001
103	0.001
104	0.000
105	0.000
106	0.000
107	0.000
108	0.000

n1= 9 n2= 9

W	p
86	0.500
87	0.466
88	0.432
89	0.398
90	0.365
91	0.333
92	0.302
93	0.273
94	0.245
95	0.218
96	0.193
97	0.170
98	0.149
99	0.129
100	0.111
101	0.095
102	0.081
103	0.068
104	0.057
105	0.047
106	0.039
107	0.031
108	0.025
109	0.020
110	0.016
111	0.012
112	0.009
113	0.007
114	0.005
115	0.004
116	0.003
117	0.002
118	0.001
119	0.001
120	0.001
121	0.000
122	0.000
123	0.000
124	0.000
125	0.000
126	0.000

n1= 9 n2=10

W	p
100	0.516
101	0.484
102	0.452
103	0.421
104	0.390
105	0.360
106	0.330

W	p
107	0.302
108	0.274
109	0.248
110	0.223
111	0.200
112	0.178
113	0.158
114	0.139
115	0.121
116	0.106
117	0.091
118	0.078
119	0.067
120	0.056
121	0.047
122	0.039
123	0.033
124	0.027
125	0.022
126	0.017
127	0.014
128	0.011
129	0.009
130	0.007
131	0.005
132	0.004
133	0.003
134	0.002
135	0.001
136	0.001
137	0.001
138	0.000
139	0.000
140	0.000
141	0.000
142	0.000
143	0.000
144	0.000
145	0.000

Table 4. Rank-sum table for $n_1, n_2 \leq 10$. (continued)

n1=10 n2= 2							
W	p	41	0.071	57	0.281	86	0.012
		42	0.053	58	0.246	87	0.009
13	0.545	43	0.038	59	0.214	88	0.007
14	0.455	44	0.027	60	0.184	89	0.005
15	0.379	45	0.018	61	0.157	90	0.003
16	0.303	46	0.012	62	0.132	91	0.002
17	0.242	47	0.007	63	0.110	92	0.002
18	0.182	48	0.004	64	0.090	93	0.001
19	0.136	49	0.002	65	0.074	94	0.001
20	0.091	50	0.001	66	0.059	95	0.000
21	0.061			67	0.047	96	0.000
22	0.030			68	0.036	97	0.000
23	0.015	n1=10 n2= 5		69	0.028	98	0.000
				70	0.021		
		W	p	71	0.016		
n1=10 n2= 3				72	0.011		
		40	0.523	73	0.008		
W	p	41	0.477	74	0.005		
		42	0.430	75	0.004		
21	0.531	43	0.384	76	0.002		
22	0.469	44	0.339	77	0.001		
23	0.406	45	0.297	78	0.001		
24	0.346	46	0.257	79	0.000		
25	0.287	47	0.220	80	0.000		
26	0.234	48	0.185	81	0.000		
27	0.185	49	0.155				
28	0.143	50	0.127				
29	0.108	51	0.103	n1=10 n2= 7			
30	0.080	52	0.082				
31	0.056	53	0.065	W	p		
32	0.038	54	0.050				
33	0.024	55	0.038	63	0.519		
34	0.014	56	0.028	64	0.481		
35	0.007	57	0.020	65	0.443		
36	0.003	58	0.014	66	0.406		
		59	0.010	67	0.370		
		60	0.006	68	0.335		
		61	0.004	69	0.300		
n1=10 n2= 4		62	0.002	70	0.268		
		63	0.001	71	0.237		
W	p	64	0.001	72	0.209		
		65	0.000	73	0.182		
30	0.527			74	0.157		
31	0.473			75	0.135		
32	0.420	n1=10 n2= 6		76	0.115		
33	0.367			77	0.097		
34	0.318	W	p	78	0.081		
35	0.270			79	0.067		
36	0.227	51	0.521	80	0.054		
37	0.187	52	0.479	81	0.044		
38	0.152	53	0.437	82	0.035		
39	0.120	54	0.396	83	0.028		
40	0.094	55	0.356	84	0.022		
		56	0.318	85	0.017		

n1=10 n2= 8

W	p
76	0.517
77	0.483
78	0.448
79	0.414
80	0.381
81	0.348
82	0.317
83	0.286
84	0.257
85	0.230
86	0.204
87	0.180
88	0.158
89	0.137
90	0.118
91	0.102
92	0.086
93	0.073
94	0.061
95	0.051
96	0.042
97	0.034
98	0.027
99	0.022
100	0.017
101	0.013
102	0.010
103	0.008
104	0.006
105	0.004
106	0.003
107	0.002
108	0.002
109	0.001
110	0.001
111	0.000
112	0.000
113	0.000
114	0.000
115	0.000
116	0.000

n1=10 n2= 9

W	p
90	0.516
91	0.484
92	0.452
93	0.421
94	0.390
95	0.360
96	0.330
97	0.302
98	0.274
99	0.248
100	0.223
101	0.200
102	0.178
103	0.158
104	0.139
105	0.121
106	0.106
107	0.091
108	0.078
109	0.067
110	0.056
111	0.047
112	0.039
113	0.033
114	0.027
115	0.022
116	0.017
117	0.014
118	0.011
119	0.009
120	0.007
121	0.005
122	0.004
123	0.003
124	0.002
125	0.001
126	0.001
127	0.001
128	0.000
129	0.000
130	0.000
131	0.000
132	0.000
133	0.000
134	0.000
135	0.000

n1=10 n2=10

W	p
105	0.515
106	0.485
107	0.456
108	0.427
109	0.398
110	0.370
111	0.342
112	0.315
113	0.289
114	0.264
115	0.241
116	0.218
117	0.197
118	0.176
119	0.157
120	0.140
121	0.124
122	0.109
123	0.095
124	0.083
125	0.072
126	0.062
127	0.053
128	0.045
129	0.038
130	0.032
131	0.026
132	0.022
133	0.018
134	0.014
135	0.012
136	0.009
137	0.007
138	0.006
139	0.004
140	0.003
141	0.003
142	0.002
143	0.001
144	0.001
145	0.001
146	0.001
147	0.000
148	0.000
149	0.000
150	0.000
151	0.000
152	0.000
153	0.000
154	0.000
155	0.000

Table 5. Individual Probabilities for the Binomial Distribution.

n = 5				p=probability of a success				
x	0.1	0.2	0.3	0.4	0.5	0.6	0.7	0.8	0.9
0	0.590	0.328	0.168	0.078	0.031	0.010	0.002	0.000	0.000
1	0.328	0.410	0.360	0.259	0.156	0.077	0.028	0.006	0.000
2	0.073	0.205	0.309	0.346	0.313	0.230	0.132	0.051	0.008
3	0.008	0.051	0.132	0.230	0.313	0.346	0.309	0.205	0.073
4	0.000	0.006	0.028	0.077	0.156	0.259	0.360	0.410	0.328
5	0.000	0.000	0.002	0.010	0.031	0.078	0.168	0.328	0.590

n = 10				p=probability of a success				
x	0.1	0.2	0.3	0.4	0.5	0.6	0.7	0.8	0.9
0	0.349	0.107	0.028	0.006	0.001	0.000	0.000	0.000	0.000
1	0.387	0.268	0.121	0.040	0.010	0.002	0.000	0.000	0.000
2	0.194	0.302	0.233	0.121	0.044	0.011	0.001	0.000	0.000
3	0.057	0.201	0.267	0.215	0.117	0.042	0.009	0.001	0.000
4	0.011	0.088	0.200	0.251	0.205	0.111	0.037	0.006	0.000
5	0.001	0.026	0.103	0.201	0.246	0.201	0.103	0.026	0.001
6	0.000	0.006	0.037	0.111	0.205	0.251	0.200	0.088	0.011
7	0.000	0.001	0.009	0.042	0.117	0.215	0.267	0.201	0.057
8	0.000	0.000	0.001	0.011	0.044	0.121	0.233	0.302	0.194
9	0.000	0.000	0.000	0.002	0.010	0.040	0.121	0.268	0.387
10	0.000	0.000	0.000	0.000	0.001	0.006	0.028	0.107	0.349

n = 15				p=probability of a success				
x	0.1	0.2	0.3	0.4	0.5	0.6	0.7	0.8	0.9
0	0.206	0.035	0.005	0.000	0.000	0.000	0.000	0.000	0.000
1	0.343	0.132	0.031	0.005	0.000	0.000	0.000	0.000	0.000
2	0.267	0.231	0.092	0.022	0.003	0.000	0.000	0.000	0.000
3	0.129	0.250	0.170	0.063	0.014	0.002	0.000	0.000	0.000
4	0.043	0.188	0.219	0.127	0.042	0.007	0.001	0.000	0.000
5	0.010	0.103	0.206	0.186	0.092	0.024	0.003	0.000	0.000
6	0.002	0.043	0.147	0.207	0.153	0.061	0.012	0.001	0.000
7	0.000	0.014	0.081	0.177	0.196	0.118	0.035	0.003	0.000
8	0.000	0.003	0.035	0.118	0.196	0.177	0.081	0.014	0.000
9	0.000	0.001	0.012	0.061	0.153	0.207	0.147	0.043	0.002
10	0.000	0.000	0.003	0.024	0.092	0.186	0.206	0.103	0.010
11	0.000	0.000	0.001	0.007	0.042	0.127	0.219	0.188	0.043
12	0.000	0.000	0.000	0.002	0.014	0.063	0.170	0.250	0.129
13	0.000	0.000	0.000	0.000	0.003	0.022	0.092	0.231	0.267
14	0.000	0.000	0.000	0.000	0.000	0.005	0.031	0.132	0.343
15	0.000	0.000	0.000	0.000	0.000	0.000	0.005	0.035	0.206

Table 6. Individual probabilities for the Sign Test. These are binomial probabilities for p=.5 . For X > 11 use P(X = x) = P(X = n-x).

x=number of successes

n	0	1	2	3	4	5	6	7	8	9	10
1	0.500	0.500									
2	0.250	0.500	0.250								
3	0.125	0.375	0.375	0.125							
4	0.063	0.250	0.375	0.250	0.063						
5	0.031	0.156	0.313	0.313	0.156	0.031					
6	0.016	0.094	0.234	0.313	0.234	0.094	0.016				
7	0.008	0.055	0.164	0.273	0.273	0.164	0.055	0.008			
8	0.004	0.031	0.109	0.219	0.273	0.219	0.109	0.031	0.004		
9	0.002	0.018	0.070	0.164	0.246	0.246	0.164	0.070	0.018	0.002	
10	0.001	0.010	0.044	0.117	0.205	0.246	0.205	0.117	0.044	0.010	0.001
11	0.000	0.005	0.027	0.081	0.161	0.226	0.226	0.161	0.081	0.027	0.005
12	0.000	0.003	0.016	0.054	0.121	0.193	0.226	0.193	0.121	0.054	0.016
13	0.000	0.002	0.010	0.035	0.087	0.157	0.209	0.209	0.157	0.087	0.035
14	0.000	0.001	0.006	0.022	0.061	0.122	0.183	0.209	0.183	0.122	0.061
15	0.000	0.000	0.003	0.014	0.042	0.092	0.153	0.196	0.196	0.153	0.092
16	0.000	0.000	0.002	0.009	0.028	0.067	0.122	0.175	0.196	0.175	0.122
17	0.000	0.000	0.001	0.005	0.018	0.047	0.094	0.148	0.185	0.185	0.148
18	0.000	0.000	0.001	0.003	0.012	0.033	0.071	0.121	0.167	0.185	0.167
19	0.000	0.000	0.000	0.002	0.007	0.022	0.052	0.096	0.144	0.176	0.176
20	0.000	0.000	0.000	0.001	0.005	0.015	0.037	0.074	0.120	0.160	0.176

Table 7. Chi-Squared distribution critical values.

upper-tail p d.f.	0.2	0.1	0.075	0.05	0.04
1	1.64	2.71	3.17	3.84	4.22
2	3.22	4.61	5.18	5.99	6.44
3	4.64	6.25	6.90	7.81	8.31
4	5.99	7.78	8.50	9.49	10.03
5	7.29	9.24	10.01	11.07	11.64
6	8.56	10.64	11.47	12.59	13.20
7	9.80	12.02	12.88	14.07	14.70
8	11.03	13.36	14.27	15.51	16.17
9	12.24	14.68	15.63	16.92	17.61
10	13.44	15.99	16.97	18.31	19.02
11	14.63	17.28	18.29	19.68	20.41
12	15.81	18.55	19.60	21.03	21.79
13	16.98	19.81	20.90	22.36	23.14
14	18.15	21.06	22.18	23.68	24.49
15	19.31	22.31	23.45	25.00	25.82
16	20.47	23.54	24.72	26.30	27.14
17	21.61	24.77	25.97	27.59	28.44
18	22.76	25.99	27.22	28.87	29.75
19	23.90	27.20	28.46	30.14	31.04
20	25.04	28.41	29.69	31.41	32.32
21	26.17	29.62	30.92	32.67	33.60
22	27.30	30.81	32.14	33.92	34.87
23	28.43	32.01	33.36	35.17	36.13
24	29.55	33.20	34.57	36.42	37.39
25	30.68	34.38	35.78	37.65	38.64
26	31.79	35.56	36.98	38.89	39.89
27	32.91	36.74	38.18	40.11	41.13
28	34.03	37.92	39.38	41.34	42.37
29	35.14	39.09	40.57	42.56	43.60
30	36.25	40.26	41.76	43.77	44.83
40	47.27	51.81	53.50	55.76	56.95
50	58.16	63.17	65.03	67.50	68.80
60	68.97	74.40	76.41	79.08	80.48
80	90.41	96.58	98.86	101.88	103.46
100	111.67	118.50	121.02	124.34	126.08

Table 7. Chi-Squared distribution critical values. (Continued)

upper-tail p d.f.	0.03	0.02	0.01	0.005	0.001
1	4.71	5.41	6.63	7.88	10.83
2	7.01	7.82	9.21	10.60	13.82
3	8.95	9.84	11.34	12.84	16.27
4	10.71	11.67	13.28	14.86	18.47
5	12.37	13.39	15.09	16.75	20.51
6	13.97	15.03	16.81	18.55	22.46
7	15.51	16.62	18.48	20.28	24.32
8	17.01	18.17	20.09	21.95	26.12
9	18.48	19.68	21.67	23.59	27.88
10	19.92	21.16	23.21	25.19	29.59
11	21.34	22.62	24.73	26.76	31.26
12	22.74	24.05	26.22	28.30	32.91
13	24.12	25.47	27.69	29.82	34.53
14	25.49	26.87	29.14	31.32	36.12
15	26.85	28.26	30.58	32.80	37.70
16	28.19	29.63	32.00	34.27	39.25
17	29.52	31.00	33.41	35.72	40.79
18	30.84	32.35	34.81	37.16	42.31
19	32.16	33.69	36.19	38.58	43.82
20	33.46	35.02	37.57	40.00	45.31
21	34.76	36.34	38.93	41.40	46.80
22	36.05	37.66	40.29	42.80	48.27
23	37.33	38.97	41.64	44.18	49.73
24	38.61	40.27	42.98	45.56	51.18
25	39.88	41.57	44.31	46.93	52.62
26	41.15	42.86	45.64	48.29	54.05
27	42.41	44.14	46.96	49.65	55.48
28	43.66	45.42	48.28	50.99	56.89
29	44.91	46.69	49.59	52.34	58.30
30	46.16	47.96	50.89	53.67	59.70
40	58.43	60.44	63.69	66.77	73.40
50	70.42	72.61	76.15	79.49	86.66
60	82.23	84.58	88.38	91.95	99.61
80	105.42	108.07	112.33	116.32	124.84
100	128.24	131.14	135.81	140.17	149.45

Table 8a. F Distribution critical values.
Degrees of Freedom in Numerator=2.

Degrees of Freedom in Denominator	0.2	0.1	0.075	0.05	0.04
1	12.00	49.50	88.39	199.50	312.00
2	4.00	9.00	12.33	19.00	24.00
3	2.89	5.46	6.93	9.55	11.32
4	2.47	4.32	5.30	6.94	8.00
5	2.26	3.78	4.55	5.79	6.56
6	2.13	3.46	4.11	5.14	5.77
7	2.04	3.26	3.84	4.74	5.28
8	1.98	3.11	3.64	4.46	4.94
9	1.93	3.01	3.50	4.26	4.70
10	1.90	2.92	3.39	4.10	4.52
11	1.87	2.86	3.31	3.98	4.37
12	1.85	2.81	3.24	3.89	4.26
13	1.83	2.76	3.18	3.81	4.17
14	1.81	2.73	3.13	3.74	4.09
15	1.80	2.70	3.09	3.68	4.02
16	1.78	2.67	3.06	3.63	3.96
17	1.77	2.64	3.03	3.59	3.91
18	1.76	2.62	3.00	3.55	3.87
19	1.75	2.61	2.98	3.52	3.83
20	1.75	2.59	2.96	3.49	3.80
21	1.74	2.57	2.94	3.47	3.77
22	1.73	2.56	2.92	3.44	3.74
23	1.73	2.55	2.91	3.42	3.71
24	1.72	2.54	2.89	3.40	3.69
25	1.72	2.53	2.88	3.39	3.67
26	1.71	2.52	2.87	3.37	3.65
27	1.71	2.51	2.86	3.35	3.64
28	1.71	2.50	2.85	3.34	3.62
29	1.70	2.50	2.84	3.33	3.60
30	1.70	2.49	2.83	3.32	3.59
40	1.68	2.44	2.77	3.23	3.49
60	1.65	2.39	2.71	3.15	3.40
100	1.64	2.36	2.66	3.09	3.32
10000	1.61	2.30	2.59	3.00	3.22

Table 8a. F Distribution critical values. (Continued)
Degrees of Freedom in Numerator=2.

Degrees of Freedom in Denominator	0.03	0.02	0.01	0.005	0.001
1	555.05	1249	4999	19997	499725
2	32.33	49.00	99.00	199.01	998.84
3	14.04	18.86	30.82	49.80	148.49
4	9.55	12.14	18.00	26.28	61.25
5	7.66	9.45	13.27	18.31	37.12
6	6.65	8.05	10.92	14.54	27.00
7	6.03	7.20	9.55	12.40	21.69
8	5.61	6.64	8.65	11.04	18.49
9	5.31	6.23	8.02	10.11	16.39
10	5.08	5.93	7.56	9.43	14.90
11	4.91	5.70	7.21	8.91	13.81
12	4.76	5.52	6.93	8.51	12.97
13	4.65	5.37	6.70	8.19	12.31
14	4.55	5.24	6.51	7.92	11.78
15	4.47	5.14	6.36	7.70	11.34
16	4.40	5.05	6.23	7.51	10.97
17	4.34	4.97	6.11	7.35	10.66
18	4.29	4.90	6.01	7.21	10.39
19	4.24	4.84	5.93	7.09	10.16
20	4.20	4.79	5.85	6.99	9.95
21	4.16	4.74	5.78	6.89	9.77
22	4.13	4.70	5.72	6.81	9.61
23	4.10	4.66	5.66	6.73	9.47
24	4.07	4.63	5.61	6.66	9.34
25	4.05	4.59	5.57	6.60	9.22
26	4.03	4.56	5.53	6.54	9.12
27	4.00	4.54	5.49	6.49	9.02
28	3.98	4.51	5.45	6.44	8.93
29	3.97	4.49	5.42	6.40	8.85
30	3.95	4.47	5.39	6.35	8.77
40	3.83	4.32	5.18	6.07	8.25
60	3.72	4.18	4.98	5.79	7.77
100	3.63	4.07	4.82	5.59	7.41
10000	3.51	3.91	4.61	5.30	6.91

Table 8b. F Distribution critical values. (Continued)
Degrees of Freedom in Numerator = 3.

Degrees of Freedom in Denominator	0.2	0.1	0.075	0.05	0.04
1	13.06	53.59	95.62	215.71	337.29
2	4.16	9.16	12.50	19.16	24.16
3	2.94	5.39	6.79	9.28	10.96
4	2.48	4.19	5.09	6.59	7.56
5	2.25	3.62	4.30	5.41	6.10
6	2.11	3.29	3.86	4.76	5.30
7	2.02	3.07	3.57	4.35	4.81
8	1.95	2.92	3.38	4.07	4.48
9	1.90	2.81	3.23	3.86	4.23
10	1.86	2.73	3.12	3.71	4.05
11	1.83	2.66	3.03	3.59	3.91
12	1.80	2.61	2.96	3.49	3.80
13	1.78	2.56	2.90	3.41	3.70
14	1.76	2.52	2.85	3.34	3.62
15	1.75	2.49	2.81	3.29	3.56
16	1.74	2.46	2.78	3.24	3.50
17	1.72	2.44	2.75	3.20	3.45
18	1.71	2.42	2.72	3.16	3.41
19	1.70	2.40	2.70	3.13	3.37
20	1.70	2.38	2.67	3.10	3.34
21	1.69	2.36	2.65	3.07	3.31
22	1.68	2.35	2.64	3.05	3.28
23	1.68	2.34	2.62	3.03	3.26
24	1.67	2.33	2.61	3.01	3.23
25	1.66	2.32	2.59	2.99	3.21
26	1.66	2.31	2.58	2.98	3.20
27	1.66	2.30	2.57	2.96	3.18
28	1.65	2.29	2.56	2.95	3.16
29	1.65	2.28	2.55	2.93	3.15
30	1.64	2.28	2.54	2.92	3.13
40	1.62	2.23	2.48	2.84	3.04
60	1.59	2.18	2.42	2.76	2.95
100	1.58	2.14	2.37	2.70	2.87
10000	1.55	2.08	2.30	2.61	2.77

Table 8b. F Distribution critical values. (Continued)
Degrees of Freedom in Numerator = 3.

Degrees of Freedom in Denominator	0.03	0.02	0.01	0.005	0.001
1	599.98	1351	5404	21614	540257
2	32.50	49.17	99.16	199.16	999.31
3	13.53	18.11	29.46	47.47	141.10
4	8.97	11.34	16.69	24.26	56.17
5	7.08	8.67	12.06	16.53	33.20
6	6.07	7.29	9.78	12.92	23.71
7	5.45	6.45	8.45	10.88	18.77
8	5.04	5.90	7.59	9.60	15.83
9	4.74	5.51	6.99	8.72	13.90
10	4.52	5.22	6.55	8.08	12.55
11	4.34	4.99	6.22	7.60	11.56
12	4.20	4.81	5.95	7.23	10.80
13	4.09	4.67	5.74	6.93	10.21
14	4.00	4.55	5.56	6.68	9.73
15	3.92	4.45	5.42	6.48	9.34
16	3.85	4.36	5.29	6.30	9.01
17	3.79	4.29	5.19	6.16	8.73
18	3.74	4.22	5.09	6.03	8.49
19	3.69	4.16	5.01	5.92	8.28
20	3.65	4.11	4.94	5.82	8.10
21	3.62	4.07	4.87	5.73	7.94
22	3.59	4.03	4.82	5.65	7.80
23	3.56	3.99	4.76	5.58	7.67
24	3.53	3.96	4.72	5.52	7.55
25	3.51	3.93	4.68	5.46	7.45
26	3.48	3.90	4.64	5.41	7.36
27	3.46	3.87	4.60	5.36	7.27
28	3.45	3.85	4.57	5.32	7.19
29	3.43	3.83	4.54	5.28	7.12
30	3.41	3.81	4.51	5.24	7.05
40	3.30	3.67	4.31	4.98	6.59
60	3.19	3.53	4.13	4.73	6.17
100	3.10	3.43	3.98	4.54	5.86
10000	2.98	3.28	3.78	4.28	5.43

Table 8b. F Distribution critical values. (Continued)
Degrees of Freedom in Numerator = 4.

Degrees of Freedom in Denominator	0.2	0.1	0.075	0.05	0.04
1	13.64	55.83	99.58	224.58	351.14
2	4.24	9.24	12.58	19.25	24.25
3	2.96	5.34	6.70	9.12	10.75
4	2.48	4.11	4.96	6.39	7.31
5	2.24	3.52	4.16	5.19	5.83
6	2.09	3.18	3.71	4.53	5.04
7	1.99	2.96	3.42	4.12	4.54
8	1.92	2.81	3.21	3.84	4.21
9	1.87	2.69	3.07	3.63	3.97
10	1.83	2.61	2.95	3.48	3.78
11	1.80	2.54	2.87	3.36	3.64
12	1.77	2.48	2.79	3.26	3.53
13	1.75	2.43	2.74	3.18	3.43
14	1.73	2.39	2.69	3.11	3.36
15	1.71	2.36	2.64	3.06	3.29
16	1.70	2.33	2.61	3.01	3.23
17	1.68	2.31	2.58	2.96	3.18
18	1.67	2.29	2.55	2.93	3.14
19	1.66	2.27	2.52	2.90	3.10
20	1.65	2.25	2.50	2.87	3.07
21	1.65	2.23	2.48	2.84	3.04
22	1.64	2.22	2.46	2.82	3.01
23	1.63	2.21	2.45	2.80	2.99
24	1.63	2.19	2.43	2.78	2.97
25	1.62	2.18	2.42	2.76	2.95
26	1.62	2.17	2.41	2.74	2.93
27	1.61	2.17	2.40	2.73	2.91
28	1.61	2.16	2.39	2.71	2.90
29	1.60	2.15	2.38	2.70	2.88
30	1.60	2.14	2.37	2.69	2.87
40	1.57	2.09	2.30	2.61	2.77
60	1.55	2.04	2.24	2.53	2.68
100	1.53	2.00	2.19	2.46	2.61
10000	1.50	1.95	2.12	2.37	2.51

Table 8b. F Distribution critical values. (Continued)
Degrees of Freedom in Numerator = 4.

Degrees of Freedom in Denominator	0.03	0.02	0.01	0.005	0.001
1	624.58	1406	5624	22501	562668
2	32.58	49.25	99.25	199.24	999.31
3	13.25	17.69	28.71	46.20	137.08
4	8.65	10.90	15.98	23.15	53.43
5	6.75	8.23	11.39	15.56	31.08
6	5.74	6.86	9.15	12.03	21.92
7	5.13	6.03	7.85	10.05	17.20
8	4.71	5.49	7.01	8.81	14.39
9	4.42	5.10	6.42	7.96	12.56
10	4.20	4.82	5.99	7.34	11.28
11	4.02	4.59	5.67	6.88	10.35
12	3.89	4.42	5.41	6.52	9.63
13	3.77	4.28	5.21	6.23	9.07
14	3.68	4.16	5.04	6.00	8.62
15	3.60	4.06	4.89	5.80	8.25
16	3.53	3.97	4.77	5.64	7.94
17	3.48	3.90	4.67	5.50	7.68
18	3.43	3.84	4.58	5.37	7.46
19	3.38	3.78	4.50	5.27	7.27
20	3.34	3.73	4.43	5.17	7.10
21	3.31	3.69	4.37	5.09	6.95
22	3.27	3.65	4.31	5.02	6.81
23	3.24	3.61	4.26	4.95	6.70
24	3.22	3.58	4.22	4.89	6.59
25	3.19	3.55	4.18	4.84	6.49
26	3.17	3.52	4.14	4.79	6.41
27	3.15	3.50	4.11	4.74	6.33
28	3.13	3.47	4.07	4.70	6.25
29	3.12	3.45	4.04	4.66	6.19
30	3.10	3.43	4.02	4.62	6.12
40	2.99	3.30	3.83	4.37	5.70
60	2.88	3.16	3.65	4.14	5.31
100	2.80	3.06	3.51	3.96	5.02
10000	2.68	2.92	3.32	3.72	4.62

Table 8b. F Distribution critical values. (Continued)
Degrees of Freedom in Numerator = 5.

Degrees of Freedom in Denominator	0.2	0.1	0.075	0.05	0.04
1	14.01	57.24	102.07	230.16	359.85
2	4.28	9.29	12.63	19.30	24.30
3	2.97	5.31	6.64	9.01	10.62
4	2.48	4.05	4.88	6.26	7.14
5	2.23	3.45	4.06	5.05	5.66
6	2.08	3.11	3.60	4.39	4.86
7	1.97	2.88	3.31	3.97	4.37
8	1.90	2.73	3.11	3.69	4.03
9	1.85	2.61	2.96	3.48	3.79
10	1.80	2.52	2.84	3.33	3.61
11	1.77	2.45	2.75	3.20	3.46
12	1.74	2.39	2.68	3.11	3.35
13	1.72	2.35	2.62	3.03	3.26
14	1.70	2.31	2.57	2.96	3.18
15	1.68	2.27	2.53	2.90	3.11
16	1.67	2.24	2.49	2.85	3.06
17	1.65	2.22	2.46	2.81	3.01
18	1.64	2.20	2.43	2.77	2.97
19	1.63	2.18	2.41	2.74	2.93
20	1.62	2.16	2.38	2.71	2.89
21	1.61	2.14	2.37	2.68	2.86
22	1.61	2.13	2.35	2.66	2.84
23	1.60	2.11	2.33	2.64	2.81
24	1.59	2.10	2.32	2.62	2.79
25	1.59	2.09	2.30	2.60	2.77
26	1.58	2.08	2.29	2.59	2.75
27	1.58	2.07	2.28	2.57	2.73
28	1.57	2.06	2.27	2.56	2.72
29	1.57	2.06	2.26	2.55	2.70
30	1.57	2.05	2.25	2.53	2.69
40	1.54	2.00	2.18	2.45	2.60
60	1.51	1.95	2.12	2.37	2.50
100	1.49	1.91	2.07	2.31	2.43
10000	1.46	1.85	2.00	2.21	2.33

Table 8b. F Distribution critical values. (Continued)
Degrees of Freedom in Numerator = 5.

Degrees of Freedom in Denominator	0.03	0.02	0.01	0.005	0.001
1	640.05	1441	5764	23056	576496
2	32.63	49.30	99.30	199.30	999.31
3	13.07	17.43	28.24	45.39	134.58
4	8.44	10.62	15.52	22.46	51.72
5	6.54	7.95	10.97	14.94	29.75
6	5.53	6.58	8.75	11.46	20.80
7	4.91	5.76	7.46	9.52	16.21
8	4.50	5.22	6.63	8.30	13.48
9	4.21	4.84	6.06	7.47	11.71
10	3.99	4.55	5.64	6.87	10.48
11	3.81	4.34	5.32	6.42	9.58
12	3.68	4.16	5.06	6.07	8.89
13	3.56	4.02	4.86	5.79	8.35
14	3.47	3.90	4.69	5.56	7.92
15	3.39	3.81	4.56	5.37	7.57
16	3.33	3.72	4.44	5.21	7.27
17	3.27	3.65	4.34	5.07	7.02
18	3.22	3.59	4.25	4.96	6.81
19	3.17	3.53	4.17	4.85	6.62
20	3.13	3.48	4.10	4.76	6.46
21	3.10	3.44	4.04	4.68	6.32
22	3.07	3.40	3.99	4.61	6.19
23	3.04	3.36	3.94	4.54	6.08
24	3.01	3.33	3.90	4.49	5.98
25	2.99	3.30	3.85	4.43	5.89
26	2.97	3.28	3.82	4.38	5.80
27	2.95	3.25	3.78	4.34	5.73
28	2.93	3.23	3.75	4.30	5.66
29	2.91	3.21	3.73	4.26	5.59
30	2.90	3.19	3.70	4.23	5.53
40	2.78	3.05	3.51	3.99	5.13
60	2.68	2.92	3.34	3.76	4.76
100	2.59	2.82	3.21	3.59	4.48
10000	2.48	2.68	3.02	3.35	4.11

Answers to Selected Exercises

Chapter 1

Exercise Set A

3. a) 75.6%

5. Because the telephone survey was done during the day, the survey may not represent those housewives that are employed outside the home.

Exercise Set C

1. a) 26.8   b) 12.9

5. 112

7. a) Yes   b) 78 inches

Exercise Set D

1. a) 24   b) 4.830

3. Dormitory 2

7. She will obtain the correct standard deviation.

Exercise Set E

4. a) 0.54 approximately   b) 0.90 approximately

6. a) roughly normal   b) 12 pounds

Exercise Set F

1. a) $X_{(1200)}$ b) $X_{(3600)}$ c) $X_{(3600)} - X_{(1200)}$

3. a) 4.25   b) 18.25

Exercise Set G

3.   a) 25th percentile = 23.5

50th percentile = 25.6

75th percentile = 29.3

Chapter 2

Exercise Set A

3.   0.7

5.   a) 474   b) 526   c) 948

7.   a) 0.167   b) 0.50   c) 0.833

Exercise Set B

1. $\dfrac{1}{36}$

3.   a) $\dfrac{1}{36}$ b) $\dfrac{2}{36}$

5.   a) 0.16   b) 0.52

Exercise Set C

1.   a) 0.00452   b) 0.00603

3.   a) 0.333   b) 0.36

5. $\dfrac{1}{52}$

Exercise Set D

1.   0.0906

3.   0.416

Chapter 3

Exercise Set A

3. a) 0.00000076   b) 0.00000076

5. "Gender" and "Apply for financial aid" are discrete. "GPA" and "Admissions test score" are continuous.

7. population mean = 3.5   population standard deviation = 1.708

Exercise Set B

1. Histogram (b).

3. Standard error of the mean = 0.71

7. a) No   b) Yes   c) 3.2

    d) 0.28   e) between 2.64 and 3.76

Exercise Set C

1. a) 0.1587   b) 0.8413   c) 0.1587   d) 0.6826

3. a) 0.8413   b) $P(Z > -1.75) = 0.9599$

5. $P(-2.07 \leq Z \leq 2.07) = 0.9616$

Exercise Set D

1. d.f. = 9 $P(t > 2.25) = 0.025$ $P(t > 3.25) = 0.005$

3. a) standard deviation of z is 1, the standard deviation of t is 1.41

    b) No, the sampling distribution of t does not approximate that of z because t is more variable than z.

5. d.f. =13 $P(-1.35 < t < 1.35) = .9 - .1 = .8$

Exercise Set E

1. a) 0.5   b) 0.05   c) Approximately 95%

3. a) $\dfrac{1}{6}$ b) No   c) 0.0162 approximately

5. Less than 0.001

Chapter 4

Exercise Set A

1. a) mean = 2.887   median = 176.5
    trimmed mean = 258.25

3. The mean.

Exercise Set B

1. Approximately 36.

3. 4.50 ± 1.05

7. $n = 181$

Exercise Set C

1. a) 235 ± 101.5
    b) The triglyceride levels must be normally distributed.

3. a) Income distributions tend to be skewed.

    b) No, the sample size is small and the distribution is probably not normally distributed.

5. 127 ± 6.2

Exercise Set D

1. a) 0.0592 ± 0.0174

    b) 0.1000 ± 0.0294

3. 0.283 ± 0.051

5. 0.163 ± 0.042

7. It is not possible because the minimum count does not exceed 5.

Chapter 5

Exercise Set A

1. $H_0 : \mu = 8$ $H_a : \mu < 8$

3. $H_0 : \mu = 0$ $H_a : \mu > 0$ where μ is the mean of the change in systolic blood pressure.

Exercise Set B

1. a) Two-tailed

    b) She would not reject the null hypothesis. She would not have sufficient information to conclude that those who lived off campus had a higher or lower GPA.

3. a) $H_0 : \mu = 20$ $H_a : \mu > 20$

    b) Yes, they would reject the null hypothesis. They would conclude that the average current exceeds 20 amperes.

    c) Chart (c).

Exercise Set C

1. a) 8

    b) $\bar{x} = -504$, $s = 218.6$

    c) $H_0 : \mu = 0$, $H_a : \mu \neq 0$

    d) $t = -6.52$, $p < 0.002$

3. Researcher A

5. a) $t = 3.81$ b) $p < 0.001$ c) Reject the null hypothesis.
    d) The mean exceeds 20.

Exercise Set D

1. a) $H_0 : \mu = 0$, $H_a : \mu > 0$

    b) $S = 25$, $p = 0.039$

    c) The experimental program was effective.

3. a) 4

    b) $S = 28$

    c) No, use the table for the signed rank statistic because $n < 10$.

    d) $p = 0.098$

    e) Do not reject the null hypothesis. We have insufficient evidence to conclude that the reading program is effective.

5. Yes, the two data sets will produce equal signed rank statistics because the rank and the signs would be equal.

Exercise Set E

1. The signed rank test would be recommended because $n \geq 10$ and the researcher was not confident that the observations came from a normal distribution.

Exercise Set F

1. $n = 34$

3. a) 14   b) 39

Chapter 6

Exercise Set A

1. a) $s_p^2 = 162859$ b) 140 ± 129.4

3. 1.6 ± 3.92

Exercise Set B

1. a) $s_p = 0.7575$

   b) $t = 1.76$ Do not reject the null hypothesis using $\alpha = 0.05$.

   c) There is not sufficient evidence to conclude that the iron supplementation is effective.

3. a) $s_p = 981.8$ b) $t = 8.58$, $p < 0.002$

   c) The histamine concentrations are greater in cardiac patients.

Exercise Set C

1. a) $W = 82$ b) $p = 0.020$ c) $p = 0.040$

   d) Yes, you would reject the null hypothesis.

3. a) 1242   b) $Z = .725$, $p = 0.234$

   c) There is insufficient evidence to conclude that the enriched course was more effective.

Exercise Set D

1. Since we are not confident that weights are normally distributed and because there are more than 10 observations in each sample we would recommend the rank-sum test.

3. a) Alcohol consumption is probably skewed to the right.

   b) No, there appears to be a great difference in variability between the groups.

   c) No, a t-test would not be the best approach because of the large difference in variability.

Exercise Set E

1. a) ratio = 1.66   A transformation would be recommended.

   b)   .061   .021   .079   0.0   -.071

            .041   .130   .021   .161   .217   .079

   c) Control st. dev. = 0.059 B_{12} st. dev. = 0.075

      ratio = 1.27   A t-test would be appropriate.

3. He should be cautious because the F-test fails to maintain its significance level with non-normal data, and this weight data may be skewed to the right.

Exercise Set F

1. a) $n_1 = n_2 = 658$

   b) $n_1 = n_2 = 30$

Chapter 7

Exercise Set A

1. a) 82   b) 77   c) ii   d) 95 on math, 60 on language

   e) 2, 8, 9   f) 1, 3, 5, 7, 10

   g) 6   h) 3, 4, 5, 7, 8

Exercise Set B

1. (a) is $r = 0.0$, (b) is $r = 0.3$, (c) is $r = -0.3$

3. No, the correlation will not change because the values of $(y_i - \bar{y})/s_y$ will not change.

5. a) 3   b) 12   c) -0.5   d) No, because the standardized values will not change.

Exercise Set C

1. It is not clear that additional expenditures will increase the college entrance exam scores because school district wealth may be a common cause.

Exercise Set D

1. c) $b = 0.75$ d) 130   e) 134.5

3. c) $b = 7.933$ d) 85   e) 61.2 f) 180.2

   g) $a = 37.4$, $\hat{y} = 37.4 + 7.933x$

Exercise Set E

1. a) line (a)

   b) line (a)

   c) line (b)

3. a) The predicted value for the first observation is 0.0242

   b) The residual for the first observation is 0.026-0.0242=0.0018

Exercise Set F

1. b) 500 c) 680 d) 320 e) Yes, there was a small regression effect.

Exercise Set G

1. a) b

    b) b is the sample slope of the least squares line whereas β is the population slope.

    c) s

    d) σ is the standard deviation of the errors in the population model and s is the root-mean-squared error that is used to estimate σ.

3. a) Yes, but the correlation is low.

    b) $t = 1.17$. Do not reject the null hypothesis. This correlation does not prove that increased advertising will increase sales.

Exercise Set H

1. b) $r_s = -0.036$

    c) The large difference between r and r_s is due to the presence of an outlier.

    d) The Spearman rank correlation would be the better statistic to describe the true relationship.

Exercise Set I

1. Pearson's correlation should be used because it is easier to interpret than the other measures. Spearman's correlation is more difficult to interpret.

Chapter 8

Exercise Set A

1. a) Categorical   b) Ordered

3. It should be analyzed as continuous data.

Exercise Set B

1. a) $p(2) = 0.375$

    b) $p(3) = 0.250$

c) $p(4) = 0.0625$

d) 0.6875

3. a) No, it is not a binomial experiment because the probability of a success does change from trial to trial.

    b) $26/52 = 0.50$

    c) $24/50 = 0.48$

5. a) 0.3928

    b) 0.304

    c) 0.304

Exercise Set C

1. a) $H_0 : p_1 = 0.5$

    b)

Outcome	Actual Count	Expected Count
Head	55	50
Tail	45	50

    c) $X^2 = 1.0$ With d.f. = 1 the critical value is 3.84 so we have insufficient evidence to conclude that the coin is not fair.

3. $X^2 = 2.1985$ With d.f. = 3 the critical value is 7.81 so we do not reject the null hypothesis.

Exercise Set D

1. a) Expected cell counts

	Drug	Violent	Property
Females	22.54	21.90	36.56
Males	120.46	117.10	195.44

    b) Drug and Property crimes.

    c) $X^2 = 14.641$ With d.f. = 2 the critical value is 5.99. We reject the null hypothesis of independence.

3. a) $X^2 = 141.528$

    b) With d.f. = 1 the critical value = 3.84 so we reject the null hypothesis.

    c) $p < 0.001$. We conclude that alcohol abuse is related to gender.

Exercise Set E

1. a) $p = 0.0139 + 0.0010 = 0.0149$

    b) Yes.

    c) The new treatment improves survival.

Exercise Set F

1. a) Yes. b) $X^2_{CMH} = 0.0267$

    c) Do not reject the null hypothesis.

    d) We have insufficient evidence to conclude that grade and residence are related.

3. a) $X^2_{CMH} = 17.93$ $p < 0.001$

    b) Yes, the FDP score appears to be related to survival.

Exercise Set G

1. a) Concordant pairs = 40

    b) Discordant pairs = 0

    c) $\gamma = 1.0$

3. a) The rows and columns are ordered.

    b) $C = 118$

    c) $D = 28$

    d) $\gamma = 0.616$

    e) Yes, there appears to be a positive association between education and job satisfaction.

Exercise Set H

1. a) Use a sign test.

   b) $p = 0.241$

   c) There is insufficient evidence to conclude that the symptoms were reduced.

Chapter 9

Exercise Set A

1. a) $\bar{x}_2 = 26.50$, $s_2 = 1.87$

   b) $SS_{treatments} = 2404.26$, $DF_{treatments} = 3$, $MS_{treatments} = 801.42$

   c) $SS_{errors} = 85.74$, $DF_{errors} = 20$, $MSE = 4.287$

   d) $F = 186.94$

   e) The measurements vary from point to point.

   f) F is so large because paint B is inferior.

3. a) Yes.   b) No.

Exercise Set B

1. a) $p = 0.005$ b) p is approximately equal to 0.15

   c) p is approximately equal to 0.035

3. a) $SS_{treatments} = 28$, $SS_{errors} = 1388$

   b)

Source	DF	SS	Mean Square
Treatments	2	28.0	14.0
Error	15	1388.0	92.533
Total	17	1416.0	

   c) $F = 0.15$, $p > 0.2$

   d) There is insufficient evidence to conclude that the treatments are different.

Exercise Set C

1. It is not possible to compute the Kruskal-Wallis test statistic because the ranks cannot be determined.

3. a) No.

b) Average ranks:

Inexpensive	Moderate	Expensive
4.1428	15.7143	13.142

c) $K = 13.424$

d) Yes, the large sample approximation is appropriate because the sample sizes exceed 5 in all groups.

e) $p = 0.001$

f) Yes, we would reject the null hypothesis.

g) The distributions of the distances vary by type of golf ball.

Exercise Set D

1. The F test is recommended because the sample sizes do not exceed 10.

Exercise Set E

1. The value of T_{13} could change because the value of the MSE could change with additional data.

3. $\alpha = 0.20$

Index